Function of Polymers in Encapsulation Process

Function of Polymers in Encapsulation Process

Editors

M. Ali Aboudzadeh
Shaghayegh Hamzehlou

MDPI • Basel • Beijing • Wuhan • Barcelona • Belgrade • Manchester • Tokyo • Cluj • Tianjin

Editors
M. Ali Aboudzadeh
CNRS, IPREM
University of Pau and Pays de l'Adour (UPPA)
Pau
France

Shaghayegh Hamzehlou
Polymat and Kimika
Aplikatua Saila
Kimika Fakultatea
University of the Basque Country UPV-EHU
Donostia-San Sebastián
Spain

Editorial Office
MDPI
St. Alban-Anlage 66
4052 Basel, Switzerland

This is a reprint of articles from the Special Issue published online in the open access journal *Polymers* (ISSN 2073-4360) (available at: www.mdpi.com/journal/polymers/special_issues/Funct_Polym_Encapsulation_Process).

For citation purposes, cite each article independently as indicated on the article page online and as indicated below:

LastName, A.A.; LastName, B.B.; LastName, C.C. Article Title. *Journal Name* **Year**, *Volume Number*, Page Range.

ISBN 978-3-0365-3776-4 (Hbk)
ISBN 978-3-0365-3775-7 (PDF)

Contents

About the Editors vii

Preface to "Function of Polymers in Encapsulation Process" ix

Abbas Rahdar, Saman Sargazi, Mahmood Barani, Sheida Shahraki, Fakhara Sabir and M. Ali Aboudzadeh
Lignin-Stabilized Doxorubicin Microemulsions: Synthesis, Physical Characterization, and In Vitro Assessments
Reprinted from: *Polymers* **2021**, *13*, 641, doi:10.3390/polym13040641 1

Bjad K. Almutairy, Abdullah Alshetaili, Amer S. Alali, Mohammed Muqtader Ahmed, Md. Khalid Anwer and M. Ali Aboudzadeh
Design of Olmesartan Medoxomil-Loaded Nanosponges for Hypertension and Lung Cancer Treatments
Reprinted from: *Polymers* **2021**, *13*, 2272, doi:10.3390/polym13142272 15

Mohammed F. Aldawsari, Md. Khalid Anwer, Mohammed Muqtader Ahmed, Farhat Fatima, Gamal A. Soliman and Saurabh Bhatia et al.
Enhanced Dissolution of Sildenafil Citrate Using Solid Dispersion with Hydrophilic Polymers: Physicochemical Characterization and In Vivo Sexual Behavior Studies in Male Rats
Reprinted from: *Polymers* **2021**, *13*, 3512, doi:10.3390/polym13203512 29

Chandrasekar Ponnusamy, Abimanyu Sugumaran, Venkateshwaran Krishnaswami, Rajaguru Palanichamy, Ravichandiran Velayutham and Subramanian Natesan
Development and Evaluation of Polyvinylpyrrolidone K90 and Poloxamer 407 Self-Assembled Nanomicelles: Enhanced Topical Ocular Delivery of Artemisinin
Reprinted from: *Polymers* **2021**, *13*, 3038, doi:10.3390/polym13183038 45

Sylwia Łukasiewicz
Development of a New Polymeric Nanocarrier Dedicated to Controlled Clozapine Delivery at the Dopamine D_2-Serotonin 5-HT_{1A} Heteromers
Reprinted from: *Polymers* **2021**, *13*, 1000, doi:10.3390/polym13071000 63

Kusha Sharma, Ze'ev Porat and Aharon Gedanken
Designing Natural Polymer-Based Capsules and Spheres for Biomedical Applications—A Review
Reprinted from: *Polymers* **2021**, *13*, 4307, doi:10.3390/polym13244307 79

Miriam Khodeir, He Jia, Alexandru Vlad and Jean-François Gohy
Application of Redox-Responsive Hydrogels Based on 2,2,6,6-Tetramethyl-1-Piperidinyloxy Methacrylate and Oligo(Ethyleneglycol) Methacrylate in Controlled Release and Catalysis
Reprinted from: *Polymers* **2021**, *13*, 1307, doi:10.3390/polym13081307 121

Michał Rudko, Tomasz Urbaniak and Witold Musiał
Recent Developments in Ion-Sensitive Systems for Pharmaceutical Applications
Reprinted from: *Polymers* **2021**, *13*, 1641, doi:10.3390/polym13101641 131

Virginija Skurkyte-Papieviene, Ausra Abraitiene, Audrone Sankauskaite, Vitalija Rubeziene and Julija Baltusnikaite-Guzaitiene
Enhancement of the Thermal Performance of the Paraffin-Based Microcapsules Intended for Textile Applications
Reprinted from: *Polymers* **2021**, *13*, 1120, doi:10.3390/polym13071120 147

Valentina Sabatini, Laura Pellicano, Hermes Farina, Eleonora Pargoletti, Luisa Annunziata and Marco A. Ortenzi et al.
Design of New Polyacrylate Microcapsules to Modify the Water-Soluble Active Substances Release
Reprinted from: *Polymers* **2021**, *13*, 809, doi:10.3390/polym13050809 . . . . **163**

Gabriel Morand, Pascale Chevallier, Cédric Guyon, Michael Tatoulian and Diego Mantovani
In-Situ One-Step Direct Loading of Agents in Poly(acrylic acid) Coating Deposited by Aerosol-Assisted Open-Air Plasma
Reprinted from: *Polymers* **2021**, *13*, 1931, doi:10.3390/polym13121931 . . . . **177**

Xiang Lai, Xuan Zhang, Shukai Li, Jie Zhang, Weifeng Lin and Longgang Wang
Polyethyleneimine-Oleic Acid Micelles-Stabilized Palladium Nanoparticles as Highly Efficient Catalyst to Treat Pollutants with Enhanced Performance
Reprinted from: *Polymers* **2021**, *13*, 1890, doi:10.3390/polym13111890 . . . . **189**

About the Editors

M. Ali Aboudzadeh

Dr. M. Ali Aboudzadeh is currently a Marie Curie research fellow in IPREM institute, which is a joint research unit attached to the CNRS and the University of Pau and Pays de l'Adour, France. He received his B.Sc. (2003) and M.Sc. (2006) in Polymer Engineering from Amir Kabir University of Technology (AUT) and Iran Polymer & Petrochemical Institute (IPPI), respectively. He obtained his Ph.D. (in 2015) in Applied Chemistry and Polymer Materials from the University of the Basque Country (UPV/EHU), Spain and so far he has done several postdoctoral fellowships in different institutes. Dr. Aboudzadeh is an author/coauthor of more than 35 professional research articles and 2 book chapters. He is also the editor of a book entitled "Emulsion-based Encapsulation of Antioxidants: Design and Performance" published by Springer Nature. His research interests include polymer synthesis and characterization, polymer physics, supramolecular assemblies, rheology, DNA nanotechnology, and encapsulation via emulsion-based systems.

Shaghayegh Hamzehlou

Dr. Shaghayegh Hamzehlou obtained her B.Sc. and M.Sc. in polymer engineering at Amir Kabir University of Technology (Poly Technique of Tehran) on 2006. Later, she worked for four years as R&D Engineer in a manufacturing company, working on polymer based insulation materials for high voltage generators. In 2010, she moved to University of Basque country to do the Ph.D. in the framework of a European project, Marie Curie training network ITN NANOPOLY. In 2014, she joined Basque Center for Macromolecular Design and Engineering as a postdoctoral fellow involving in a European project RECOBA for three years. Currently, she is working as a researcher at University of Basque Country. Her research is focused on polymer reaction engineering, modelling, and simulation of kinetics, topology, microstructure, and morphology of the complex polymerization systems. She has (co)authored more than 30 scientific articles, 4 book chapters and had oral presentations in more than 20 national and international conferences and was a keynote lecturer at the Polymer Reaction Engineering X (PRE 10) May 20-25, 2018.

Preface to "Function of Polymers in Encapsulation Process"

Encapsulation technology comprises enclosing active agents (core materials) within a homogeneous/heterogeneous matrix (wall material) at the micro/nano scale. In the last few years encapsulation has gained a lot of interest. Using this process, a physical barrier is developed between the inner substance and the environment which, on one hand, prevents its degradation and facilitates its handling and transportation and, on the other hand, allows the controlled release of the core material in a certain ambiance. Polymers may be used to trap the material of interest inside the micro/nano-capsules. Such encapsulated systems have many applications in the fields of the food industry, drug delivery, agriculture, cosmetics, coatings, adhesives, and so forth. Various biopolymers, such as alginate, chitosan, carrageenan, gums, gelatin, whey protein, or starch, act as a barrier against external conditions. Encapsulation in biodegradable polymers can also enhance the permeability and stability of the active agent and, thus, its bioavailability. Choosing the right polymer is very important in this process due to its impact on target delivery and controlled release, and, therefore, on the bioavailability of active agents. It should have the necessary properties, such as being non-reactive with the active agent, flexibility, stability, strength, and impermeability. If the active agent has application in the food industry, the used polymer should be "generally recognized as safe" (GRAS), biodegradable, and capable of preserving the encapsulated material from the atmosphere.

There are a number of chemical, physical or mechanical processes available for encapsulation such as emulsion-solvent evaporation/extraction methods, coacervation-phase separation, spray drying, interfacial and in situ polymerization. The choice of a particular technique depends on the attributes of the polymer and the active agent. There are still many aspects to be developed in this field, which offer new challenges and breakthrough opportunities. The main objective of this interdisciplinary book is to bring together, at an international level, a high-quality collection of reviews and original research articles dealing with the importance of natural or synthetic polymers in encapsulation processes and their applications. A deep understanding and relevant theoretical calculations for exploring the functions of the materials (involved in the formulations) have also been obtained by fundamental investigations. We believe that the present book has explored the latest research on the function of polymers in encapsulation technology including fundamental theory and experiments together with reviews and articles. More efficient designs and preparation processes, as well as further understandings of the interfacial chemistry of encapsulated materials within the polymeric systems, are still needed.

The main aim of the book is to inspire and to guide scientists in this field. For the industrial establishments, the book also presents easy-to-achieve approaches that have been developed so far and could create a platform for industrial material production.

The Editors express their appreciation to all contributors from different parts of the world that have cooperated in the preparation of this book. In this context, this international book gives the active reader different perspectives on the subject and encourages him/her to read the entire book.

M. Ali Aboudzadeh and Shaghayegh Hamzehlou
Editors

polymers

MDPI

Article

Lignin-Stabilized Doxorubicin Microemulsions: Synthesis, Physical Characterization, and In Vitro Assessments

Abbas Rahdar [1,*], Saman Sargazi [2], Mahmood Barani [3], Sheida Shahraki [2], Fakhara Sabir [4] and M. Ali Aboudzadeh [5,6,*]

1 Department of Physics, University of Zabol, Zabol 98613-35856, Iran
2 Cellular and Molecular Research Center, Resistant Tuberculosis Institute, Zahedan University of Medical Sciences, Zahedan 98167-43463, Iran; sgz.biomed@gmail.com (S.S.); sheida.shahraki@gmail.com (S.S.)
3 Department of Chemistry, Shahid Bahonar University of Kerman, Kerman 76169-14111, Iran; mahmoodbarani7@gmail.com
4 Faculty of Pharmacy, Institute of Pharmaceutical Technology and Regulatory Affairs, University of Szeged, H-6720 Szeged, Hungary; fakhra.sabir@gmail.com
5 Centro de Física de Materiales, CSIC-UPV/EHU, Paseo Manuel Lardizábal 5, 20018 Donostia-San Sebastián, Spain
6 Donostia International Physics Center (DIPC), Paseo Manuel Lardizábal 4, 20018 Donostia-San Sebastián, Spain
* Correspondence: a.rahdar@uoz.ac.ir (A.R.); mohammadali.aboudzadeh@ehu.eus (M.A.A.)

Abstract: Encapsulation of the chemotherapy agents within colloidal systems usually improves drug efficiency and decreases its toxicity. In this study, lignin (LGN) (the second most abundant biopolymer next to cellulose on earth) was employed to prepare novel doxorubicin (DOX)-loaded oil-in-water (O/W) microemulsions with the aim of enhancing the bioavailability of DOX. The droplet size of DOX-loaded microemulsion was obtained as $\approx$ 7.5 nm by dynamic light scattering (DLS) analysis. The entrapment efficiency (EE) % of LGN/DOX microemulsions was calculated to be about 82%. In addition, a slow and sustainable release rate of DOX (68%) was observed after 24 h for these microemulsions. The cytotoxic effects of standard DOX and LGN/DOX microemulsions on non-malignant (HUVEC) and malignant (MCF7 and C152) cell lines were assessed by application of a tetrazolium (MTT) colorimetric assay. Disruption of cell membrane integrity was investigated by measuring intracellular lactate dehydrogenase (LDH) leakage. *In vitro* experiments showed that LGN/DOX microemulsions induced noticeable morphological alterations and a greater cell-killing effect than standard DOX. Moreover, LGN/DOX microemulsions significantly disrupted the membrane integrity of C152 cells. These results demonstrate that encapsulation and slow release of DOX improved the cytotoxic efficacy of this anthracycline agent against cancer cells but did not improve its safety towards normal human cells. Overall, this study provides a scientific basis for future studies on the encapsulation efficiency of microemulsions as a promising drug carrier for overcoming pharmacokinetic limitations.

Keywords: microemulsion; doxorubicin; *in vitro*; cytotoxicity; lignin (LGN)

Citation: Rahdar, A.; Sargazi, S.; Barani, M.; Shahraki, S.; Sabir, F.; Aboudzadeh, M.A. Lignin-Stabilized Doxorubicin Microemulsions: Synthesis, Physical Characterization, and In Vitro Assessments. *Polymers* **2021**, *13*, 641. https://doi.org/10.3390/polym13040641

Academic Editor: Iolanda De Marco

Received: 3 February 2021
Accepted: 18 February 2021
Published: 21 February 2021

Publisher's Note: MDPI stays neutral with regard to jurisdictional claims in published maps and institutional affiliations.

1. Introduction

Doxorubicin (DOX) is an anthracycline and active anticancer drug that contains a naphthacenequinone nucleus with a glycosidic bond to an amino sugar [1]. DOX is structurally similar to daunorubicin except that the former contains a hydroxyl acetyl group instead of an acetyl group at the 8-position. DOX is slightly soluble in normal saline and sparingly soluble in alcohol, and has broad-spectrum activity against neoplasms, lymphomas, solid tumors, and breast tumors [2]. It has been widely used as a first-line therapy in testicular, breast, and hepatocellular carcinoma. DOX has different biomedical applications in various chemotherapeutic regimens. It was previously applied in combination with bleomycin, dacarbazine, cyclophosphamide vincristine, and prednisone for the treatment of non-Hodgkin's and Hodgkin's lymphomas. Another application of DOX and cyclophosphamide can be used as adjuvant therapy with or without including fluorouracil followed by paclitaxel for breast cancer. The combination of DOX with a greater dose of cisplatin and methotrexate has been successfully applied to treat osteogenic sarcoma [3,4]. DOX attachment to DNA via intercalation could inhibit the function of topoisomerase II, resulting in disruption of DNA and RNA. The quinone group of DOX is reduced by cytochrome P450 reductase to produce semiquinone oxygen free radicals that can attack the cells' DNA.

Moreover, DOX binds to cell membranes and modifies the fluidity of ion transport [5,6]. The cell membranes are permeable to the lipid-soluble anthracycline molecules with an unprotonated sugar amino group like DOX with a pKa value of about 8.2. These compounds have direct access to the intracellular sites in all cells, including tumor cells.

The limitations in the application of conventional DOX are its bone marrow suppression, nephrotoxicity, and cardiotoxicity. The toxicity issues related to DOX limit its long-term use for clinical purposes [7,8]. P-glycoprotein (P-gp) also shows multidrug-resistance-associated protein-1 (MRP1)-mediated efflux that makes the tumor less responsive towards DOX. The other challenges are its short half-life, poor solubility, lack of oral dosage formulations, instability of the drugs under gastric conditions, and hepatic first-pass effect [9,10]. Several approaches have been used to reduce the toxicity and enhance the oral efficacy of DOX. Some of them employ prolonged infusion along with simultaneous administration with other cardioprotective agents (dexrazoxane) [11]. Furthermore, anthracycline analogs and other novel drug delivery systems can be applied to modify its distribution and reduce its accumulation in the heart and lower its toxicity. DOX can be incorporated within different carrier systems, for example, DOX-loaded liposomes for efficient targeting against tumors [6]. Furthermore, encapsulation of DOX in dendrimers, nanocrystals, nanogels, nanotubes, and nanoemulsions has been studied [12].

Microemulsion, as a potential drug delivery system, allows controlled or sustained release of drugs for oral, topical, ocular, and percutaneous administration. In comparison to other dosage forms, microemulsions offer the advantages of easy formation, high scale-up, greater stability, the higher drug solubilization of hydrophobic drugs, and enhanced bioavailability [13,14]. A microemulsion system composed of oil, surfactant (non-ionic or anionic), co-surfactants, and water can form a large number of configurations and phases via mixing different proportions of components to design the formulation [14,15]. Non-ionic surfactants are preferable because of their high tolerability, lower irritation and toxicity, and they have the potential to enhance the biocompatibility of the colloidal system. Microemulsions can be developed by applying single or double chain surfactants. Single chains do not lower the interfacial tension, and that is why co-surfactants are needed. The co-surfactants may exhibit toxicity in the formulation of microemulsions [16]. In this context, the selection of surfactant and co-surfactant is of great significance. Phospholipids-based microemulsions are preferred over other synthetic surfactants from a toxicity point of view [17]. Green surfactants (GSs) provide benefits over other synthetic surfactants in many drug delivery methods [18]. The most relevant properties of GSs are biodegradability, economic production, environmental tolerance, specificity, and structural diversity [4,19,20]. Despite the many advantages of GSs such as the commercialization and economic availability in industries, they are not productive enough due to the high operational and material costs needed for their synthesis. Hence, the use of nanotechnological methods to increase the production of GSs via microbial induction has been evaluated. A lot of research studies have been carried out on decreasing the cost sources for manufacturing GSs [21].

Lignin (LGN) (a phenolic polymer derived from phenylpropanoid units,) is the type of GS that we used in this study. This biodegradable natural polymer, which is the result of industrial wood processing, has widespread implementations as a renewable bioresource for producing different end-products including detergents, surfactants, and dispersants [22]. LGNs are developed from different methods, including chemical modifications using alkylation, destruction, sulfonation, etc. [23]. LGN is also a potential material for biomedical applications because of its eco-friendly properties such as biodegradability, biocompatibility, and low toxicity [24,25]; due to its specific aromatic structure, it can be easily adsorbed at various interfaces. That is why among the applications proposed for lignin particles, interfacial stabilization is highly promising, especially for the encapsulation of sensitive drugs such as DOX. For example, Zhou et al. developed LGN-based hollow nanoparticles (NPs) to deliver DOX. The results showed that the folic magnetic functionalized LGN hollow NPs could enhance the cellular uptake of NPs into HeLa cells [26]. In another study, the same research group applied these LGN hollow NPs as useful vehicles for the antineoplastic drug DOX. The authors related the enhanced encapsulation efficiency and drug loading of the NPs to the surface area and pore volume. In this study, the focus was on the mechanism of encapsulation of DOX into LGN hollow NP. The results of loading and release behavior of DOX were due to the various structures, stabilities, and sizes of LGN hollow NPs. The controlled release behavior of LGN hollow NPs, the access of DOX-loaded LGN hollow NPs into cells, and the low pH environment of lysosomes can support the release of DOX [27]. Whereas these pioneering reports made use of complex nanocarriers such as hollow NPs, here, we report the facile encapsulation of DOX into microemulsion systems which are more stable, easy-to-formulate, along with the greater potential for industrial scale-up. This novel strategy formulates a novel green carrier (DOX-loaded microemulsion) that has not been implemented before for these types of cancer cell lines (MCF7 and C152). In addition, in this formulation, we used a low amount of surfactant for the preparation of microemulsions, which can overcome the challenges and issues related to the toxicity of synthetic surfactants. These findings may be further sculpted into new cancer treatment strategies if this newly synthesized formulation induces desirable cytotoxic activity.

2. Materials and Methods

2.1. Chemicals

Standard laboratory-grade chemicals, including DOX, lignin (alkali) with CAS 8068-05-1, sodium caprylate, and ethyl butyrate, were provided by Sigma Chemical Co (St. Louis, MI, USA). Fetal bovine serum (FBS) and culture media (RPMI1640 and Dulbecco's modified Eagle's medium (DMEM)) were all obtained from GIBCO (Grand Island, NY, USA). Dimethyl sulfoxide (DMSO), 1% penicillin/streptomycin solution, and phosphate-buffered saline (PBS) were purchased from KalaZist company (KalaZist Co., Tehran, Iran). Amphotericin B and 3-(4,5-Dimethylthiazol-2-yl)-2,5 diphenyltetrazolium bromide (MTT) were procured from Sigma-Aldrich Co. (St. Louis, MI, USA). All plastic materials were provided by Sorfa Medical Plastic Co. (Hangzhou, China).

2.2. Cell Lines

Human oral squamous carcinoma (C152) and human umbilical vein endothelial (HUVEC) cell lines were obtained from the cell bank of Pasteur Institute of Iran (Tehran, Iran) and cultured in DMEM medium. Michigan Cancer Foundation-7 (MCF7) human breast cancer cells were obtained from the cell repository of the Research Institute of Biotechnology, Ferdowsi University of Mashhad (Mashhad, Iran) and maintained in RPMI1640 medium. MCF7 and C152 were selected as appropriate *in vitro* models for solid tumors. Both cell lines are well-suited for *in vitro* cytotoxic assessments and morphology evaluations since they can be cultured easily. HUVEC was chosen as a widely studied non-cancerous cell line for further assessments. Culture media were supplemented with 250 μg/mL of amphotericin B, 50 U/mL of penicillin, 50 μg/mL of streptomycin, and 10% heat deactivated FBS. Cells were maintained under standard culture conditions, described in our previous work [28].

2.3. Formulation of DOX-Loaded Microemulsions

DOX-loaded O/W microemulsions were prepared according to a reported procedure [29], as 1% *w/w* solutions of oil by dissolving appropriate quantities of LGN, fatty acid (sodium caprylate), and finally, ethyl butyrate in PBS (pH 7.4) via vigorous stirring at an oil-to-LGN molar ratio of 1. All the preparation steps were carried out in a room-temperature environment. The microemulsions were subsequently allowed to equilibrate for at least 1 day prior to use. A schematic depiction of the contents of the microemulsions prepared in this work is shown in Scheme 1.

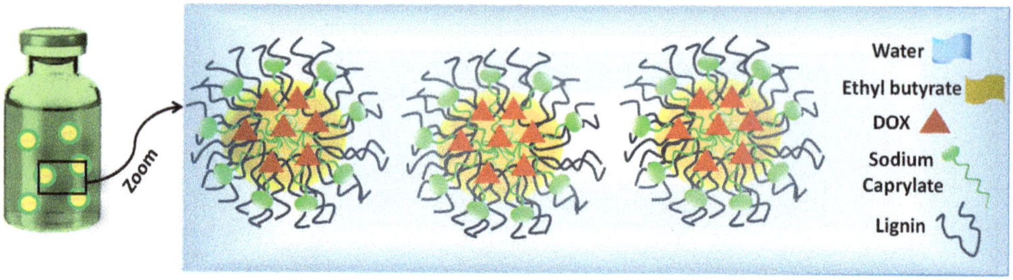

Scheme 1. Schematic representation of DOX-loaded O/W microemulsion and its contents prepared in this study.

2.4. Characterization of DOX-Loaded Microemulsions by DLS

DLS characterization of DOX-loaded LNG microemulsions was carried out using an ALV-5000F Goniometer System (Sartorius, Germany) coupled with a diode-pumped solid-state laser. These measurements were performed at an angle of θ = 90° to the incident ray through calibrating the intensity scale by toluene against scattering. The system was also integrated with a digital correlator (ALV SP-86, Sartorius, Germany) with a sample range of 25 ns to 100 ms. A description of DLS and the mechanism of data analysis via this technique are presented in the Supporting Information. The polydispersity index was calculated by the average size of the droplets divided by the average number of measured droplets.

2.5. Entrapment Efficiency of DOX

To study the entrapment efficiency (EE), DOX was loaded into LGN-stabilized microemulsions, and the percentage of EE was determined by a UV spectrophotometric approach (Agilent Technologies, Cary 50, Pittsburgh, USA) [30–32]. Stock solutions of DOX with a concentration of 2 to 250 μg/mL were prepared with ethanol/PBS 7.4 (1:1). The absorbance peak of DOX was recorded in the range 250 to 750 nm, where DOX showed a characteristic peak at a wavelength of 480 nm. Based on this characteristic peak, the calibration curve of standard DOX was achieved with an R^2 of 0.9894. LGN/DOX were centrifuged at 20,000 rpm for 60 min (model MC-20,000, Medline, UK). The centrifuging procedure was continued until a clear supernatant was obtained. The absorbance of the supernatant was measured at 480, and the EE of the DOX in the LGN/DOX was determined using Equation (1).

$$\text{Entrapment effeciency\%} = \frac{(\text{Total Dox} - \text{Free Dox})}{\text{Total Dox}} \times 100 \quad (1)$$

2.6. Release Study

The release rate was evaluated using the dialysis technique with a 6000 Dalton pore size dialysis membrane [30,31,33]. First, the dialysis bags were immersed in the PBS buffer for 24 h (24 h). Then, 1 mL of standard DOX solution or LGN/DOX microemulsion was placed in the dialysis bag and immersed in 50 mL PBS 7.4: ethanol (in 1:1 ratio as the buffer solution). The dialysis bag was left to stir at 90 rpm at 37 °C using a heater stirrer. At different time intervals up to 24 h, 1 mL of the buffer solution was removed and then replaced by adding the same volume of fresh buffer (preheated at 37 °C). The UV spectrophotometer recorded the absorbance of the samples at 480 nm to quantify released DOX.

To investigate the nature of the release mechanism, based on our previous studies, we have fitted the *in vitro* DOX release with different kinetic methods [30–32,34].

2.7. Cell Viability Assay and Evaluation of Cell Morphology

The target cells used in this experiment were C152 and MCF7, the two well-studied cell lines for analyzing the *in vitro* growth-inhibitory effects of anticancer agents. Endothelial HUVEC cells were used as target non-malignant cells. The cytotoxic effects of standard DOX and LGN/DOX microemulsions were evaluated using the MTT reduction assay [35].

Cells (5×10^3 cell/well) were seeded in 96-well microplates and incubated for 24 h to obtain a monolayer culture. Next, cells were treated with increasing DOX concentrations in standard or encapsulated state from 0.0625 to 1 μg/mL, and incubated for 48 h. Then, the supernatant was removed from each well, and 200 μL MTT reagent (5 mg/mL) was placed into each well and kept for another 3 h in a humidified incubator. The MTT solution was then replaced with 200 μL of DMSO, and the plates were placed on a shaker to dissolve the formazan crystals completely. The optical density (OD) of each well was measured at a test wavelength of 570 nm using a microplate reader (Stat Fax 2100, Awareness, Technologies Inc, Palm City, FL, USA). The percentage of viable cells was calculated using the following formula:

$$\text{Cell viability (\%)} = \text{OD sample}/\text{OD control} \times 100 \quad (2)$$

The half-maximal inhibitory concentration (IC50) of LGN/DOX microemulsions and standard DOX was calculated via GraphPad Prism software version 7.0.

For morphology assessment, cells were plated in 6-well microplates at a density of 1×10^5 cell/well. Following overnight incubation, cells were treated with given MTT concentrations for 48h, and untreated cells served as controls. Changes in cell morphology were monitored by an inverted phase-contrast microscope (IX71, Olympus Inc., Japan) at 20× magnification and imaged using a digital camera.

2.8. Membrane Integrity

Lactic dehydrogenase (LDH) leakage was measured in the medium of cultivated cells using a cytotoxicity assay kit (Cayman item no 601170) (Cayman Chemical Company, Ann Arbor, MI, USA) to evaluate membrane integrity. Cells were inserted in 96-well microplates (5×10^3 cell/well) and kept in an incubator for 24 h. Next, the culture medium was removed, and cells were exposed to a concentration equal to IC50 concentrations of both agents (standard DOX and LGN/DOX microemulsions). After treatment for 48 h, 100 μL of supernatant was placed into each well of a 96-well microplate for assessment. The LDH leakage (% of total) as a marker of cytotoxicity was determined as percentage of LDH in medium compared with total LDH in cell lysate for each microwell using the following formula:

$$\text{LDH leakage (\%)} = (OD_{test} - OD_{spontaneous\ release})/(OD_{maximum\ release} - OD_{spontaneous\ release})\times100 \quad (3)$$

where OD_{test} is the absorbance of cells treated with our microemulsion, $OD_{spontaneous\ release}$ is the absorbance of microwells containing assay buffer, and $OD_{maximum\ release}$ represents the absorbance of microwells containing cells.

Absorbance was read at 490 nm via a microplate reader (Spectra Max Gemini ®, Molecular Devices Cooperation, Sunnyvale, CA, USA).

2.9. Statistical Analysis

Results were analyzed using SPSS software version 20 (SPSS Inc., Chicago, IL, USA). Differences among treated and untreated cells were determined using the sample T-test and non-parametric one-way analysis of variance (ANOVA) test. $p < 0.05$ was considered statistically significant.

3. Results

3.1. Characterization of DOX-Loaded Microemulsions by DLS

Figure S1 (in supplementary materials) shows an autocorrelation function (ACF) against time for the prepared microemulsions. Fitting a curve to the ACF yielded a decay rate from which the droplet size of DOX-based microemulsions was measured to be approximately 7.4 nm based on the diffusion data (3.1938×10^{-11} m^2/s) [30–32]. For encapsulating drugs, without any doubt, a small particle would be an advantage since it would carry only a small drug quantity, which will help in minimizing the onset of abrupt toxic response [36]. The PDI obtained from DLS analysis showed values (0.1–0.2), indicating a good size distribution of the oil droplet in the microemulsion system. This parameter directly reflects the size homogeneity of the droplets in the total microemulsion. The stability of the microemulsion was confirmed visually after passing three months of preparation and no aggregation was observed (Figure S1b).

3.2. Entrapment Efficiency

Generally, an ideal drug carrier should have high encapsulation efficiency (EE) [33,37–39]. High EE (above 70%) can increase the efficacy of the drug delivery system and decrease the side effects of the drug [33,39]. The EE% of the LGN/DOX microemulsion prepared in this study was calculated to be about 82%. Interactions between DOX and LGN microemulsions can lead to this high EE. LGN molecules have a similar polyphenolic structure to DOX, thus hydrogen bonding and π–π stacking enhance this interaction between them as well. The microemulsion core is hydrophobic, and it could be concluded that DOX was mainly entrapped in the core of the microemulsion. Moreover, the EE of our LGN/DOX microemulsion system is well above the EE of LGN-based NPs that have been designed for encapsulating different compounds [40]. For example, the EE of the pesticide avermectin (AVM) in hollow lignin azo colloidal spheres was determined to be 61.49% (*w/w*) [41]. In another similar study, Li et al. synthesized lignosulfonate-based colloidal spheres from a mixture of sodium lignosulfonate and cetyltrimethylammonium bromide through self-assembly for the encapsulation of AVM, and the EE of AVM reached up to 62.58% [42]. In contrast, when (similar to our study) DOX was used as the active compound, its encapsulation efficiency in LGN-based hollow nanoparticles (NPs) was obtained as 67.5% [26].

3.3. In Vitro Release Experiment

One of the crucial characteristics of drug delivery systems in biomedicine is to impart the sustained release of a drug. Thus, the sustained drug release improves the accumulation of DOX at the tumor site while enhancing its anticancer efficiency. The *in vitro* release of DOX solution (standard DOX) and LGN/DOX microemulsions was performed using dialysis methods at PBS 7.4/ethanol (1:1) and 37 °C up to 24 h. As shown in Figure 1, the release rate of standard DOX was significantly faster than LGN/DOX microemulsions. The DOX release reached 68% after 24 h for the LGN/DOX microemulsion that showed a slow and sustainable release rate.

In a similar study, Heggannavar et al. prepared DOX-loaded magnetic silica-pluronic F-127 nanocarriers and conjugated them with transferrin (Tf) to treat glioblastoma. The release of DOX in these thermosensitive NPs was predominant during the first 24 h at 37 °C and reached around 17.2%. However, when the temperature was raised to 42 °C, DOX release was significantly increased from 17.2 to 32.2% for 24 h and reached up to 51.4% for 120 h [43].

After matching the first-order, zero-order, Higuchi, and Korsmeyer–Peppas models with the DOX release profile, it was found that the kinetics of release followed the Higuchi model with an R^2 of 0.965 (Figure S2 in supplementary materials). This describes that DOX release as a diffusion process is based on Fick's law, which is square root time-dependent [30].

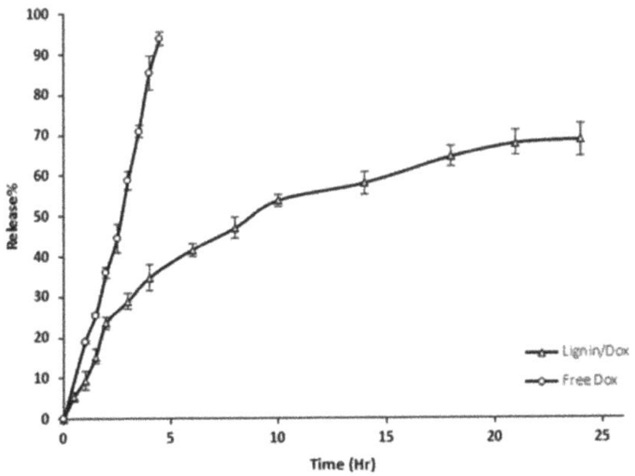

Figure 1. *In vitro* release behavior of free DOX and LGN/DOX microemulsions at PBS 7.4/ethanol (1:1) and 37 °C after 24 h. The error bars indicate the uncertainty.

The effect of different parameters on the release of active ingredients from nanocarriers has been evaluated in several studies [37–39]. Thus, the release of DOX from the NPs was found to follow an anomalous type of behavior. For example, Heggannavar et al. investigated the nature of the DOX release mechanism and obtained values of n (a parameter that represents the nature of the release mechanism) varying from 0.6175 to 0.6589 [43]. Figure S2 shows profiles of different kinetic models for release of DOX from LGN/DOX microemulsions.

3.4. *Effects of* Standard and Encapsulated DOX *on Cells Viability and Morphology*

Compared with untreated cells, incubation with different concentrations of standard DOX significantly inhibited the proliferation of MCF7, C152, and HUVEC cells in a dose-dependent fashion (Figure 2). Likewise, exposing these cells to gradually increasing concentrations of LGN/DOX microemulsions exhibited a dose-dependent inhibitory effect and a significant reduction in the number of MCF7, C152, and HUVEC cells compared with untreated cells ($p < 0.05$), presenting cell viability that ranged from 6.14 to 76.23% (for MCF7 cells), 6.16 to 60.95% (for C152 cells), and 9.82 to 75.78% (for HUVEC cells) (Figure 2). IC50 values for 48 h treatment of MCF7, C152, and HUVEC cells with standard DOX were 0.798, 0.244, and 0.716 µg/mL, respectively. At the same time, the IC50 concentrations of LGN/DOX microemulsions were 0.197 µg/mL (for HUVEC cells), 0.094 µg/mL (for C152 cells), and 0.208 µg/mL (for MCF7 cells). Between the studied cell lines, C152 cells were most susceptible to encapsulated DOX. Our MTT assay results showed that encapsulated DOX induced higher toxicity than standard DOX on normal HUVEC cells, indicated by a lower IC50.

All three cell lines were characterized by altered morphology and proliferation rate. MCF7 cells grew faster, and the cells were larger and more branched than the other two cell lines. Among the studied cell lines, the HUVEC cells were relatively smaller and were more compact. Gradual increasing concentrations of standard DOX could moderately decrease the viability of C152 (Figure 3), MCF7 (Figure 4), and HUVEC (Figure 5) cells without causing noticeable morphologic changes. Contrastingly, treatment with LGN/DOX microemulsions caused evident concentration-dependent morphological changes, specifically in C152 cells (Figure 3). In this regard, treatment with 0.0625 µg/mL nano-DOX decreased the number of living C152 cells, while higher concentrations of nano-DOX caused progressive nuclear shrinkage and formation of apoptotic bodies. Upon exposing MCF7 cells to nano-DOX at a concentration of 0.5 µg/mL, cells were shrunk, and apoptotic bodies were similarly formed (Figure 3). The highest concentration of LGN/DOX microemulsions (1 µg/mL) led to the adhesion of MCF7, C152, and HUVEC cells to the plate, which was not observed when cells were treated with standard DOX at given concentrations (Figure 3, Figure 4, and Figure 5).

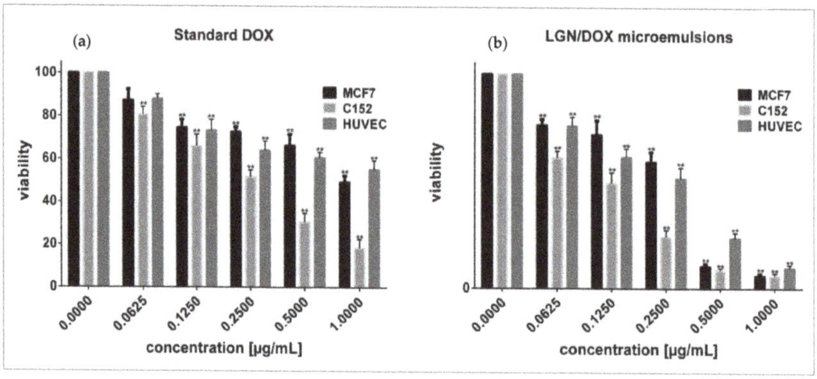

Figure 2. Cytotoxicity assessment of (**a**) standard DOX and (**b**) LGN/DOX microemulsions on malignant (MCF7, C152) and non-malignant (HUVEC) cells following 48 h treatment. (** $p < 0.05$ compared with untreated cells).

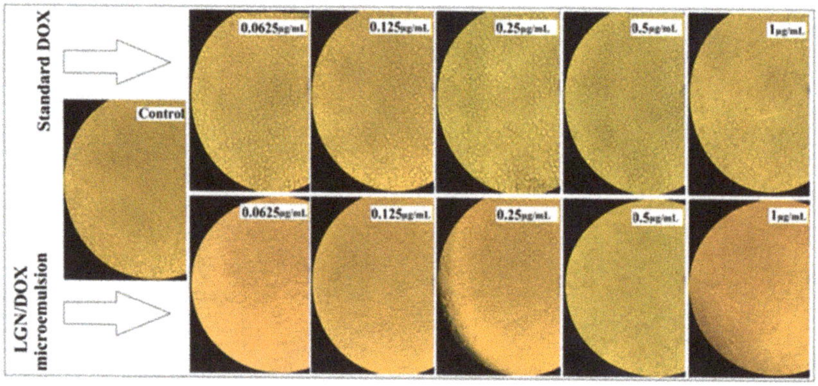

Figure 3. Optical microscopy images of malignant C152 cells treated with standard DOX and LGN/DOX microemulsions for 48 h.

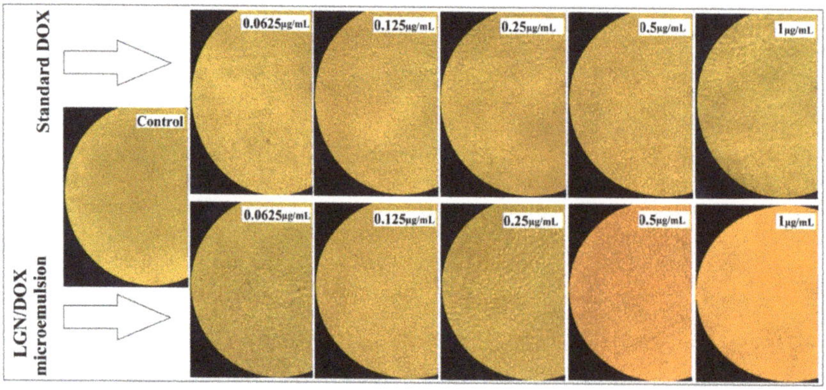

Figure 4. Optical microscopy images of malignant MCF7 cells treated with standard DOX and LGN/DOX microemulsions for 48 h.

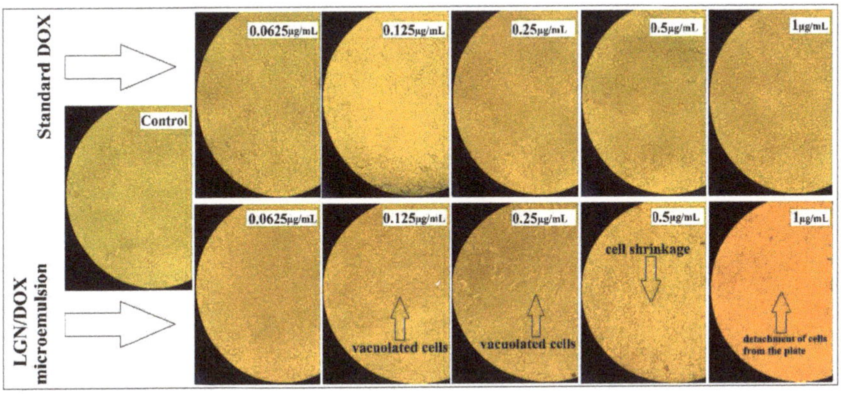

Figure 5. Optical microscopy images of non-malignant HUVEC cells treated with standard DOX and LGN/DOX microemulsions for 48 h.

The clinical success of nanomedicine relies on the synthesis of stable nanocarriers that can expeditiously encapsulate chemotherapeutics and quickly release them into target malignant cells. Anthracyclines entrapped in small-sized nanocarriers have the advantage of reduced systematic toxicity and gradual release of the payload [44]. In the current study, we observed that LGN/DOX microemulsions showed higher cytotoxic activity compared to standard DOX, indicating that the different level of cytotoxic activity is due to the encapsulation of DOX. It has been demonstrated that the encapsulation process boosts the therapeutic efficacy and the intracellular uptake of DOX [45]. The incorporation of DOX into long-circulating liposomes resulted in increased DOX accumulation in tumor tissue associated with higher toxicity than unencapsulated DOX [46].

Lately, technological advancements have been made in controlling the release of anticancer drugs and enhancing their solubility and stability by applying innovative drug delivery systems, such as microemulsions [47]. Multiple surfactant-based delivery systems, such as microemulsions, have recently been designed for efficient localization or systematic delivery of antitumor drugs to tumor cells [48]. Shakeel and coworkers showed that 5-fluorouracil (5-FU) in an optimized nanoemulsion inhibits the proliferation of SK-MEL-5 tumor cells more effectively than free 5-FU [49]. Chen et al. showed how Pluronic-based functional polymer mixed nanomicelles could serve as a potential nanodrug delivery system for efficient co-delivery of DOX and paclitaxel, two conventional chemotherapeutic agents [50]. In our study, the enhanced antitumor activity of LGN/DOX microemulsions might be due to the prolonged and controlled release of DOX by using this new formulation.

Similarly, few studies reported the desirable antitumor efficacy of DOX-loaded nanocarriers. In 2019, Abbasian et al. investigated the growth-inhibitory effects of newly designed LA/chitosan/NaX/Fe_3O_4/DOX nanofibers. They observed that these nanofibers decreased the number of viable H1355 cancer cells by 82% following a seven-day treatment [51]. Hence, as a beneficial carrier for hydrophobic antitumor agents, LGN-based microemulsions can be applied to treat solid tumors.

On the other hand, alkylating agents may cause adverse effects on normal cells; therefore, their application for chemotherapy is limited [52]. Compared with standard DOX treatment, we noticed a lower IC50 concentration for nano-DOX treated cells derived from human umbilical vein endothelial cells. This might not be a desirable effect for our newly developed encapsulated DOX since these microemulsions were cytotoxic at high concentrations toward normal human cells. More studies are needed to fully elucidate the cytotoxic activity of LGN-based microemulsions on normal cells derived from different human tissues.

3.5. LDH-Based Cytotoxicity Assay

If the cell membrane is disrupted, intracellular LDH is released from the cytosol into the culture medium. Therefore, the percentage of LDH leakage is an indicator of cell membrane damage as a result of necrosis or late-stage apoptosis. In comparison to control cells, exposure to standard DOX and LGN/DOX microemulsions enhanced LDH leakage by 1.48- and 1.71- (for MCF7 cells), 1.32- and 2.31- (for C152 cells), and 1.24- and 1.28- (for HUVEC cells) fold, respectively (Figure 6). Interestingly, compared to standard DOX, a significant increase in LDH leakage was observed following treatment of C152 cells with LGN/DOX microemulsions ($p < 0.05$). This shows that encapsulated DOX induced greater damage on cell membrane and might trigger necrosis as an alternative cell death mechanism. Compared to standard DOX, LGN/DOX microemulsions did not augment the leakage of cytosolic LDH in HUVEC and MCF7 cells ($p > 0.05$).

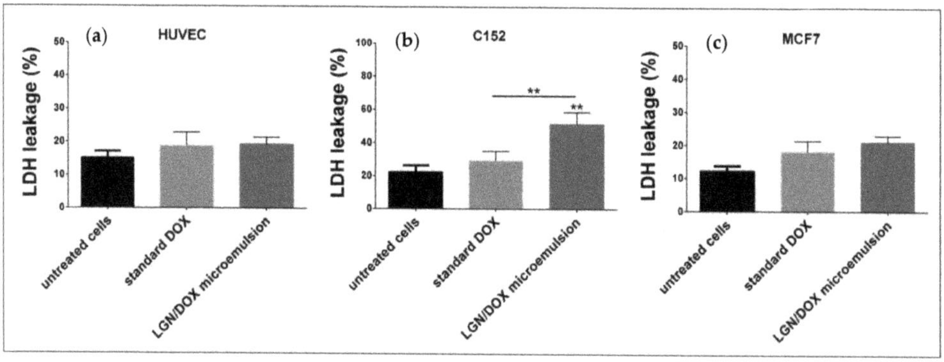

Figure 6. Effect of standard DOX and LGN/DOX microemulsions on the membrane integrity of (**a**) HUVEC, (**b**) C152, (**c**) MCF7, and cells after 48 h of exposure. ** $p < 0.05$.

Former studies demonstrated that apoptotic cell death could be the principal mechanism of DOX-induced cell death [53]. Recently, it has been shown that Bcl-2-like 19kDa-interacting protein 3 [54] and poly-ADP-ribose polymerase 1 [55] could mediate DOX-induced necrosis as well. In the current study, cells were treated with standard DOX and LGN/DOX microemulsions, and some hallmarks of apoptotic cell death, including the formation of membrane-bound apoptotic bodies, cell shrinkage, and roundness, appeared as evident morphological alterations. On the other hand, loss of membrane integrity and LDH leakage are indications of necrosis [56], a type of cell-death that lacks features of apoptosis and is generally considered to be uncontrolled [57]. In our study, changes in LDH leakage were different for the investigated cell lines. In comparison to untreated cells, encapsulated DOX did not enhance LDH leakage in HUVEC cells, which is a promising outcome. In contrast, both agents did not significantly enhance LDH leakage in malignant MCF7 cells, recommending that apoptosis might be the primary mechanism of cell death in these cells. This is in agreement with the findings of Pilco-Ferreto et al., suggesting that DOX activating apoptosis in MCF7 cells is the main mechanism of cell death through inducing proteolytic processing of anti-apoptotic proteins, i.e., B cell lymphoma 2 (BCL2) family members [58]. Additionally, Sharifi and coworkers confirmed that DOX induces mitochondrial-dependent apoptosis by up-regulating *caspase-9* and proapoptotic protein (*Bax*), and down-regulating an anti-apoptotic protein (*Bcl-xL*) [59]. However, compared to untreated cells, we observed enhanced LDH leakage like treated C152 cells with encapsulated DOX. To the best of our knowledge, no study has already established the principal mechanism for DOX-induced cell death in C152 oral carcinoma cells. Another study also reported that DOX treatment causes necrotic cell death, mostly in cells that had floated from culture plates [60]. This suggests that two modes of cell death might be induced by DOX in different cell lines. Thus, changes in LDH leakage between MCF7 and C152 cells might be due to the activation of distinct cell death pathways within these cells. Furthermore, LGN/DOX microemulsions induced more toxicity in C152 cells compared with standard DOX. This might explain the significant enhancement in LDH leakage in C152 cells exposed to this formulation.

Overall, these findings suggest that LGN/DOX microemulsions might induce cell death through activation of both apoptosis and necrosis mechanisms. Further studies are needed to discover the underlying cell death mechanisms by which these microemulsions kill cancer cells.

3.6. Practical Applications and Future Research Perspectives

Novel methods for preparing drug-loaded microemulsion systems and further investigation with the aim of expanding their practical applications are worth exploration. However, in drug delivery technology, employing these systems is limited because of their drug loading capacity and the used level of excipients [61]. In this context, surfactants and co-surfactants can be toxic at high concentrations and may be limited in their uptake levels. In the present study, we aimed to maximize DOX loading capacity while using the minimum amount of surfactant for preparing LGN/DOX microemulsions. A substantial amount of work on physicochemical O/W formulations and *in vitro* assessments are required to be performed before they can live up to their potential as widely used drug carriers.

The diversity in lignin structure, chemical reactivity, and its safety profile can be further exploited in different applications of lignin-based materials. Advances in genetics, bio-, and analytical chemistry have resulted in a deeper understanding of lignin biosynthesis and the structure of nanoparticles in the field of nanomedicines.

4. Conclusions

Our results indicated that novel synthesized LGN/DOX microemulsions exert cytotoxic effects on oral and breast carcinoma cells while inducing unfavorable toxic effects on normal human cells. Performing more studies is encouraged to investigate the growth-inhibitory activity of this novel formulation on other malignant and non-malignant cell lines. Due to the toxic effects of LGN/DOX microemulsions on normal human cells, multiple considerations should be undertaken before using these microemulsions to treat cancer patients. The take-home message of the current research is that DOX encapsulation in O/W microemulsions enhanced its therapeutic efficacy against solid tumors. This could be further sculpted into new approaches for treating different types of cancers.

Supplementary Materials: The following are available online at https://www.mdpi.com/2073-4360/13/4/641/s1, Figure S1: Dynamic light scattering autocorrelation function of microemulsions Figure S2: Profiles of different kinetic models for release of DOX from LGN/DOX microemulsions.

Author Contributions: A.R., M.B., S.S. (Saman Sargazi), F.S. wrote the manuscript, performed the main analysis and data collection. A.R., M.B., S.S. (Sheida Shahraki) performed data interpretations. S.S. performed and interpreted the *in vitro* analysis, A.R. and M.A.A. supervised and performed data interpretations. A.R. performed the DLS data analysis. A.R and M.A.A. supervised all the tasks and finalized the manuscript. All authors have read and agreed to the published version of the manuscript.

Funding: This research received no external funding.

Institutional Review Board Statement: The study was conducted according to the guidelines of the Declaration of Helsinki and approved by the Institutional Review Board (or Ethics Committee) of Zahedan University of Medical Sciences (protocol code 10267).

Informed Consent Statement: Not applicable.

Data Availability Statement: The data presented in this study are available on request from the corresponding author.

Conflicts of Interest: The authors declare no conflict of interest.

References

1. Zhao, N.; Woodle, M.C.; Mixson, A.J. Advances in delivery systems for doxorubicin. *J. Nanomed. Nanotechnol.* **2018**, *9*, 519. [CrossRef]
2. Renu, K. Molecular mechanism of doxorubicin-induced cardiomyopathy–An update. *Eur. J. Pharmacol.* **2018**, *818*, 241–253. [CrossRef]
3. Shafei, A. A review on the efficacy and toxicity of different doxorubicin nanoparticles for targeted therapy in metastatic breast cancer. *Biomed. Pharmacother.* **2017**, *95*, 1209–1218. [CrossRef] [PubMed]
4. Fernandez, A.M. New green surfactants for emulsion polymerization. *Progr. Organ. Coat.* **2005**, *53*, 246–255. [CrossRef]
5. Trouet, A. Extracellularly tumor-activated prodrugs for the selective chemotherapy of cancer: Application to doxorubicin and preliminary *in vitro* and *in vivo* studies. *Cancer Res.* **2001**, *61*, 2843–2846. [PubMed]
6. Rahmani, S. Synthesis of mesoporous silica nanoparticles and nanorods: Application to doxorubicin delivery. *Solid State Sci.* **2017**, *68*, 25–31. [CrossRef]
7. Ghosh, S. Biomedical application of doxorubicin coated hydroxyapatite—poly (lactide-co-glycolide) nanocomposite for controlling osteosarcoma therapeutics. *J. Nanosci. Nanotechnol.* **2020**, *20*, 3994–4004. [CrossRef] [PubMed]
8. Iwamoto, T. Clinical application of drug delivery systems in cancer chemotherapy: Review of the efficacy and side effects of approved drugs. *Biol. Pharm. Bullet.* **2013**, *36*, 715–718. [CrossRef]
9. Pugazhendhi, A. Toxicity of Doxorubicin (Dox) to different experimental organ systems. *Life Sci.* **2018**, *200*, 26–30. [CrossRef] [PubMed]
10. Tan, M.L.; Choong, P.F.; Dass, C.R. Doxorubicin delivery systems based on chitosan for cancer therapy. *J. Pharm. Pharmacol.* **2009**, *61*, 131–142. [CrossRef]
11. Alavi, M.; Nokhodchi, A. Microformulations and Nanoformulations of Doxorubicin for Improvement of Its Therapeutic Efficiency. *Crit. Rev. Ther. Drug Carrier Syst.* **2020**, *37*. [CrossRef]
12. Formariz, T. Doxorubicin biocompatible O/W microemulsion stabilized by mixed surfactant containing soya phosphatidylcholine. *Colloids Surf. B Biointerf.* **2006**, *51*, 54–61. [CrossRef]
13. Sedyakina, N.E. Formulation, drug release features and *in vitro* cytotoxic evaluation of nonionic mixed surfactant stabilized water-in-oil microemulsion loaded with doxorubicin. *Mendel. Commun.* **2019**, *29*, 320–322. [CrossRef]
14. Aboudzadeh, M.A. Low-Energy Encapsulation of α-Tocopherol Using Fully Food Grade Oil-in-Water Microemulsions. *ACS Omega* **2018**, *3*, 10999–11008. [CrossRef]
15. Candido, C.D. Biocompatible microemulsion modifies the tissue distribution of doxorubicin. *J. Pharm. Sci.* **2014**, *103*, 3297–3301. [CrossRef] [PubMed]
16. Bera, A.; Mandal, A. Microemulsions: A novel approach to enhanced oil recovery: A review. *J. Petrol. Exp. Product. Technol.* **2015**, *5*, 255–268. [CrossRef]
17. Muzaffar, F.; Singh, U.; Chauhan, L. Review on microemulsion as futuristic drug delivery. *Int. J. Pharm. Pharm. Sci.* **2013**, *5*, 39–53.
18. Grampurohit, N.; Ravikumar, P.; Mallya, R. Microemulsions for topical use–a review. *Ind. J. Pharm. Ed. Res.* **2011**, *45*, 100–107.
19. Rebello, S. Surfactants: Toxicity, remediation and green surfactants. *Environ. Chem. Lett.* **2014**, *12*, 275–287. [CrossRef]
20. Wenan, L. Application and research advance of green surfactants. *J. Anhui Agric. Sci.* **2007**, *35*, 5691.
21. Kandasamy, R. New Trends in the Biomanufacturing of Green Surfactants: Biobased Surfactants and Biosurfactants, in Next Generation Biomanufacturing Technologies. *Am. Chem. Soc.* **2019**, *1329*, 243–260.
22. Saratale, R.G. A comprehensive overview and recent advances on polyhydroxyalkanoates (PHA) production using various organic waste streams. *Bioresour. Technol.* **2021**, *325*, 124685. [CrossRef] [PubMed]
23. Shulga, G.; Shakels, V.; Skudra, S.; Bogdanovs, V. Modified lignin as an environmentally friendly surfactant. In Proceedings of the 8th International Scientific and Practical Conference, Rēzekne, Latvia, 20–22 June 2011; pp. 276–281.
24. Figueiredo, P. In vitro evaluation of biodegradable lignin-based nanoparticles for drug delivery and enhanced antiproliferation effect in cancer cells. *Biomaterials* **2017**, *121*, 97. [CrossRef]
25. Li, H. Preparation of nanocapsules via the self-assembly of kraft lignin: A totally green process with renewable resources. *ACS Sustain. Chem. Eng.* **2016**, *4*, 1946. [CrossRef]
26. Zhou, Y. Preparation of targeted lignin–based hollow nanoparticles for the delivery of doxorubicin. *Nanomaterials* **2019**, *9*, 188. [CrossRef]
27. Zhou, Y. Effects of Lignin-Based Hollow Nanoparticle Structure on the Loading and Release Behavior of Doxorubicin. *Materials* **2019**, *12*, 1694. [CrossRef]
28. Sargazi, S. Hydro-alcoholic Extract of Achillea Wilhelmsii, C. Koch Reduces the Expression of Cell Death-Associated Genes while Inducing DNA Damage in HeLa Cervical Cancer Cells. *Ir. J. Med. Sci.* **2020**, *45*, 359.
29. Varshney, M. Pluronic microemulsions as nanoreservoirs for extraction of bupivacaine from normal saline. *J. Am. Chem. Soc.* **2004**, *126*, 5108–5112. [CrossRef]
30. Rahdar, A. Dynamic light scattering and zeta potential measurements: Effective techniques to characterize therapeutic nanoparticles. *J. Nanoanal.* **2019**. [CrossRef]
31. Rahdar, A. Effect of ion exchange in NaAOT surfactant on droplet size and location of dye within Rhodamine B (RhB)-containing microemulsion at low dye concentration. *J. Mol. Liq.* **2018**, *252*, 506–513. [CrossRef]

32. Rahdar, A. Deferasirox-loaded pluronic nanomicelles: Synthesis, characterization, *in vitro* and *in vivo* studies. *J. Mol. Liq.* **2020**, *323*, 114605. [CrossRef]
33. Barani, M. A new formulation of hydrophobin-coated niosome as a drug carrier to cancer cells. *Mater. Sci. Eng. C* **2020**, *113*, 110975. [CrossRef]
34. Hajizadeh, M.R. In vitro cytotoxicity assay of D-limonene niosomes: An efficient nanocarrier for enhancing solubility of plant-extracted agents. *Res. Pharm. Sci.* **2019**, *14*, 448.
35. Gerlier, D.; Thomasset, N. Use of MTT colorimetric assay to measure cell activation. *J. Immunol. Methods* **1986**, *94*, 57. [CrossRef]
36. Acosta, E. Bioavailability of nanoparticles in nutrient and nutraceutical delivery. *Curr. Opin. Colloid Interf. Sci.* **2009**, *14*, 3. [CrossRef]
37. Rahdar, A. Effect of tocopherol on the properties of Pluronic F127 microemulsions: Physico-chemical characterization and *in vivo* toxicity. *J. Mol. Liquids* **2019**, *277*, 624–630. [CrossRef]
38. Pandey, M. Hyaluronic acid-modified betamethasone encapsulated polymeric nanoparticles: Fabrication, characterisation, *in vitro* release kinetics, and dermal targeting. *Drug Deliv. Trans. Res.* **2019**, *9*, 520–533. [CrossRef]
39. Anwer, M.K. Preparation of sustained release apremilast-loaded PLGA nanoparticles: In vitro characterization and *in vivo* pharmacokinetic study in rats. *Int. J. Nanomed.* **2019**, *14*, 1587. [CrossRef] [PubMed]
40. Li, Y. pH-responsive lignin-based complex micelles: Preparation, characterization and application in oral drug delivery. *Chem. Eng. J.* **2017**, *327*, 1176. [CrossRef]
41. Deng, Y.H. Hollow lignin azo colloids encapsulated avermectin with high anti-photolysis and controlled release performance. *Ind. Crop. Prod.* **2016**, *87*, 191. [CrossRef]
42. Li, Y. Lignin-based microsphere: Preparation and performance on encapsulating the pesticide avermectin. *ACS Sustain. Chem. Eng.* **2017**, *5*, 3321. [CrossRef]
43. Heggannavar, G.B. Development of doxorubicin-loaded magnetic silica–pluronic F-127 nanocarriers conjugated with transferrin for treating glioblastoma across the blood–brain barrier using an *in vitro* model. *ACS Omega* **2018**, *3*, 8017. [CrossRef] [PubMed]
44. Siegal, T.; Horowitz, A.; Gabizon, A. Doxorubicin encapsulated in sterically stabilized liposomes for the treatment of a brain tumor model: Biodistribution and therapeutic efficacy. *J. Neurosurg.* **1995**, *83*, 1029–1037. [CrossRef]
45. Mohan, P.; Rapoport, N. Doxorubicin as a molecular nanotheranostic agent: Effect of doxorubicin encapsulation in micelles or nanoemulsions on the ultrasound-mediated intracellular delivery and nuclear trafficking. *Mol. Pharm.* **2010**, *7*, 1959–1973. [CrossRef]
46. Working, P.K.; Newman, M.S.; Huang, S.K.; Mayhew, E.; Vaage, J.; Lasic, D.D. Pharmacokinetics, biodistribution and therapeutic efficacy of doxorubicin encapsulated in Stealth®liposomes (Doxil®). *J. Lip. Res.* **1994**, *4*, 667–687. [CrossRef]
47. Saini, A.; Panesar, P.S.; Bera, M.B. Valorization of fruits and vegetables waste through green extraction of bioactive compounds and their nanoemulsions-based delivery system. *Bioresour. Bioproc.* **2019**, *6*, 26. [CrossRef]
48. Kaur, P.; Garg, T.; Rath, G.; Murthy, R.S.; Goyal, A.K. Surfactant-based drug delivery systems for treating drug-resistant lung cancer. *Drug Deliv.* **2016**, *23*, 717–728. [CrossRef]
49. Shakeel, F.; Haq, N.; Al-Dhfyan, A.; Alanazi, F.K.; Alsarra, I.A. Chemoprevention of skin cancer using low HLB surfactant nanoemulsion of 5-fluorouracil: A preliminary study. *Drug Deliv.* **2015**, *22*, 573–580. [CrossRef]
50. Chen, Y.; Zhang, W.; Huang, Y.; Gao, F.; Sha, X.; Fang, X. Pluronic-based functional polymeric mixed micelles for co-delivery of doxorubicin and paclitaxel to multidrug resistant tumor. *Int. J. Pharm.* **2015**, *488*, 44–58. [CrossRef] [PubMed]
51. Abasian, P.; Radmansouri, M.; Jouybari, M.H.; Ghasemi, M.V.; Mohammadi, A.; Irani, M.; Jazi, F.S. Incorporation of magnetic NaX zeolite/DOX into the PLA/chitosan nanofibers for sustained release of doxorubicin against carcinoma cells death *in vitro*. *Int. J. Biol. Macromol.* **2019**, *121*, 398–406. [CrossRef]
52. Hassett, M.J.; O'Malley, A.J.; Pakes, J.R.; Newhouse, J.P.; Earle, C.C. Frequency and cost of chemotherapy-related serious adverse effects in a population sample of women with breast cancer. *J. Nat. Cancer Inst.* **2006**, *98*, 1108–1117. [CrossRef]
53. Lee, T.; Lau, T.; Ng, I. Doxorubicin-induced apoptosis and chemosensitivity in hepatoma cell lines. *Cancer Chemother. Pharmacol.* **2002**, *49*, 78–86. [CrossRef]
54. Dhingra, R.; Margulets, V.; Chowdhury, S.R.; Thliveris, J.; Jassal, D.; Fernyhough, P.; Dorn, G.W.; Kirshenbaum, L.A. Bnip3 mediates doxorubicin-induced cardiac myocyte necrosis and mortality through changes in mitochondrial signaling. *Proc. Nat. Acad. Sci. USA* **2014**, *111*, E5537–E5544. [CrossRef]
55. Shin, H.J.; Kwon, H.K.; Lee, J.H.; Gui, X.; Achek, A.; Kim, J.H.; Choi, S. Doxorubicin-induced necrosis is mediated by poly-(ADP-ribose) polymerase 1 (PARP1) but is independent of p53. *Sci. Rep.* **2015**, *5*, 15798. [CrossRef]
56. Krysko, D.V.; Berghe, T.V.; Parthoens, E.; D'Herde, K.; Vandenabeele, P. Methods for distinguishing apoptotic from necrotic cells and measuring their clearance. *Methods Enzymol.* **2008**, *442*, 307–341. [PubMed]
57. Golstein, P.; Kroemer, G. Cell death by necrosis: Towards a molecular definition. *Trends Biochem. Sci.* **2007**, *32*, 37–43. [CrossRef]
58. Pilco-Ferreto, N.; Calaf, G.M. Influence of doxorubicin on apoptosis and oxidative stress in breast cancer cell lines. *Int. J. Oncol.* **2016**, *49*, 753–762. [CrossRef] [PubMed]
59. Sharifi, S.; Barar, J.; Hejazi, M.S.; Samadi, N. Doxorubicin changes Bax/Bcl-xL ratio, caspase-8 and 9 in breast cancer cells. *Adv. Pharm. Bullet.* **2015**, *5*, 351. [CrossRef]

60. Eom, Y.W.; Kim, M.A.; Park, S.S.; Goo, M.J.; Kwon, H.J.; Sohn, S.; Kim, W.H.; Yoon, G.; Choi, K.S. Two distinct modes of cell death induced by doxorubicin: Apoptosis and cell death through mitotic catastrophe accompanied by senescence-like phenotype. *Oncogene.* **2005**, *24*, 4765–4777. [CrossRef]
61. Narang, A.S.; Delmarre, D.; Gao, D. Stable drug encapsulation in micelles and microemulsions. *Int. J. Pharm.* **2007**, *345*, 9–25. [CrossRef]

polymers

Article

Design of Olmesartan Medoxomil-Loaded Nanosponges for Hypertension and Lung Cancer Treatments

Bjad K. Almutairy [1], Abdullah Alshetaili [1], Amer S. Alali [1], Mohammed Muqtader Ahmed [1], Md. Khalid Anwer [1,*] and M. Ali Aboudzadeh [2,*]

1 Department of Pharmaceutics, College of Pharmacy, Prince Sattam Bin Abdulaziz University, Alkharj 11942, Saudi Arabia; b.almutairy@psau.edu.sa (B.K.A.); a.alshetaili@psau.edu.sa (A.A.); a.alali@psau.edu.sa (A.S.A.); mo.ahmed@psau.edu.sa (M.M.A.)

2 Institut des Sciences Analytiques et de Physico-Chimie pour l'Environnement et les Matériaux, University Pau & Pays Adour, 64000 Pau, France

* Correspondence: mkanwer2002@yahoo.co.in (M.K.A.); m.aboudzadeh-barihi@univ-pau.fr (M.A.A.)

Abstract: Olmesartan medoxomil (OLM) is one of the prominent antihypertensive drug that suffers from low aqueous solubility and dissolution rate leading to its low bioavailability. To improve the oral bioavailability of OLM, a delivery system based on ethylcellulose (EC, a biobased polymer) nanosponges (NSs) was developed and evaluated for cytotoxicity against the A549 lung cell lines and antihypertensive potential in a rat model. Four OLM-loaded NSs (ONS1-ONS4) were prepared and fully evaluated in terms of physicochemical properties. Among these formulations, ONS4 was regarded as the optimized formulation with particle size (487 nm), PDI (0.386), zeta potential ($\zeta P = -18.1$ mV), entrapment efficiency (EE = 91.2%) and drug loading (DL = 0.88%). In addition, a nanosized porous morphology was detected for this optimized system with NS surface area of about 63.512 m^2/g, pore volume and pore radius Dv(r) of 0.149 cc/g and 15.274 Å, respectively, measured by nitrogen adsorption/desorption analysis. The observed morphology plus sustained release rate of OLM caused that the optimized formulation showed higher cytotoxicity against A549 lung cell lines in comparison to the pure OLM. Finally, this system (ONS4) reduced the systolic blood pressure (SBP) significantly ($p < 0.01$) as compared to control and pure OLM drug in spontaneously hypertensive rats. Overall, this study provides a scientific basis for future studies on the encapsulation efficiency of NSs as promising drug carriers for overcoming pharmacokinetic limitations.

Keywords: ethylcellulose; encapsulation; lung cancer; nanosponge; oral bioavailability; systolic blood pressure

Citation: Almutairy, B.K.; Alshetaili, A.; Alali, A.S.; Ahmed, M.M.; Anwer, M..K.; Aboudzadeh, M.A. Design of Olmesartan Medoxomil-Loaded Nanosponges for Hypertension and Lung Cancer Treatments. *Polymers* **2021**, *13*, 2272. https://doi.org/10.3390/polym13142272

Academic Editor: Florian J. Stadler

Received: 14 June 2021
Accepted: 7 July 2021
Published: 11 July 2021

1. Introduction

Globally, lung cancer is one of the leading causes of the cancer-related mortality and palliative care could bring about up the survival rate to 5 years for 15% of patients suffering from it [1]. Female breast cancer, colorectal, stomach and lung cancer make the 40% of the total cancer cases reported. Life-style modification shows an increase in the number of female smokers and smoking habits are relatively correlated for lung cancer due to the presence of mutagens in the inhaled smoke [2]. The prevalence of lung cancer is classically categorized into small cell lung cancer (SCLC) and non-small cell lung cancer (NSCLC) which is typically based on the histological features [3]. Among lung cancer patients, 85% exhibit non-small cell lung cancer (NSCLC) while the rest (15%) of the patients have small cell lung cancer (SCLC). Due to the fast metastasis and quick metabolism of SCLC, if left untreated, the mean survival rate would be about 120 days [4]. On the other hand, NSCLC is any type of epithelial lung cancer other than SCLC, which can be further, divided into lung adenocarcinoma (LUAD), broncho-alveolar, lung squamous cell carcinoma (LUSC), large cell carcinoma and bronchial carcinoid tumor [5]. Generally, cancer can be treated by surgery, radiation therapy, chemotherapy, immunotherapy, targeted therapy, hormone therapy, stem cell transplant, precision medicine and biomarker testing as suggested by the National Institutes of Health (NIH). Due to numerous limitations associated with these conventional methods, several researchers have exploited nanotechnology-based approaches for the efficient diagnosis and delivery of therapeutic agents [6]. In this context, nanomedicine has been applied successfully in developing novel nanocarriers, such as gold-nanoparticles, silver-nanoparticles, polymeric nanoparticles, nanoemulsion, self-emulsifying nanoemulsion and nanosponges [7].

Nanosponges (NSs) are porous nanocarriers with a particle size ≤1000 nm which is favorable for increasing the solubility, dissolution rate and sustained drug release action for temporal and targeted purposes [8]. The porous nature of the NSs could accommodate more drugs and diffusion of the solvent will be easy [9]. NS carriers are implicated in numerous disease conditions and early trials considered this nanotechnology five-folds more effective in comparison to conventional drug delivery systems, especially in malignant cancer treatment [10–12]. For example, it was shown in a study that the therapeutic efficacy of the anticancer drug was increased by the fabrication of ethylcellulose-based NSs containing brigatinib [13]. In another report, Zhang et al. introduced NSs as an antagonistic actuator to a viral mutation that successfully helped to neutralize the virus and inhibit SARS-CoV-2 infectivity [14]. Recently, we developed apremilast-loaded NSs as an efficient nanocarrier for the effective treatment of psoriasis and psoriatic arthritis condition. We observed that the pharmacokinetic profile of these carriers was increased 1.64-fold in terms of bioavailability compared to the pure apremilast suspension [8]. Moreover, in another study, we demonstrated that butenafine-loaded NSs could enhance the therapeutic efficacy through channeling the drug deeper into the skin layers at the target site to completely eradicate fungal infections [15]. In addition, starch derivatives and especially, cyclodextrin-based NSs have recently emerged due to the excellent properties owing to their distinct structure [16,17].

Drugs such as olmesartan medoxomil (OLM), which have been used in this study, have some intrinsic biopharmaceutical drawbacks that could be overcome by nanomedicine technology. Fast metabolism and poor water solubility of this drug result in its poor bioavailability (~28%). Along this portentous dissociation of OLM causes enteropathy due to direct facing of intestinal villi to free olmesartan that stimulate the cell-mediated immune reaction on prolong usage of this medicine [18,19]. Due to numerous therapeutic applications of OLM over commercially available antihypertensive drugs, formulators are exploring novel nanocarriers to improve the intestinal permeability, oral bioavailability and therapeutic effectiveness of OLM through its encapsulation in nanocarriers with the aim of achieving a sustained drug release pattern. PLGA nanoparticles [19,20], self-emulsifying drug delivery systems (SEDDS) [21,22], nanosuspension [23], nanocrystal [24] and other conventional and novel drug delivery systems [25,26] have been developed in this regard. It is worth remarking here that OLM exhibits anti-proliferative and anti-metastatic effects on tumors too [27]. For example, a study conducted in a mouse model by Vassiliou et.al [28] reported that OLM could reduce levels of plasminogen activator inhibitor-1 (PAI-1), whose high levels could cause oral cancer. Besides, another study reports the effectiveness of OLM against the A549 cell lines indicating lung cancer [29].

OLM, chemically named as 2,3-dihydroxy-2-butenyl4-(1-hydroxy-1-methylethyl)-2-propyl-1-[p-(o-1Htetrazol-5-ylphenyl)benzyl]imidazole-5-carboxylate, cyclic-2,3-carbonate, is a potent orally active angiotensin II receptor blocker. It is a prodrug that is hydrolyzed to the active olmesartan throughout absorption from the gastrointestinal tract (GIT) [30]. Angiotensin II receptor blockers are major regulators of blood pressure that competitively inhibit the binding of angiotensin II to its receptor. OLM is an angiotensin-II receptor antagonist (AIIRA) that acts selectively by binding angiotensin receptor 1 (AT1), thereby, preventing the protein angiotensin-II from binding, responsible for vasoconstriction. Physiologically, olmesartan reduces blood pressure (BP), cardiac activity, and aldosterone level and increases sodium excretion [31–33]. Hypertension is a high pressure exerted by circulating blood against the walls of the body's arteries. If the measured pressure is ≥140 mmHg in the heart contract (systolic) and ≥90 mmHg for diastolic pressure at the heart rest, is considered as hypertension (HPT) also called as high blood pressure (HBP). HBP often referred as the silent killer with the prevalence of 972 million people globally accounted for about 26% of the world's population which is expected to grow up to 29% by 2025. The global mortality of this disease is about 7.6 million deaths per year (13.5% of the total). The global antihypertensive drugs accounted for $26.3 billion in 2018 to $27.8 billion by 2023 with a compound annual growth rate (CAGR) of 1.1% for the period of 2018–2023 [34].

The primary aim of this work is to promote the oral bioavailability of OLM through its encapsulation in nanosponge carriers and secondly to investigate the ability of OLM-loaded ethylcellulose-based nanosponges to improve the antihypertensive and cytotoxicity activity against A549 lung cancer cell lines. The specific A549 cell lines were selected because OLM represents established in vitro experimental model to study the cytotoxicity potential of OLM [29]. Therefore, four different OLM-loaded nanosponges (ONS1-ONS4) were developed by varying the content of ethylcellulose polymer. Analyzing the formulations in terms of physicochemical properties and the encapsulation efficiency allowed us to select the optimized nanosponge carrier which was investigated further for antihypertensive activity and cytotoxicity against A549 lung cancer cell lines. The findings in this study may be further sculpted into new cancer and hypertension treatment strategies if this newly synthesized formulation induces desirable cytotoxic activity.

2. Materials and Methods

2.1. Materials

Olmesartan medoxomil (OLM) was purchased from "Mesochem Technology Co. Ltd., Beijing, China". Ethylcellulose (EC), polyvinyl alcohol (PVA) and dichloromethane (DCM) were purchased from Sigma-Aldrich, St. Louis, MO, USA. All the other chemicals were of analytical grade.

2.2. Preparation of OLM-Loaded Nanosponges

OLM-loaded NSs were prepared by emulsion-solvent evaporation method [8]. Four NSs developed by dissolving EC (50–200 mg) and OLM (40 mg) in 5 mL of DCM and ultra-sonicated (Ultrasonic-Water Bath; Daihan Scientific, Model: WUG-D06H, Gangwon, Korea) for 3 min, after which this organic phase was emulsified by adding dropwise into 100 mL of aqueous phase (PVA, 0.2%, *w/v*) solution using probe-sonication (SONIC DISMEMBRATOR Model-FB120, Fisher scientific, Waltham, MA, USA) attached with probe (435 A) working at 60% power for 3 min (Figure 1). Thereafter, the organic solvent was evaporated by keeping the dispersion under stirring (Isotemp®, Fisher scientific, Model-DLM 1886×3, Waltham, MA, USA) over night at 700 rpm under atmospheric condition. Once the solvent was completely evaporated, the dispersion was centrifuged at 12,000 rpm, 25 °C for 15 min (Hermle Labortechnik GMBH, Model-Z216MK, Wehingen, Germany). The pellet of OLM-loaded NSs were then washed several times with milli-Q water to avoid adsorption of PVA on the surface of NSs. Stabilizer-free NSs were lyophilized (Millrock Technology, Kingston, NY, USA) and preserved for particle analysis and evaluations. The composition and precursors of the four NSs composition are tabulated in Table 1.

2.3. Measurement of Particle Size, Polydispersity Index (PDI) and Zeta-Potential (ζP)

The size, PDI and ζP of OLM-loaded NSs were evaluated by dynamic light scattering (DLS) technique using Malvern-Nano ZS-Zetasizer, Malvern, UK. This instrument determines the particle size from intensity–time fluctuations of a laser beam (633 nm) scattered from a sample at an angle of 173°. To avoid multiple scattering, the samples under investigation were dispersed into Milli-Q water in (1:200) dilution concentration and ultra-sonicated (Ultrasonic-Water Bath; DaihanbScientific, Model: WUG-D06H, Gangwon, Korea) for three min in order to break the agglomerates and separate adherents. The sample is then filled in the disposable cuvettes (Folded capillary zeta cell—DTS1070) and kept in the machine with calibrated temperature of 25 °C [35]. Three measurements were performed with 100 cycles in each measurement for both size and PDI together followed by ζP.

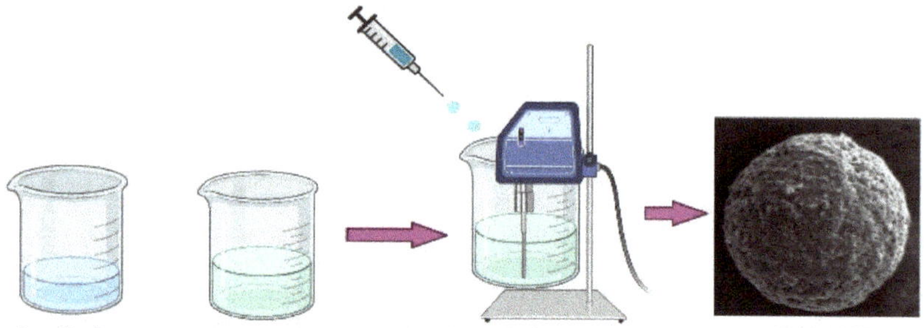

Figure 1. Schematic representation of nanosponge preparation.

Table 1. Composition of OLM-loaded nanosponges.

Nanosponges	OLM (mg)	EC (mg)	PVA (*w*/*v*%)
ONS1	40	50	0.2
ONS2	40	100	0.2
ONS3	40	150	0.2
ONS4	40	200	0.2

2.4. Measurement of Percent Entrapment Efficiency (%EE) and Drug Loading (%DL)

An indirect method was followed in order to measure the %EE and %DL. The aqueous dispersion of NSs were centrifuged (Hermle Labortechnik GMBH, Model-Z216MK, Wehingen, Germany) in order to separate nanosponge particles, and supernatant was analyzed for free OLM drug by UV-spectroscopy (Jasco-UV-visible spectrophotometer, Model: V-630, Tokyo, Japan) at wavelength 255 nm. Each sample was analyzed chemically for drug content in supernatant (n = 3). The %EE and %DL of the OLM-loaded NSs were calculated using Equations (1) and (2), respectively [36].

$$\%EE = \frac{\text{Amount of drug added} - \text{amount of drug in supernatant}}{\text{Amount of drug added}} \times 100 \quad (1)$$

$$\%DL = \frac{\text{Amount of drug in nanosponge}}{\text{Total amount of polymer added}} \times 100 \quad (2)$$

2.5. Fourier Transform Infra-Red (FTIR) Spectroscopy

The FTIR spectra of pure OLM and OLM-loaded NSs (ONS1-ONS4) were recorded using FTIR spectrometer (Jasco FTIR Spectrophotometer, Tokyo, Japan). Each samples were diluted with crystalline potassium bromide (sample:KBr, 1:10 *wt*/*wt*), and pressed into transparent film. The film was kept on sample holder and spectra was recorded using spectra manager software [37].

2.6. Differential Scanning Calorimetry (DSC) Studies

The DSC thermal spectra of pure OLM and their developed nanosponges (ONS1-ONS4) were recorded using a DSC instrument (N-650, Scinco, Seoul, Korea). Each sample (approx. 5 mg) was pressed in aluminum pan covered with a lid. The pressed pan was placed in DSC sample holder and heated in the temperature range of 25 to 200 °C at a rate of 20 °C/min with continuous purge of nitrogen gas during the analysis [37].

2.7. Powder X-ray Diffraction (PXRD) Studies

PXRD of pure OLM and their developed nanosponges (ONS1-ONS4) were recorded using a X-ray Diffractometer (Ultima-IV, Rigaku, Japan) in the range of 0–90° (2θ) at a scan rate of 4°/min. The PXRD spectra of each sample were taken at voltage and current 30 kV and 25 mA, respectively [37].

2.8. In Vitro Release and Kinetic Studies

In vitro release study was performed as reported based on a previous study [8]. Briefly, pure OLM (50 mg) and the optimized NS drug carrier (ONS4) were suspended in 5 mL of phosphate buffer (pH-6.8) and enclosed in diffusion semipermeable cellophane membrane (Hi-media Mol. 12,000 Dalton). The dialysis bag is then dipped into 50 mL of buffer medium, maintained at 37 °C under stirring at 50 rpm on thermostatically controlled magnetic stirrer (Isotemp®, Fisher scientific, Model-DLM 1886×3, Waltham, MA, USA). The samples were withdrawn (1 mL) and replaced with equal amount with the freshly prepared buffer at predetermined time intervals. The aliquots were analyzed spectrophotometrically at 255 nm (Jasco-UV-visible spectrophotometer, Model: V-630, made in Japan) [20]. The concentration was determined, and cumulative (%) release curve was plotted against time intervals (0, 0.5, 1, 2, 3, 4, 5, 6, 12 and 24 h). The release data were fitted to various release kinetic models, viz., zero order, first order, Higuchi and Krosmayer–Peppas kinetic models by imposing relation between % drug release vs. time, % log cumulative drug release vs. time, % log cumulative drug release vs. square root time and % log cumulative drug release vs. log time, respectively [36].

2.9. Scanning Electron Microscopy (SEM) Studies

Based on the physicochemical parameter and in vitro release studies, the NS (ONS4) was optimized and further examined for the surface morphology and nanocarrier diameter using SEM imaging (Joel JSM-SEM, model: JSM6330 LV, Tokyo, Japan). The lyophilized NSs (ONS4) was cautiously mounted on the SEM stubs and coated with Au (gold). Scanning was carried out at different magnifications and zones were captured and further processed to elucidate the surface property [37].

2.10. Nitrogen Adsorption/Desorption Characterization of OLM-Loaded NSs

The surface area, pore volume and pore radius of the optimized nanosponge drug carrier (ONS4) were analyzed by nitrogen sorption isotherm using Brunauer Emmett Teller (BET) method [38] (Quantachrome Instruments-version 5.0, Anton Paar, FL, USA), whereby the under investigation powdered sample was first put in the glass-bulb sample holder followed by heating overnight at 50 °C under negative pressure of 0.1 MPa in order to remove the moisture. Surface area was calculated by BET analysis from the nitrogen sorption data over a relative pressure (*P/PO*) of 1.003–5.047 range. However, the pore volume and pore radius were also obtained from the Barrett–Joiner–Halenda (BJH) summary data.

2.11. Cell Viability MTT Test

Lung cell lines A549 cells were seeded in 96-well plates (Thermo Scientific™ PCR Plate, 96-well, Waltham, MA USA) at a density of 5000 cell/well and were treated at different concentrations (6.25, 12.5, 25, and 50 µg/mL) of OLM-pure, ONS4 as well as the blank-NSs for 24 h. The cells added with the respective samples were treated with 10.0 µL of a 5.0 mg/mL MTT solution, incubated at 37 °C in presence of CO_2. Precipitate of pre-washed PBS (phosphate buffer saline) was dissolved in 150 µL DMSO for 20 min and analyzed at 563 nm [39]. Cell viability (%) of the blank NS and the pure OLM and ONS4 carrier were compared for each concentration followed by processing the data for calculation of IC50.

2.12. *In Vivo Antihypertensive Studies*

The antihypertensive efficacy of the developed optimized nanosponge system (ONS4) was evaluated on spontaneous hypertensive rats (SHRs). The study protocol was reviewed and approved by "Animal Ethics Committee (Approval number: BERC 005-05-19), College of Pharmacy, Prince Satam Bin Abdulaziz University, Alkharj, Saudi Arabia". The rats were divided into three groups (n = 6 per group) and they were kept for 7 days in controlled conditions, temperature (22 °C), RH (55 ± 5%), 12 h light/dark cycle and fed with standard diet with water ad libitum. Group I (control) received 0.5% *w/v*, sodium carboxy methylcellulose, group II received pure OLM drug suspension (1 mg/kg) and group III received the optimized nanosponge drug carrier (ONS4) (equivalent to 1 mg/kg of OLM). The rats were kept in a holder, pre-warmed and systolic blood pressure (SBP) of animals were measured using non-invasive tail cuff device (Kent Scientific Corp, Torrington, CT, USA) at 0 h, before the treatment and then at 0.5, 1, 2, 4, 6, 8, 10 and 12 h after treatment with drug and formulation [40].

2.13. *Statistical Analysis*

Results of physicochemical characterizations were expressed as the mean ± standard error of the mean (SEM). For in vivo antihypertensive studies, statistical variations of different treatment groups were analyzed according to one-way analysis of variance (ANOVA) followed by post-hoc Tukey's test. $p < 0.05$ was considered statistically significant. Statistical analysis was performed using the GraphPad Prism program (version 4) (GraphPad Software, San Diego, CA, USA).

3. Results and Discussions

3.1. *Measurement of Size, PDI and ζP*

The particle size (PS), PDI and ζP of the developed OLM-loaded NSs (ONS1-ONS4) are presented in Table 2. The particle size of ONS1-ONS4 was measured in the range of 381–487 nm. The smaller (381 nm) and bigger size (487 nm) of OLM-loaded NS were measured for ONS1 and ONS4, respectively. It was observed that the size of NS was increased with the increase in amount of EC polymer. For encapsulating drugs, a small particle undoubtedly would be an advantage as it would carry only a small drug quantity, which will help in minimizing the onset of abrupt toxic response [41]. The PDI values of the NSs were measured in the range of 0.312–0.446, which is considered to be acceptable in drug delivery applications and indicates a homogenous population of particles. The aqueous dispersion was used to determine zeta potential in order to get stability of prepared NSs. The ζP of the OLM-loaded NSs were found in the range of −15.7 to −19.0 mV. It is believed that the values of ζP ≥ ±30 mV indicate formation of a stable dispersion, however it has to be remarked that the usual Smoluchowski method to calculate ζP is only valid for hard spheres. In this case, due to the soft nature of NSs, ζP calculated by conventional analysis does not reflect the state of agglomeration or stability. Our soft NS particles were stable despite ζP < ±30 mV [42].

Table 2. Characterization of developed OLM-loaded nanosponges.

Nanosponges	Physicochemical Properties *				
	PS ± SD (nm)	PDI	ζP ± SD (mV)	%EE ± SD	%DL ± SD
ONS1	381 ± 8.2 *	0.312 ± 0.02 *	−15.7 ± 1.6 *	87.1 ± 4.1	1.89 ± 0.15
ONS2	387 ± 9.5 *	0.446 ± 0.02 *	−19.0 ± 2.1 *	87.9 ± 5.2	1.61 ± 0.28
ONS3	404 ± 12.4 *	0.355 ± 0.05 *	−18.2 ± 1.2 *	88.6 ± 4.6	1.52 ± 0.16
ONS4	487 ± 12.8 *	0.386 ± 0.03 *	−18.1 ± 1.8 *	91.2 ± 1.9	0.88 ± 0.35

* Data are not statistically different ($p > 0.05$). Values are presented in mean ± SD (n = 3) in each column.

3.2. Measurement of %EE and %DL

Generally, an ideal drug carrier should have high entrapment efficiency (EE). High EE (above 70%) can increase the efficacy of the drug delivery system and decrease the side effects of the drug [35,43,44]. The %EE and %DL of developed OLM-loaded NSs (ONS1-ONS4) were measured in the range of 86.6–91.2% and 0.88–1.89%, respectively (Table 2). Nanosponges can encapsulate drug up to three times to their weight due higher porosity [45]. The highest drug encapsulation was detected in ONS4, probably due to presence of large amount of EC polymer that prevents the leakage of drug from NSs, therefore, we selected this formulation as the optimized carrier system. Due to poor aqueous solubility, the OLM and ehylcellulose polymer were solubilized in DCM forming a viscous solution, which prevents the drug molecules from coming out to aqueous phase. That resulted in more drug encapsulation in polymeric system [46]. It is revealed that the more EC polymer content, the higher drug entrapment. As polymer content increased, the binding capacity or matrix forming ability of polymer with the drug also increased, due to this the maximum amounts of drug get entrapped in NSs producing more percentage of EE in higher drug to polymer ratio than lower ratio [47]. In this context, Maji et al. found that the drug %EE of EC microparticles surronding metformin HCl prepared by solvent-evaporation technique increased with increasing polymer concentration, when the surfactant content and the stirring speed were constant, increasing polymer content induced better coating onto the drug particles [48].

3.3. FTIR Studies

The interactions between drug and excipients were analyzed by comparing the FTIR spectra of the pure OLM and the obtained OLM-loaded nanosponges (ONS1-ONS4) (Figure S1). The FTIR spectra of pure OLM showed its sharp characteristic peaks assigned at 3289 cm^{-1} for O–H stretching vibrations, 2954 cm^{-1} for C–H stretching of aromatic rings, 1764 cm^{-1} for C=O stretching of the carboxylic group and 1173 cm^{-1} for C–N stretching vibrations, all these peaks were identical to the OLM confirming its purity [20]. Compared to the pure OLM, the characteristic peaks of OLM were present, weakened or shifted in OLM-loaded nanosponges (ONS1-ONS4). The FTIR interpretation of pure OLM and nanosponges (ONS1-ONS4) confirmed no significant interaction of OLM with excipient and therefore suggested successful entrapment and dispersion of OLM inside EC polymer [20]. From this spectral analysis, the OH and carbonyl groups were affected by possible host/guest hydrogen bond formation. In fact, when a carbonyl group and/or hydroxyl groups connect to a hydroxylic compound by hydrogen bonds, the stretching band shifts or weakens in intensity. This appears clearly in the spectrums of ONS1-ONS4 where there are a clear shift and fall in the intensity of the characteristic OH and C=O stretchings of the hydrate, respectively. Moreover, the characteristic peaks of OLM in the wavelenght range 1000 to 2000 cm^{-1} were weakened due to intermolecular interaction between C=O and C–N groups, resulting from higher electronegativity of oxygen and nitrogen. These findings stated that in different spectra of nanosponges no new peaks appeared which indicate that no chemical bonds were created in the developed formulations.

3.4. DSC Studies

DSC analysis confirmed the structural changes observed by FTIR measurements. DSC thermal behavior of the pure OLM and the developed nanosponges drug carriers (ONS1-ONS4) were presented in Figure 2. A sharp endothermic peak at 188 °C was observed in the free OLM drug, which was corresponded to the melting temperature of the drug [20]. A reduced intensity in endothermic peak could be seen in ONS1 formulation compared to the pure OLM, probably due to the less amount of EC used in formulating ONS1 nanosponges. However, in the systems ONS2, ONS3 and ONS4, complete disappearance of endothermic peak corresponding to 188 °C was observed, which was probably due to conversion of crystalline into amorphous OLM or molecularly dispersed into polymer matrix that could confirm complete encapsulation of drug inside EC polymer.

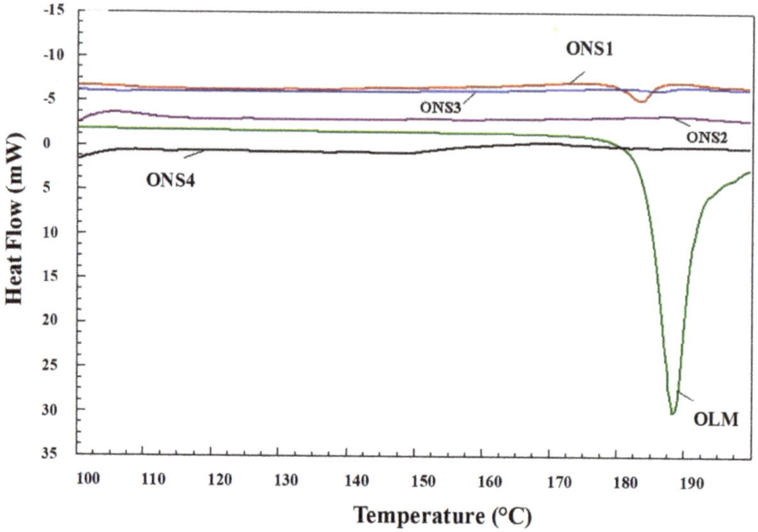

Figure 2. Comparative DSC spectra of OLM-loaded nanosponges.

3.5. *PXRD Studies*

X-ray diffractograms of pure OLM and the developed OLM-loaded nanosponges (ONS1-ONS4) were presented in Figure S2. The XRD results strongly supported the data of DSC analysis. The XRD pattern of the pure OLM drug showed various intense peaks at 2θ values of 7.20°, 9.20°, 10.60°, 11.60°, 12.80°, 14.50°, 16.60°, 18.50°, 19.70°, 21.90°, 22.10°, 23.40°, 24.70°, 25.20°, 38.10°, 44.30° and 77.5°, confirming the crystalline nature of the drug [20]. However, the OLM-loaded nanosponges (ONS1-ONS4) were evaluated by weakening or disappearance of intense peaks when compared to the pure OLM drug. This clearly reveals conversion of crystals to amorphous form due to the dispersion of OLM drug in EC polymer. The polymeric encapsulation layers confine the drug film, which is then unable to crystallize at the solid–air interface. As a consequence, the coating layer introduces another solid–solid boundary. This process is called amorphous solid dispersion and is certainly the result of disrupting intermolecular interactions in the drug's crystal lattice and forming drug–polymer interactions [49].

3.6. *In Vitro Release and Kinetic Studies*

One of the important characteristics of drug delivery systems in biomedicine is to impart the sustained release of a drug [50]. Thus, the sustained drug release improves the accumulation of OLM at the tumor site while enhancing its anticancer performance. In vitro release profile of OLM and the optimized NS drug carrier (ONS4) are presented in Figure 3. The rate of OLM drug release depends on the content of EC polymer. ONS4 carrier showed the sustained release pattern, 89.5% of OLM was released in 24 h, probably due to viscous and swelling nature of EC polymer. In contrast, free OLM drug released 97.36% in the first 4 h. The decreased release rate of OLM-loaded NSs might be due to reduction of drug crystallinity (confirmed by PXRD studies), carrier solubilization effect, reduction of drug agglomeration, conversion of amorphous state and increased drug wettability. In addition, EC guarantees drug dissolution in entire gastrointestinal tract that maintains the constant release of drug for longer period of time and eliminates multiple dosing in a day; hence it improves the efficacy of drug [51]. Furthermore, EC is a non-toxic, hydrophobic, viscous and biocompatible polymer with good compressibility that make it suitable for developing sustained drug release formulations [51,52]. The in vitro release data were fitted to various kinetic models and highest correlation of coefficient (R^2) was found for

Higuchi model (R^2 = 0.9908) (Figure S3), this reveals that the drug release mechanism is governed by diffusion process [53].

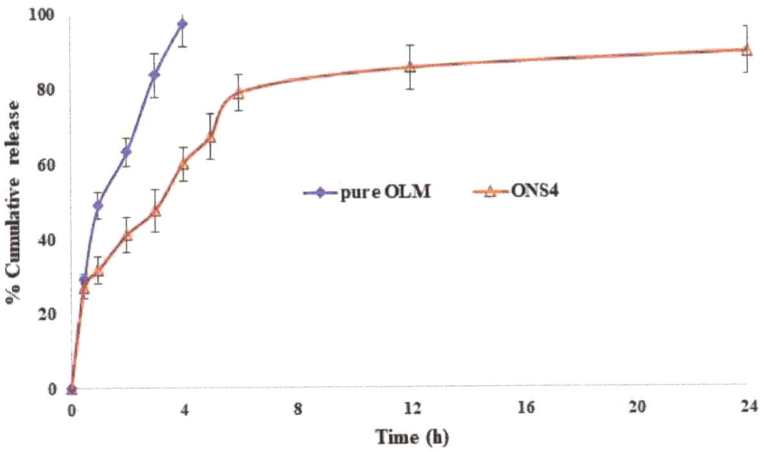

Figure 3. Comparative in vitro release profile of the pure OLM and the optimized nanosponges drug carrier (ONS4) at pH-6.8 and 37 °C after 24 h.

3.7. SEM Studies

The SEM images of optimized NS (ONS4) are shown in Figure 4, which revealed a nanosized, spongy spherical NSs with a porous surface. Numerous tiny pores could be seen on the surface of nanosponge. The formation of pore was probably due to inward diffusion of DCM solvent [54]. The size of particles observed from SEM confirmed the findings observed by DLS method.

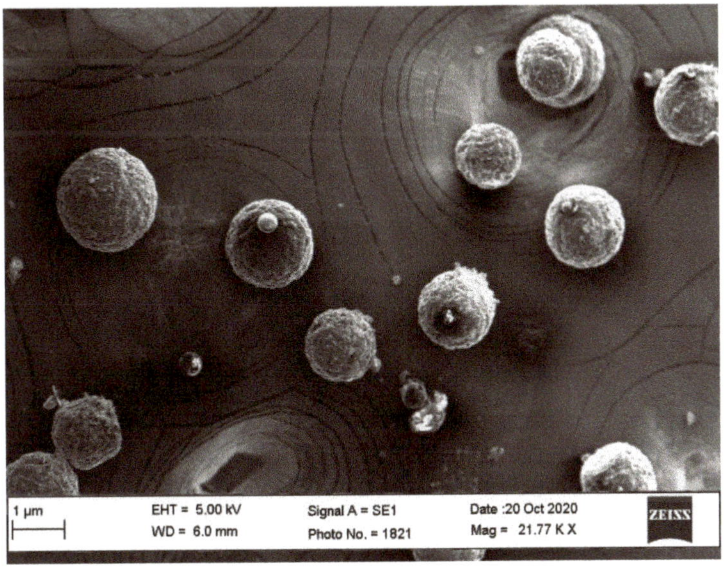

Figure 4. SEM images of the optimized nanosponge system (ONS4).

3.8. Nitrogen Adsorption/Desorption Characterization of OLM-Loaded NSs

Nitrogen adsorption/desorption analysis of the optimized NS fomulation (ONS4) revealed a surface area of 63.512 m^2/g, with pore volume and pore radius Dv(r) of 0.149 cc/g, 15.274 Å, respectively. The nitrogen sorption curve (Figure 5) demonstrates hysteresis loop due to the mono and multilayer adsorption of nitrogen on the ONS4. The ramified hysteresis loop was observed from relative pressure (p/p_0-value) of 0.5–0.9 which reflects condensation of nitrogen in different size inter-connected nanopores of NS. The pore size distribution curve exhibits that most of the pores were found to be in the range between 3.05 and 441.80 nm. In our results, ONS4 showed very low surface area, less pore size and volume probably due to use of ethylcellulose polymer in comparison to previously reported work on cyclodextrin nanosponges [55].

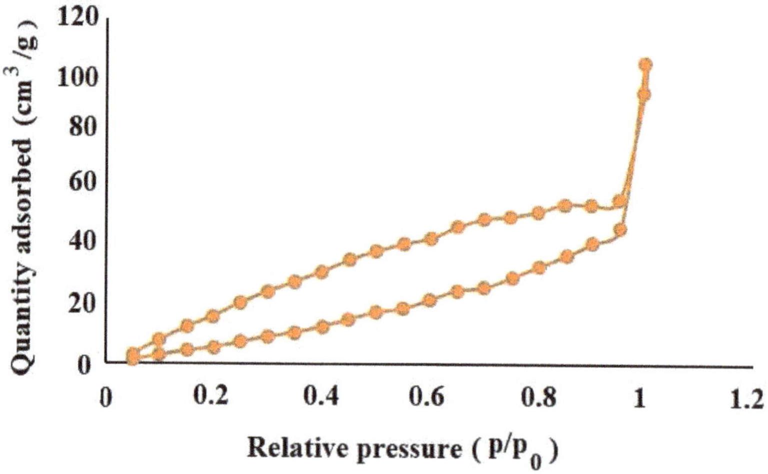

Figure 5. The Nitrogen adsorption–desorption isotherms for the optimized ONS4 system.

3.9. Cell Viability MTT Test

The chemical-colorimetric assay based on the MTT test exhibited concentration dependent reduction in the cell viability. This test measures the cellular metabolic activity, correlated with the darkness of the solution indicates greater metabolic activity and cell viability. The concentration versus percent cell viability data for the cells incubated with blank NSs, free OLM drug and the optimized NS carrier are presented in Figure 6. The optimized ONS4 carrier produced the lowest cell viability (31.44 ± 0.40%, 48.10 ± 1.21%, 59.91 ± 1.13% and 83.27 ± 1.10%) at concentrations of 50, 25, 12.5 and 6.25 µg/mL, respectively in comparison to free OLM (51.25 ± 1.12%, 58.42 ± 1.25%, 69.92 ± 1.36% and 90.39 ± 0.87%) and (72.29 ± 1.43%, 81.71 ± 1.67%, 92.07 ± 0.87% and 97.43 ± 1.76%) for the blank NSs prepared without any drug. The efficacy between ONS4 and the free OLM was much more marked at higher concentrations. The IC50 values for ONS4, OLM, blank-NSs were found to be 14.80 ± 1.21, 35.96 ± 0.45, 40.29 ± 1.17 µg/mL, respectively. IC50 is a half maximal inhibitory concentration, which represents 50% inhibition of the cell growth in vitro indicating the effectiveness of the cytotoxic entity. The optimized ONS4 carrier showed the significant cytotoxicity due to the sustained release of OLM and nano-range of NSs. It has been reported that RAS (renin-angiotensin system) and NF-κB (nuclear factor kappa-light-chain-enhancer of activated B cells) signaling is key factor for the survival of cancer cells, which causes therapeutic intervention. OLM blocks the RAS and NF-κB pathway [56] that could be the reason for significant cytotoxicity activity against A549 cell lines in our studies.

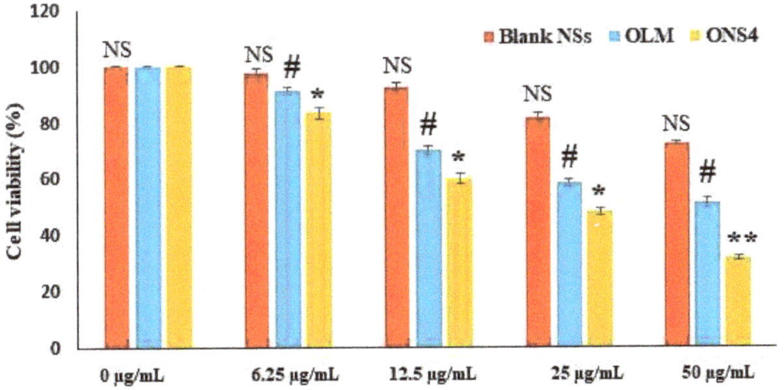

Figure 6. Comparative cell viability of blank NSs, pure OLM and OLM-loaded nanosponges (ONS4). * $p < 0.005$ highly significant—Compared between ONS4 vs. blank NSs and pure OLM at concentration of 50 µg/mL. ** $p < 0.01$ highly significant—Compared between the pure OLM and ONS4 carrier. # $p < 0.05$ significant—Compared between pure OLM vs. blank NSs at all concentrations. NS—Non-significant at all concentrations.

3.10. *In Vivo* Antihypertensive Studies

The systolic blood pressure (SBP) after oral administration of control, pure OLM and ONS4 was shown in Figure 7. The SBP was decreased significantly ($p < 0.05$) at each time point as compared to the control. The oral administration of the optimized nanosponge carrier (ONS4) reduced SBP highly significantly ($p < 0.01$) as compared to the control and the pure OLM drug. A sustained antihypertensive effect was observed in nanosponge carrier (ONS4) administered group during 12 h of the experiment, probably due to drug release retardation by ethylcellulose polymer. The maximum SBP lowering of pure OLM and ONS4 were observed as 180.3 ± 2.07 mmHg and 169.3 ± 1.37 mmHg at 10 h and 6 h, respectively. The OLM-loaded NSs (ONS4) controlled the SBP in hypertensive rats for prolong period of time by minimizing the limitation of oral delivery of OLM. The sustained release and reduced first pass metabolism by ethylcellulose polymer could be the reason of reduction in SBP.

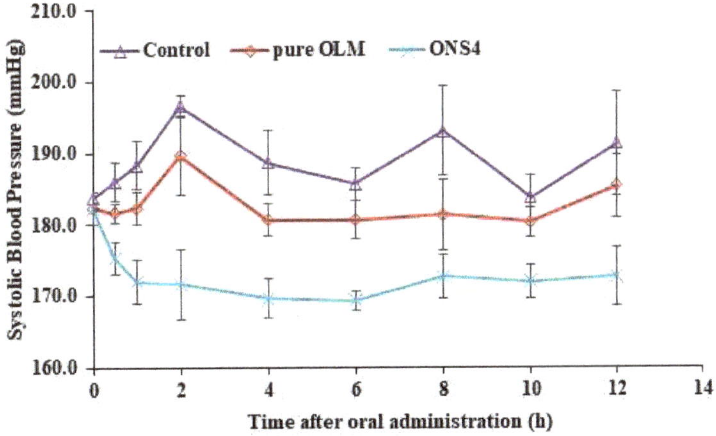

Figure 7. Systolic blood pressure lowering effect of the control, pure OLM and ONS4 samples after oral administration of in SHRs.

These data suggest that when this formulation is administered orally, it can sustain the release of OLM in human body and will produce effective therapeutic action in hypertension and lung cancer, and simultaneously will reduce the side effects of OLM drug.

4. Conclusions

Ehtylcellulose-based OLM-loaded nanosponges (ONS1-ONS4) have been successfully developed using emulsion-solvent evaporation method and was further evaluated by DSC, FTIR, XRD, SEM and in vitro release studies. Based on preliminary characterization, OLM-loaded nanosponges (ONS4) was considered as the optimized formulation and was evaluated for morphology, porosity, in vitro MTT assay against A549 cell line and in vivo antihypertensive activity in rat model. The optimized nanosponges (ONS4) presented a better sustained release with improved antihypertensive efficacy and antitumor activity against A549 cell lines. Hence, it was concluded that the developed nanosponges benefits from its nanosize, porous nature and promises better therapeutic efficacy. The proposed study proved that OLM-loaded nanosponges could have promising application for the treatment of systolic blood pressure and lung cancer; however, further discoveries need to be done in this regard which deserves to be deeply explored.

Supplementary Materials: The following are available online at https://www.mdpi.com/article/10.3390/polym13142272/s1. Figure S1: FTIR spectrum of the pure OLM and OLM-loaded nanosponge carriers. The main characteristic peaks of the pure OLM are shown with dotted vertical lines. These peaks are present, weakened or shifted in OLM-loaded nanosponges (ONS1-ONS4); Figure S2: Comparative XRD diffractograms of the pure OLM and OLM-loaded nanosponge; Figure S3: Profiles of different kinetic models for release mechanism of OLM from OLM-loaded nanosponges.

Author Contributions: Conceptualization, B.K.A. and M.K.A.; methodology, M.M.A. and A.S.A.; formal analysis, B.K.A. and A.A.; investigation, A.S.A.; resources, B.K.A. and A.S.A.; writing—original draft preparation, M.K.A. and M.M.A.; writing—review and editing, M.K.A. and M.A.A.; supervision, B.K.A. and M.A.A. All authors have read and agreed to the published version of the manuscript.

Funding: This research received no external funding.

Institutional Review Board Statement: The animal study protocol was reviewed and approved by "Animal Ethics Committee (Approval number: BERC 005-05-19), College of Pharmacy, Prince Satam Bin Abdulaziz University, Alkharj, Saudi Arabia".

Informed Consent Statement: Not applicable.

Data Availability Statement: The data presented in this study are available on request from the corresponding author.

Acknowledgments: This publication was supported by the Deanship of Scientific Research at Prince Sattam Bin Abdulaziz University, Al-Kharj Saudi Arabia.

Conflicts of Interest: The authors declare no conflict of interest.

References

1. Sung, G.; Ferlay, J.; Siegel, R.L.; Laversanne, M.; Soerjomataram, I.; Jemal, A.; Bray, F. Global cancer statistics 2020: GLOBOCAN estimates of incidence and mortality worldwide for 36 cancers in 185 countries. *CA Cancer J. Clin.* **2021**. [CrossRef]
2. De Lima, M.P.B.; Ramos, D.; Freire, A.P.C.F.; Uzeloto, J.S.; de Silva, B.L.; Ramos, E.M.C. Quality of life of smokers and its correlation with smoke load. *Fisioter. Pesqui.* **2017**, *24*, 273–279. [CrossRef]
3. Sabir, F.; Qindeel, M.; Zeeshan, M.; Ul Ain, Q.; Rahdar, A.; Barani, M.; González, E.; Aboudzadeh, M.A. Onco-Receptors Targeting in Lung Cancer via Application of Surface-Modified and Hybrid Nanoparticles: A Cross-Disciplinary Review. *Processes* **2021**, *9*, 621. [CrossRef]
4. Molina, J.R.; Yang, P.; Cassivi, S.D.; Schild, S.E.; Adjei, A.A. Non-small cell lung cancer: Epidemiology, risk factors, treatment, and survivorship. *Mayo Clin. Proc.* **2008**, *83*, 584–594. [CrossRef]
5. Saab, S.; Zalzale, H.; Rahal, Z.; Khalifeh, Y.; Sinjab, A.; Kadara, H. Insights into Lung Cancer Immune-Based Biology, Prevention, and Treatment. *Front. Immunol.* **2020**, 11. [CrossRef] [PubMed]

6. Jin, C.; Wang, K.; Oppong-Gyebi, A.; Hu, J. Application of nanotechnology in cancer diagnosis and therapy - A mini-review. *Int. J. Med. Sci.* **2020**, *17*, 2964–2973. [CrossRef] [PubMed]
7. Prasad, M.; Lambe, U.P.; Brar, B.; Shah, I.; Ranjan, M.J.K.; Rao, R.; Kumar, S.; Mahant, S.; Khurana, S.K.; Iqbal, H.M.N.; et al. Nanotherapeutics: An insight into healthcare and multi-dimensional applications in medical sector of the modern world. *Biomed. Pharmacother.* **2018**, *97*, 1521–1537. [CrossRef]
8. Ahmed, M.M.; Anwer, M.K.; Fatima, F.; Iqbal, M.; Ezzeldin, E.; Alalaiwe, A.; Aldawsari, M.F. Development of ethylcellulose based nanosponges of apremilast: In vitro and in vivo pharmacokinetic evaluation. *Latin Am. J. Pharm.* **2020**, *39*, 1292–1299.
9. Bano, N.; Ray, S.K.; Shukla, T.; Upmanyu, N.; Khare, R.; Pandey, S.P.; Jain, P. Multifunctional nanosponges for the treatment of various diseases: A review. *Asian J. Pharm. Pharmacol.* **2019**, *5*, 235–248. [CrossRef]
10. Bayda, S.; Adeel, M.; Tuccinardi, T.; Cordani, M.; Rizzolio, F. The history of nanoscience and nanotechnology: From chemical-physical applications to nanomedicine. *Molecules* **2020**, *25*, 112. [CrossRef]
11. Sutradhar, K.B.; Amin, M.L. Nanotechnology in Cancer Drug Delivery and Selective Targeting. *ISRN Nanotechnol.* **2014**, *2014*, 1–12. [CrossRef]
12. Navya, P.N.; Kaphle, A.; Srinivas, S.P.; Bhargava, S.K.; Rotello, V.M.; Daima, H.K. Current trends and challenges in cancer management and therapy using designer nanomaterials. *Nano Converg.* **2019**, *6*. [CrossRef] [PubMed]
13. Ahmed, M.M.; Fatima, F.; Anwer, M.K.; Ansari, M.J.; Das, S.S.; Alshahrani, S.M. Development and characterization of ethyl cellulose nanosponges for sustained release of brigatinib for the treatment of non-small cell lung cancer. *J. Polym. Engn.* **2020**, *40*, 823–832. [CrossRef]
14. Zhang, Q.; Honko, A.; Zhou, J.; Gong, H.; Downs, S.N.; Vasquez, J.H.; Fang, R.H.; Gao, W.; Griffiths, A.; Zhang, L. Cellular Nanosponges Inhibit SARS-CoV-2 Infectivity. *Nano Lett.* **2020**, *20*, 5570–5574. [CrossRef]
15. Ahmed, M.M.; Fatima, F.; Anwer, M.K.; Ibnouf, E.O.; Kalam, M.A.; Alshamsan, A.; Aldawsari, M.F.; Alalaiwe, A.; Ansari, M.J. Formulation and in vitro evaluation of topical nanosponge-based gel containing butenafine for the treatment of fungal skin infection. *Saudi Pharm J.* **2021**, *29*, 467–477. [CrossRef] [PubMed]
16. Krabicová, I.; Appleton, S.L.; Tannous, M.; Hoti, G.; Caldera, F.; Rubin Pedrazzo, A.; Cecone, C.; Cavalli, R.; Trotta, F. History of Cyclodextrin Nanosponges. *Polymers* **2020**, *12*, 1122. [CrossRef] [PubMed]
17. Anwer, M.K.; Iqbal, M.; Ahmed, M.M.; Aldawsari, M.F.; Ansari, M.N.; Ezzeldin, E.; Khalil, N.Y.; Ali, R. Improving the Solubilization and Bioavailability of Arbidol Hydrochloride by the Preparation of Binary and Ternary β-Cyclodextrin Complexes with Poloxamer 188. *Pharmaceuticals* **2021**, *14*, 411. [CrossRef] [PubMed]
18. Théophile, H.; David, X.R.; Miremont-Salamé, G.; Haramburu, F. Five cases of sprue-like enteropathy in patients treated by olmesartan. *Dig. Liver Dis.* **2014**, *46*, 465–469. [CrossRef] [PubMed]
19. Si, S.; Li, H.; Han, X. Sustained release olmesartan medoxomil loaded PLGA nanoparticles with improved oral bioavailability to treat hypertension. *J. Drug. Deliv. Sci. Technol.* **2020**, *55*, 101422. [CrossRef]
20. Anwer, M.K.; Jamil, S.; Ansari, M.J.; Iqbal, M.; Imam, D.; Shakeel, F. Development and evaluation of olmesartan medoxomil loaded PLGA nanoparticles. *Mat. Res. Innov.* **2016**, *20*, 193–197. [CrossRef]
21. Gorain, B.; Choudhury, H.; Kundu, A.; Sarkar, L.; Karmakar, S.; Jaisankar, P.; Pal, T.K. Nanoemulsion strategy for olmesartan medoxomil improves oral absorption and extended antihypertensive activity in hypertensive rats. *Colloids Surf. B* **2014**, *115*, 286–294. [CrossRef]
22. Kang, M.J.; Kim, H.S.; Jeon, H.S.; Park, J.H.; Lee, B.S.; Ahn, B.K.; Moon, K.Y.; Choi, Y.W. In situ intestinal permeability and in vivo absorption characteristics of olmesartan medoxomil in self-microemulsifying drug delivery system. *Drug Dev. Ind. Pharm.* **2012**, *38*, 587–596. [CrossRef]
23. Thakkar, H.P.; Patel, B.V.; Thakkar, S.P. Development and characterization of nanosuspensions of olmesartan medoxomil for bioavailability enhancement. *J. Pharm. Bioall. Sci.* **2011**, *3*, 426–434. [CrossRef]
24. Jain, S.; Patel, K.; Arora, S.; Reddy, V.A.; Dora, C.P. Formulation, optimization, and in vitro-in vivo evaluation of olmesartan medoxomil nanocrystals. *Drug Deliv. Transl. Res.* **2017**, *7*, 292–303. [CrossRef]
25. Prajapati, S.T.; Bulchandani, H.H.; Patel, D.M.; Dumaniya, S.K.; Patel, C.N. Formulation and evaluation of liquisolid compacts for olmesartan medoxomil. *J. Drug Deliv. Sci. Technol.* **2013**, 870579. [CrossRef] [PubMed]
26. Gunda, R.K.; Manchineni, P.R.; Dhachinamoorthi, D. Design, development, and in vitro evaluation of sustained release tablet formulations of olmesartan medoxomil. *MOJ Drug Des. Dev. Ther.* **2018**, *7*, 164–169. [CrossRef]
27. Abd-Alhaseeb, M.M.; Zaitone, S.A.; Abou-El-Ela, S.H.; Moustafa, Y.M. Olmesartan potentiates the anti-angiogenic effect of sorafenib in mice bearing Ehrlich's ascites carcinoma: Role of angiotensin (1-7). *PLoS ONE* **2014**, *9*, e85891. [CrossRef]
28. Vassiliou, S.; Nkenke, E.; Lefantzis, N.; Ioannidis, A.; Yapijakis, C.; Zoga, M.; Papakosta, V.; Derka, S.; Nikolaou, C.; Vairaktaris, E. Effect of Olmesartan on the Level of Oral Cancer Risk Factor PAI1. *Anticancer Res.* **2016**, *36*, 6093–6096. [CrossRef] [PubMed]
29. Gayathri, E.; Punnagai, K.; Chellathai, D.D. Evaluation of Anticancer Activity of Olmesartan and Ramipril on A549 Cell. *Biomed. Pharmacol. J.* **2018**, *11*, 1351–1357. [CrossRef]
30. Rote, A.R.; Bari, P.D. Spectrophotometric estimation of olmesartan medoxomil and hydrochlorothiazide in tablet. *Indian. J. Pharm. Sci.* **2010**, *72*, 111–113. [CrossRef]
31. Approved Product of OLMETEC®(Olmesartan Medoxomil). 2013. Available online: https://www.tga.gov.au/sites/default/files/auspar-olmesartan-medoxomil-130226-pi.pdf (accessed on 10 July 2021).

32. Brunner, H. The new oral angiotensin II antagonist olmesartan medoxomil: A concise overview. *J. Hum. Hypertens.* **2002**, *16*, S13–S16. [CrossRef] [PubMed]
33. Kreutz, R. Olmesartan/amlodipine: A review of its use in the management of hypertension. *Vasc. Health Risk Manag.* **2011**, *7*, 183–192. [CrossRef] [PubMed]
34. Alexander, M.R.; Madhur, M.S.; Harrison, D.G.; Dreisbach, A.W.; Riaz, K. What is the global prevalence of hypertension (high blood pressure)? *Medscape* **2019**. Available online: https://www.medscape.com/answers/241381-7614/what-is-the-global-prevalence-of-hypertension-high-blood-pressure (accessed on 10 July 2021).
35. Anwer, M.K.; Mohammad, M.; Ezzeldin, E.; Fatima, F.; Alalaiwe, A.; Iqbal, M. Preparation of sustained release apremilast-loaded PLGA nanoparticles: In vitro characterization and in vivo pharmacokinetic study in rats. *Int. J. Nanomed.* **2019**, *14*, 1587–1595. [CrossRef] [PubMed]
36. Anwer, M.K.; Iqbal, M.; Aldawsari, M.F.; Alalaiwe, A.; Ahmed, M.M.; Muharram, M.M.; Ezzeldin, E.; Mahmoud, M.A.; Imam, F.; Ali, R. Improved antimicrobial activity and oral bioavailability of delafloxacin by self-nanoemulsifying drug delivery system (SNEDDS). *Drug. Deliv. Sci. Technol.* **2021**, *64*, 102572. [CrossRef]
37. Anwer, M.K.; Iqbal, M.; Muharram, M.M.; Mohammad, M.; Ezzeldin, E.; Aldawsari, M.F.; Alalaiwe, A.; Imam, F. Development of Lipomer Nanoparticles for the Enhancement of Drug Release, Anti-microbial Activity and Bioavailability of Delafloxacin. *Pharmaceutics* **2020**, *12*, 252. [CrossRef]
38. Sinha, P.; Datar, A.; Jeong, C.; Deng, X.; Chung, Y.G.; Lin, L. Surface Area Determination of Porous Materials Using the Brunauer–Emmett–Teller (BET) Method: Limitations and Improvements. *J. Phy. Chem. C* **2019**, *123*, 20195–20209. [CrossRef]
39. Alshetaili, A.S. Gefitinib loaded PLGA and chitosan coated PLGA nanoparticles with magnified cytotoxicity against A549 lung cancer cell lines. *Saudi J. Biol. Sci.* **2021**. [CrossRef]
40. Michalowski, C.B.; Arbo, M.D.; Altknecht, L.; Anciuti, A.N.; Abreu, A.; Alencar, L.; Pohlmann, A.R.; Garcia, S.C.; Guterres, S.S. Oral Treatment of Spontaneously Hypertensive Rats with Captopril-Surface Functionalized Furosemide-Loaded Multi-Wall Lipid-Core Nanocapsules. *Pharmaceutics* **2020**, *12*, 80. [CrossRef] [PubMed]
41. Acosta, E. Bioavailability of nanoparticles in nutrient and nutraceutical delivery. *Curr. Opin. Colloid Interf. Sci.* **2009**, *14*, 3. [CrossRef]
42. Lerche, D.; Sobisch, T. Evaluation of particle interactions by in situ visualization of separation behavior. *Colloids Surf. A* **2014**, *440*, 122–130. [CrossRef]
43. Rahdar, A.; Sargazi, S.; Barani, M.; Shahraki, S.; Sabir, F.; Aboudzadeh, M.A. Lignin-Stabilized Doxorubicin Microemulsions: Synthesis, Physical Characterization, and In Vitro Assessments. *Polymers* **2021**, *13*, 641. [CrossRef] [PubMed]
44. Rahdar, A.; Taboada, P.; Hajinezhad, M.R.; Barani, M.; Beyzaei, H. Effect of tocopherol on the properties of Pluronic F127 microemulsions: Physico-chemical characterization and in vivo toxicity. *J. Mol. Liq.* **2019**, *277*, 624–630. [CrossRef]
45. Aloorkar, N.H.; Kulkarni, A.S.; Ingale, D.J.; Patil, R.A. Microsponges as Innovative Drug Delivery Systems. *Int. J. Pharm. Sci. Nanotechnol.* **2012**, *5*, 1597–1606. [CrossRef]
46. Sharma, N.; Madan, P.; Lin, S. Effect of process and formulation variables on the preparation of parenteral paclitaxel-loaded biodegradable polymeric nanoparticles: A co-surfactant study. *Asian. J. Pharm Sci.* **2016**, *11*, 404–416. [CrossRef]
47. Pandav, S.; Naik, J. Preparation and In Vitro Evaluation of Ethylcellulose and Polymethacrylate Resins Loaded Microparticles Containing Hydrophilic Drug. *J. Pharm.* **2014**, 904036. [CrossRef]
48. Maji, R.; Ray, S.; Das, B.; Nayak, A. Ethyl Cellulose Microparticles Containing Metformin HCl by Emulsification-Solvent Evaporation Technique: Effect of Formulation Variables. *Int. Schol. Res. Not.* **2012**, 1–7. [CrossRef]
49. Pandi, P.; Bulusu, R.; Kommineni, N.; Khan, W.; Singh, M. Amorphous solid dispersions: An update for preparation, characterization, mechanism on bioavailability, stability, regulatory considerations and marketed products. *Int. J. Pharm.* **2020**, *586*, 119560. [CrossRef]
50. Perrin, J.H. *Sustained and Controlled Release Drug Delivery Systems*; Robinson, J. Dekker: New York, NY, USA, 1978; Volume 6, p. 773. [CrossRef]
51. Wasilewska, K.; Winnicka, K. Ethylcellulose–A Pharmaceutical Excipient with Multidirectional Application in Drug Dosage Forms Development. *Materials* **2019**, *12*, 3386. [CrossRef]
52. Trofimiuk, M.; Wasilewska, K.; Winnicka, K. How to modify drug release in paediatric dosage forms? Novel technologies and modern approaches with regard to children's population. *Int. J. Mol. Sci.* **2019**, *20*, 3200. [CrossRef]
53. Mathew, S.T.; Devi, S.G.; KV, S. Formulation and evaluation of ketorolac tromethamine-loaded albumin microspheres for potential intramuscular administration. *AAPS PharmSciTech.* **2007**, *8*, 14. [CrossRef] [PubMed]
54. Sharma, R.; Walker, R.B.; Pathak, K. Evaluation of the Kinetics and Mechanism of Drug Release from Econazole nitrate Nanosponge Loaded Carbapol Hydrogel. *Indian J. Pharm. Edu. Res.* **2011**, *45*, 25–31.
55. Sadjadi, S.; Heravi, M.M.; Daraie, M. Cyclodextrin nanosponges: A potential catalyst and catalyst support for synthesis of xanthenes. *Res. Chem. Intermed.* **2017**, *43*, 843–857. [CrossRef]
56. Bakhtiari, E.; Hosseini, A.; Boroushaki, M.B.; Mousavi, S.H. Synergistic, cytotoxic and apoptotic activities of olmesartan with NF-κB inhibitor against HeLa human cell line. *Toxicol. Mech. Methods* **2015**, *25*, 614–621. [CrossRef]

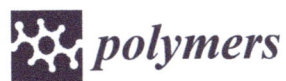

polymers

MDPI

Article

Enhanced Dissolution of Sildenafil Citrate Using Solid Dispersion with Hydrophilic Polymers: Physicochemical Characterization and In Vivo Sexual Behavior Studies in Male Rats

Mohammed F. Aldawsari [1,*], Md. Khalid Anwer [1], Mohammed Muqtader Ahmed [1], Farhat Fatima [1], Gamal A. Soliman [2,3], Saurabh Bhatia [4,5], Ameeduzzafar Zafar [6] and M. Ali Aboudzadeh [7]

1 Department of Pharmaceutics, College of Pharmacy, Prince Sattam Bin Abdulaziz University, Alkharj 11942, Saudi Arabia; mkanwer2002@yahoo.co.in (M.K.A.); mo.ahmed@psau.edu.sa (M.M.A.); s.soherwardi@psau.edu.sa (F.F.)
2 Department of Pharmacology, College of Pharmacy, Prince Sattam Bin Abdulaziz University, P.O. Box 173, Al-Kharj 11942, Saudi Arabia; g.soliman@psau.edu.sa
3 Department of Pharmacology, College of Veterinary Medicine, Cairo University, Giza 12211, Egypt
4 Natural and Medical Sciences Research Center, University of Nizwa, Birkat Al Mauz 616, Nizwa P.O. Box 33, Oman; sbsaurabhbhatia@gmail.com
5 School of Health Science, University of Petroleum and Energy Studies, Dehradun 248007, Uttarakhand, India
6 Department of Pharmaceutics, College of Pharmacy, Jouf University, Sakaka 72341, Aljouf Region, Saudi Arabia; azafar@ju.edu.sa
7 Institut des Sciences Analytiques et de Physico-Chimie pour l'Environnement et les Matériaux, IPREM, UMR5254, CNRS, University Pau & Pays Adour, E2S UPPA, 64000 Pau, France; m.aboudzadeh-barihi@univ-pau.fr
* Correspondence: moh.aldawsari@psau.edu.sa

Citation: Aldawsari, M.F.; Anwer, M.K.; Ahmed, M.M.; Fatima, F.; Soliman, G.A.; Bhatia, S.; Zafar, A.; Aboudzadeh, M.A. Enhanced Dissolution of Sildenafil Citrate Using Solid Dispersion with Hydrophilic Polymers: Physicochemical Characterization and In Vivo Sexual Behavior Studies in Male Rats. *Polymers* **2021**, *13*, 3512. https://doi.org/10.3390/polym13203512

Academic Editors: Dimitrios Bikiaris and Shaghayegh Hamzehlou

Received: 3 September 2021
Accepted: 11 October 2021
Published: 13 October 2021

Abstract: Sildenafil citrate (SLC) is a frequently used medication (Viagra®) for the treatment of erectile dysfunction (ED). Due to its poor solubility, SLC suffers from a delayed onset of action and poor bioavailability. Hence, the aim of the proposed work was to prepare and evaluate solid dispersions (SDs) with hydrophilic polymers (Kolliphor®P188, Kollidon®30, and Kollidon®-VA64), in order to enhance the dissolution and efficacy of SLC. The SLC-SDs were prepared using a solvent evaporation method (at the ratio drug/polymer, 1:1, *w*/*w*) and characterized by Differential Scanning Calorimetry (DSC), Fourier-transform infrared spectroscopy (FTIR), X-ray diffraction (XRD), Scanning electron microscope (SEM), drug content, yield, and in vitro release studies. Based on this evaluation, SDs (SLC-KVA64) were optimized, with a maximum release of drug (99.74%) after 2 h for all the developed formulas. The SDs (SLC-KVA64) were further tested for sexual behavior activity in male rats, and significant enhancements in copulatory efficiency (81.6%) and inter-copulatory efficiency (44.9%) were noted in comparison to the pure SLC drug, when exposed to the optimized SLC-KVA64 formulae. Therefore, SD using Kollidon®-VA64 could be regarded as a potential strategy for improving the solubility, in vitro dissolution, and therapeutic efficacy of SLC.

Keywords: erectile dysfunction; Kolliphor®P188; Kollidon®30; Kollidon®-VA64; polymer; sildenafil citrate

1. Introduction

Erectile dysfunction (ED) is the inability of a male to achieve and maintain erection for a sufficient period of time for satisfactory intercourse with a counterpart female partner [1]. It is also referred to as male impotence. ED is common medical problem that directly affects sexual wellbeing and quality of life. Presently, millions of men around the world have some degree of ED, and more than twice that number are anticipated to be affected by 2025 [2,3]. Men suffer from ED due to the rise in synthetic hormone levels present in our diet/environment and a nutritionally poor and imbalanced diet, resulting in low levels of testosterone formation in the body.

Sildenafil citrate (SLC) is a potent phophodiasterase-5 inhibitor, marketed under the brand name of Viagra. SLC is an orally administered medication, selectively used to treat ED and pulmonary hypertension (PH). SLC absorbs quickly and acts within 1 h of oral administration, but due to a low aqueous solubility and hepatic first pass metabolism (~ 80% of administered dose), its relative bioavailability is 41% [4,5]. The solubility and bioavailability of SLC can improved by various means, such as, cyclodextrin complex [6,7], orodissovable films [8], dry foam tablets [9], salts and co-crystals [10], Self-nanoemulsifying drug delivery systems (SNEDDS) [11], and spray dried amorphous solid dispersions [12]. The advantages of SD have been mentioned in many investigations, including the improvement of dissolution rate and efficacy of poorly water insoluble drugs [13,14].

Solid dispersion (SD) is an efficient approach to improve the solubility and bioavailability of Biopharmaceutical classification system (BCS) II and IV class drugs, which involves dispersion of active ingredients within an inert carrier in a solid state [15,16]. The selection of carrier for the preparation of SDs is very important and directly affects the efficacy and stability of the formulation. Thermodynamically unstable SDs have the tendency to recrystallize into amorphous drug during storage, even with traces of crystalline drug left during preparation and long storage periods [17]. Therefore, complete amorphous SD formation is important, to avoid recrystallization and hence improve the physical properties of the drug. The encapsulation of drugs within hydrophilic polymeric carriers induces better wettability and particle micronization; the main procedure by which SDs improve the solubility and bioavailability of poorly soluble drugs. Hydrophilic polymer carriers play a vital role in increasing the dissolution and bioavalibility of poorly soluble drugs. The function of polymers in formulating SDs is to impart stability and solubility, and modify the dissolution rate. Various polymeric carriers, notably water soluble drug carriers such as polyethylene glycol (PEG) and polyvinylpyrrolidone (PVP), with different molecular weight grades have been used for the preparation of SDs and solid solutions. Cellulose derived natural polymers such as hydroxy propyl methyl cellulose acetate succinate (HPMCAS), ethyl cellulose (EC), and hydroxypropyl methyl cellulose (HPMC) have the desired physicochemical properties, and, hence, they are extensively used in formulating SDs. PEG enables a disordered crystalline state of the drug and forms amorphous interstitial solid solutions by encompassing the drug entity in the interstitial spaces of the polymeric carrier. Glass solution is the term devised for SDs in which hydrophilic polymeric carriers are used to increase the solubility; especially for BCS class II and IV. In order to molecularly dissolve the drug, a large quantity of the hydrophilic polymer is used, which in turn increases the physical instability of the dispersions due to phase separation. The mechanisms involved in the improved solubility and dissolution include the detachment of drug molecules as the hydrophilic carrier dissolves, subsequently forming a supersaturated solution of the drug [18–22]. A study on SDs using a ibuprofen model drug revealed that polymers under the trade names "poloxamer 407" and "poloxamer 188" could increase the solubility and reduce the crystal growth of the drug with their coexistence in the polymeric dispersion, which could have been due to the disruption of the ibuprofen and formation of hydrogen bonds between the drug and polymer [23].

Another polymeric carrier, produced under the trade name Kolliphor®P 188 and commercially available through the BASF corporation, acts as a co-emulsifier in creams, an emulsifier for skin delivery applications, and solid and liquid dispersions [24]. It is a synthetic tri-block copolymer, containing a central hydrophobic chain of polyoxypropylene linked by two hydrophilic chains of polyoxyethylene. The generic name of this nonionic linear copolymer is poloxamer 188 (P188), the letter P indicates the state of the polymer (as a powder). Poloxamer 188 (P188) copolymer has been approved by the FDA as a blood thinner and is used pharmaceutically as a surfactant in toothpastes and mouth washes. The nature of the conforming blocks gives Kolliphor®P 188 amphiphilic and surface active properties, which vary depending on its poly(propylene oxide) and poly(ethylene oxide) contents. Poloxamer 188 containing 80% ethylene oxide acts as a water soluble polymeric carrier, used in solid dispersions to improve the solubility and dissolution rate of poorly water soluble active pharmaceutical ingredients (APIs). Kolliphor®P 188 facilitates solubilization process by micelle formation, in which the drug is enclosed in a hydrophobic core externally covered by a polar hydrophilic head. SDs prepared with ploxamer are reported to enhance the solubility and dissolution rate of the hydrophobic drug, Ebastine [25].

Kollidon®30 is another water soluble drug carrier (from BASF Germany) that is a polyvinyl-pyrrolidone derivative with a molecular weight of 44,000–54,000 g/mol, transition temperature of 149 °C, and soluble in both aqueous and organic solvents. Commonly called povidon(e), poly(1-vinyl-2-pyrrolidone), povidonum, and polyvidone. Kollidon®30 is widely used in SDs to improve the solubility and dissolution rate of the drugs by forming a water soluble complex with insoluble drugs. Co-precipitation and co-milling technologies are reported to increase the dissolution rate and bioavailability of water insoluble drugs with the usage of Kollidon®30. Amorphous solid dispersions (ASDs) prepared by Kollidon®30 were reported to improve the efficiency of nifedipine [26,27].

A copovidone with an exceptionally high binding capacity, traded as Kollidon®VA 64 and produced by BASF, Germany, has applications as a dry binder for direct compression tableting and as a soluble binder for granulation. These properties make it an attractive and cost-effective alternative to natural binders. Kollidon VA 64 was first used to prepare Lopinavir-Ritonavir combination SDs by Abbott laboratories; thereafter, the solubility of many APIs was increased thanks to copovidone. This copolymer of vinyl-pyrrolidone and vinyl-acetate in a ratio of 6:4 possess a transition temperature (100 °C) and degrades at temperature (230 °C) that allow using APIs of varied polarity and with a wide melting temperature range, and it is extensively employed in hot melt extrusion (HME) and spray drying solid dispersions (SDSDs). Single-phase glassy solutions formed by copovidone, in which API was amorphously lodged and dissolved along with the water soluble carrier, controlling the process polymer, improved the dissolution rate [28,29]. Moreover, copovidone-based SDs have also been found to generate nanoparticles during the process of mass conversion from solid to liquid (dissolution), contributing to the improved solubility and bioavailability of the drug. Recently, Moseson et al. [30] reported that copovidone act as crystal nucleation and growth inhibitor by polymer adsorption on the crystalline drug surface, thus improving the dissolution profile.

This study focuses on the influence of hydrophilic polymers in the dissolution enhancement of SLC. Here, the prepared SDs transformed the crystalline SLC drug into an amorphous state. The interaction of polymers and SLC in SDs was evaluated and optimized by DSC, FTIR, XRD, SEM, and in vitro release studies. The optimized SD (SLC-K64) showed enhanced dissolution, due to the improve wetting properties of the drug. The SLC-K64 formulae significantly improved the sexual behavior in male rats. The goal of the current study was to develop and optimize a SD that could extend the penile erection in male rats and be a potential approach for the treatment of erectile dysfunction.

2. Materials and Methods

2.1. Materials

Sildenafil citrate (SLC) was obtained as a gift sample from Jazeera Pharmaceutical Industry (JPI), Riyadh, Saudi Arabia. Kolliphor®P188 (K188), Kollidon®30 (K30) and Kollidon®-VA64 (KVA64) were received as a gift sample from BASF Co., Ltd. (Ludwigshafen, Germany). All solvents and chemicals used for the study were pure and analytical grade.

2.2. Preparation of SLC Solid Dispersion by Solvent Evaporation

The solid dispersions of SLC with each of the polymers (Kolliphor®P188, Kollidon®30 or Kollidon®-VA64) were prepared (at the ratio drug/polymer, 1:1, *w*/*w*) using a solvent evaporation method [31]. Briefly, an accurately weighed amount of SLC and polymer was dissolved in 60 mL of ethanol and water mixture (1:1, v/v). The resultant solution was transferred into round bottom flask and evaporated on a rotary evaporator "(Buchi Rotavapor R-215, Essen, Germany)" at 60 °C and 50 rpm for 4 h (Figure 1). The solids retained in the flask were dried under a vacuum overnight to remove residual solvent. The final powder was ground into fine particles and stored for further use.

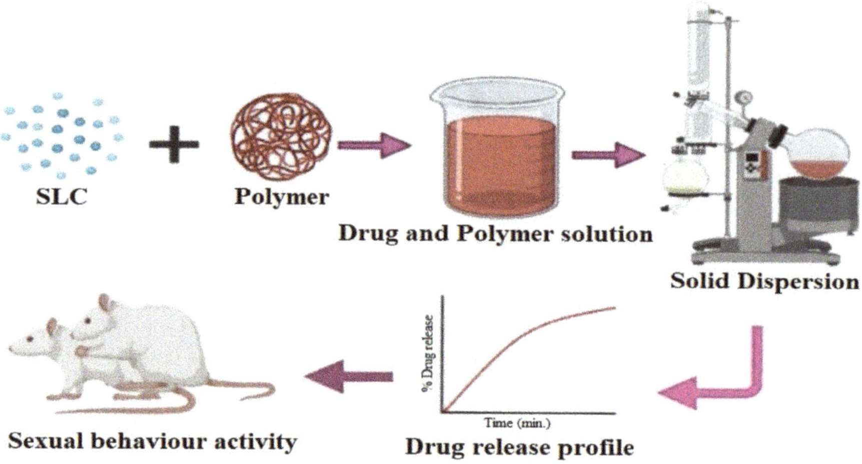

Figure 1. Schematic diagram of preparation of solid dispersion.

2.3. Practical Percentage Yield

The yield of the process was calculated to determine the efficiency of the preparation process. SDs were collected and the percentage yield was estimated using the following equation.

Yield (%) = (Practical weight of the SD)/(Theoretical weight of SLC + Polymer) × 100

2.4. Drug Content Estimation

Drug content was estimated by dissolving 50 mg equivalent weight of SLC in methanol. The solution was then filtered through a syringe filter (0.45 μm), the filtrate was suitably diluted with distilled water and analyzed for the quantity of drug using a UV spectrophotometer at 291 nm against a blank using distilled water (UV-visible spectrophotometer, Jasco 645, Tokyo, Japan).

2.5. Differential Scanning Calorimetry (DSC) Studies

DSC spectra of pure SLC, and their solid dispersions (SLC-K188, SLC-K30, and SLC-KVA64) were recorded using a DSC instrument (SCINCO, DSC N-650, Seoul, Korea) at the temperature range of 50.0–250.0 °C, at a heating rate of 10 °C/min. The instrument was purged with nitrogen gas at a flow rate of 20 mL/min. The DSC apparatus was connected with a sample holder and cooling chamber [32]. Each sample was weighed accurately (approx. 5 mg) and pressed into a hermetically sealed aluminum pan.

2.6. Fourier Transform Infra-Red (FTIR) Spectroscopy

The FTIR spectra of pure SLC and their solid dispersions (SLC-K188, SLC-K30, and SLC-KVA64) were recorded using an FTIR spectrometer (Jasco FTIR Spectrophotometer, Tokyo, Japan). Each sample was ground with crystalline potassium bromide using a glass mortar and pestle into a very fine particles, and pressed into transparent film. The transparent film was kept on a sample holder and the spectra was recorded using spectra manager software (Jasco, Tokyo, Japan) [33].

2.7. Powder X-ray Diffraction (PXRD) Studies

The PXRD pattern of pure SLC and solid dispersions (SLC-K188, SLC-K30, and SLC-KVA64) were recorded with an Ultima IV Diffractometer (Rigaku Inc. Tokyo, Japan at College of Pharmacy, King Saud University, Riyadh, KSA). The set parameters for PXRD were 0–60° (2θ) at a 10°/min scan speed. The anode tube of the instrument used was "Cu with Ka = 0.1540562 nm with mono-chromatized graphite crystal". The spectra was recorded using a voltage and current of 40 kV and 40 mA, respectively [34].

2.8. Scanning Electron Microscopy (SEM)

The morphology of pure SLC, SLC-K188, SLC-K30, and SLC-KVA64 was observed under SEM equipment (JEOL JSM-5900-LV, Tokyo, Japan) operated at 15 KV. The samples were coated with gold using a sputter coater under a vacuum and analyzed using SEM [33].

2.9. In Vitro Release Studies

In vitro release studies of SCL from the prepared solid dispersions (SLC-K188, SLC-K30, and SLC-KVA64) compared to pure SLC drug were performed using an USP-2 dissolution apparatus (Fiber optic dissolution system, Model Distek 2500i, Software Rev 1.02, North Brunswick, NJ, USA). Briefly, an accurately weighed sample (equivalent to 50 mg of SLC) was dispersed in a dissolution basket containing 900 mL of phosphate buffer (pH 6.8). The dissolution apparatus was set to run at 100 rpm at 37 ± 0.5 °C. At a predetermined time interval, 5 mL of sample was withdrawn, compensated with fresh media, and analyzed for drug content using UV spectroscopy at 291 nm [35]. Each sample was analyzed in triplicate.

2.9.1. Mean Dissolution Time (MDT)

Model independent approaches such as mean dissolution time (MDT) were assessed to study the effect of different polymers within the SDs [36]. The rate of drug dissolved of each SD was expressed by MDT and was calculated using the following equation:

$$MDT = \frac{\sum_{j=1}^{n} t_j^* \Delta M_j}{\sum_{j=1}^{n} \Delta M_j}$$

where, (j) is the number of the sample (SDs), (n) is the number of samples in the dissolution study, (t^*_j) is the midpoint time between t and t ($j-1$), and (ΔMj) is the additional amount of drug dissolved between t and t ($j-1$).

2.9.2. Similarity Index (F2)

The similarity index or fit factor of the prepared SDs was assessed, as suggested by Moore and Flanner, comparing the dissolution profiles of the test (SDs) with the reference (SLC). The following equation was used to calculate the f_2 value.

$$f_2 = 50\ X\ log\{[1 + (1/n) \sum_{t-1}^{n} (Rt - Tt)^2]^{-0.5} \times 100\}$$

where, f_2 means similarity index, n stands for dissolution time, and Rt and Tt denote reference (pure drug) and test (SDs) dissolution values at time t. If f_2 values were <50, this suggests a significant difference between the dissolution profiles under comparative study [34,36].

2.9.3. Drug Release Kinetics

To study the release kinetics, dissolution data were then fit to the drug release kinetic models, and the correlation coefficient (R^2) was calculated using regression analysis. The zero-order rate describes the concentration independent release kinetics from the SDs, first-order specifies the concentration dependent release rate, the Higuchi' model depicts the release of drug based on Fickian diffusion, whereas the Korsmeyer–Peppas model equation demonstrates a relationship between the drug release from the polymeric SDs.

$$Qt = Q_0 + k_0 t \quad \text{(Zero-order)}$$

$$logQ_t = logQ_0 - k_1 t \ / \ 2.303 \ \text{(First-order)}$$

$$Qt = k_H t^{1/2} \quad \text{(Matrix diffusion)}$$

$$Mt/M\infty = kt^n \quad \text{(Korsmeyer-Peppas)}$$

where Q_t (dissolution of drug over time t), Q_0 (amount of drug dissolved in diffusion medium at zero time), k_0 (zero order constant), k_1 (first-order constant), and k_H (Higuchi model constant). Mt and $M\infty$ are the cumulative drug release at time t and infinite time, respectively; k is the rate constant of drug-polymer particle's feature, t is the release time. Diffusional exponent (n) indicates the drug release mechanism. When $n = 0.45$ (Case I or Fickian diffusion), $0.45 < n < 0.89$ (anomalous behavior or non-Fickian transport), $n = 0.89$ (Case II transport) and $n > 0.89$ (Super Case II), based on the exponent value release mechanisms reported.

2.10. *In Vivo Sexual Behavior Studies*

2.10.1. Animals

Male (250–300 g) and female albino rats (150–200 g) were used for the sexual behavior study [37]. All animals were bred in the lab care unit at the College of Pharmacy, Prince Sattam Bin Abdulaziz University, Alkharj, Saudi Arabia. Rats were kept in separate cages with access to food and water ad libitum. The study protocol was reviewed and approved by the "Animal Ethics Committee (Approval number: BERC 005-05-19), College of Pharmacy, Prince Satam Bin Abdulaziz University, Alkharj, Saudi Arabia".

2.10.2. Preparation of Male and Female Rats

The male rats were trained sexually with receptive females three times a day for four days before commencement of the experiment. The male rats that did not show any sexual activity during training were excluded from experiment. Eighteen sexually active male rats were selected for the sexual activity of sildenafil citrate and its optimized solid dispersion. The female rats were made receptive by administering estradiol benzoate (10 mg/kg body weight) and progesterone (1.5 mg/kg body weight) subcutaneously, 48 h and 4 h prior to pairing with male rats, respectively. The sexual activity of the female rats was confirmed prior to the test by exposing them to male rats. The most receptive female rats were marked and selected for the study.

2.10.3. Experimental Procedure

The aphrodisiac experiments were performed as per a previously reported method [35]. Eighteen healthy and sexually active male rats (250–300 g) were selected for the study. They were divided into three groups of 6 animals, each group was isolated alone in separate cages during the study. Group I (Control) received 1% w/v sodium carboxymethyl cellulose (Na-CMC) as a vehicle at rate of 5 mL/kg. Group II (reference) received SLC at a dose of 5 mg/kg, and, finally, group III (optimized formulation) received formulation (equivalent to 5 mg/kg SLC pure). The control vehicle, SLC and formulation were administered as a single dose by orogastric cannula one hour before the start of the study.

2.10.4. Monitoring of Sexual Behavior

The most sexually active female rats were selected for the study. The experiment was performed at 19:00 h in a noiseless room under dim red light in transparent cages. The single female rat was introduced into the cage of single male rat for 15 min, considered as an adaptive period, and after this period, the females were separated from the male cages, then control (Na-CMC), SLC suspension, and formulation were administered orally.

The treated female rat was again paired with same male rat in the cage, and the sexual behavior of the male rat was immediately started and continued for the first two matings. The following sexual behavior parameters were monitored and noted as described in a previous study [35,38].

The following definitions were considered for this test: mounting latency (ML), the time from the pairing of the female and male in one cage and first mount; intromission latency (IL), the time from the pairing of a female and male in one cage and first intromission (vaginal penetration) by the male; mount frequency (MF), the number of mounts before ejaculation, that is lifting of the male's fore body over the hind body of the female and clasping her flanks with his forepaw; intromission frequency (IF), the number of vaginal penetrations before ejaculation; ejaculation latency (EL), the time from the first vaginal penetration of a series to the ejaculation; post-ejaculatory interval (PEI), the time from ejaculation to the first vaginal penetration of the next copulatory series. In the second mating only the EL was recorded. Percentage copulatory efficiency (%CE) was calculated using the following equation:

$$\%CE = IF/MF \times 100$$
$$\%ICE = IF/IF + MF \times 100$$

2.11. Statistical Evaluation

The significance of difference between the means was determined by one-way analysis of variance (ANOVA) with a post-hoc test. A p-value < 0.05 was considered significant.

3. Results and Discussion

3.1. Practical Percentage Yield

The percentage yield of prepared SDs was determined, to ascertain the loss during the solvent evaporation process. SDs prepared by solvent evaporation showed percentage yield values ranging between 93 and 95.8%. The high percentage yield indicates the minimum loss, homogeneity, and accuracy of the process, and hence the suitability for scale-up.

3.2. Drug Content Estimation of Solid Dispersions

Drug content estimation determined the uniformity of the drug in the polymeric dispersion, the value of drug estimation was found to be in the range of 97–98.99%.

3.3. DSC Studies

DSC is a thermal analytical technique used in formulation development to understand the physicochemical properties of pure drugs and their formulations. The presence of crystallinity of pure SLC was detected by DSC spectra, because of a sharp endothermic peak at 207 °C (Figure 2), which was close to the previous reported data [39]. The DSC peaks indicate that SLC does not have any exothermic/degradation peaks, confirming its stability up to a temperature of 250 °C. The solid dispersions SLC-K188 and SLC-K30 showed a broad endothermic peak at 198 °C and 192 °C with reduced intensity, respectively, indicating their partial crystallinity. However, SLC-KVA64 showed the complete disappearance of endothermic peaks corresponding to SLC drug peaks, indicating the transformation of crystalline phase to amorphous phase, due to dissolution into the KVA64 polymer matrix. The SLC drug did not crystallize in the SLC-KVA64 system, due to complete dispersion in the polymer matrix.

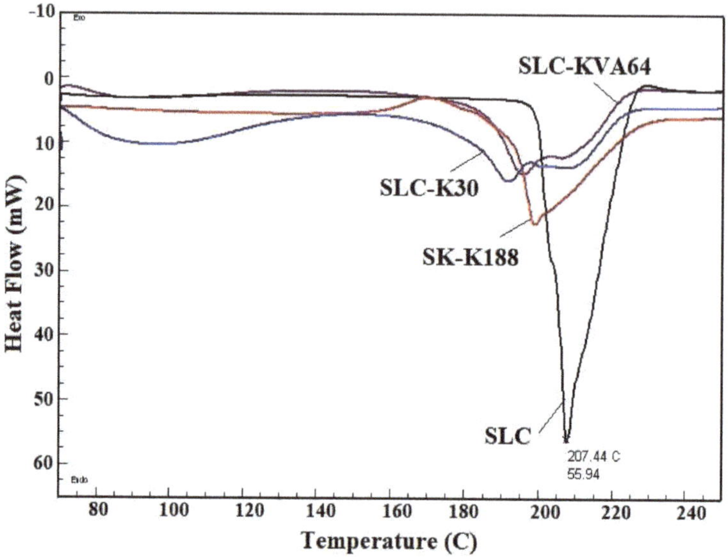

Figure 2. Comparative DSC thermogram of pure SLC, SLC-K188, SLC-K30, and SLC-KVA64.

3.4. FTIR Studies

FTIR spectra of pure SLC and solid dispersions SLC-K188, SLC-K30, and SLC-KVA64 are presented in Figure 3. The pure SLC showed characteristic peaks in frequency at 1170 cm^{-1} and 1265 cm^{-1} for asymmetric and symmetric SO_2 bands. A strong peak at 1495 cm^{-1} could be assigned to the –COOH group present in citric acid. The two strong peaks could be attributed to the –N–H bend and –N–H stretching at frequency 1582 cm^{-1} and 3301 cm^{-1}; these assigned peaks confirmed the purity of the drug [35,40]. Compared to the pure SLC, the characteristic peak of SLC was absent or weakened in the prepared solid dispersions (SLC-K188, SLC-K30, and SLC-KVA64), confirming the successful dispersion of polymers with drug [41].

3.5. PXRD Studies

PXRD spectral analysis is a useful tool for identifying the crystalline/amorphous nature of a solid state powder. The PXRD patterns of pure SLC, SLC-K188, SLC-K30, and SLC-KVA64 are shown in Figure 4. The PXRD pattern of pure SLC revealed several diffraction peaks between 0 and 60° (2θ), which confirmed the crystalline nature of the drug [37,41]. The solid dispersions SLC-K188 and SLC-K30 showed a few diffraction peaks corresponding to SLC with reduced intensity, indicating partial crystallinity. However, the PXRD pattern of the SD, SLC-KV64, showed a typical profile of an amorphous compound, suggesting that the polymer KVA64 inhibited the drug crystallization by reordering of the crystal lattice; this findings strongly supports the DSC analysis [42]. The amorphous powder influenced the faster dissolution of the drug, due to increased internal energy and molecular motion, which improved the thermodynamic property as compared to the pure SLC crystalline drug. The reduction of crystallinity of the drug could have been due to dispersion of the polymers (K188, K30, and KVA64) with the SLC drug [43].

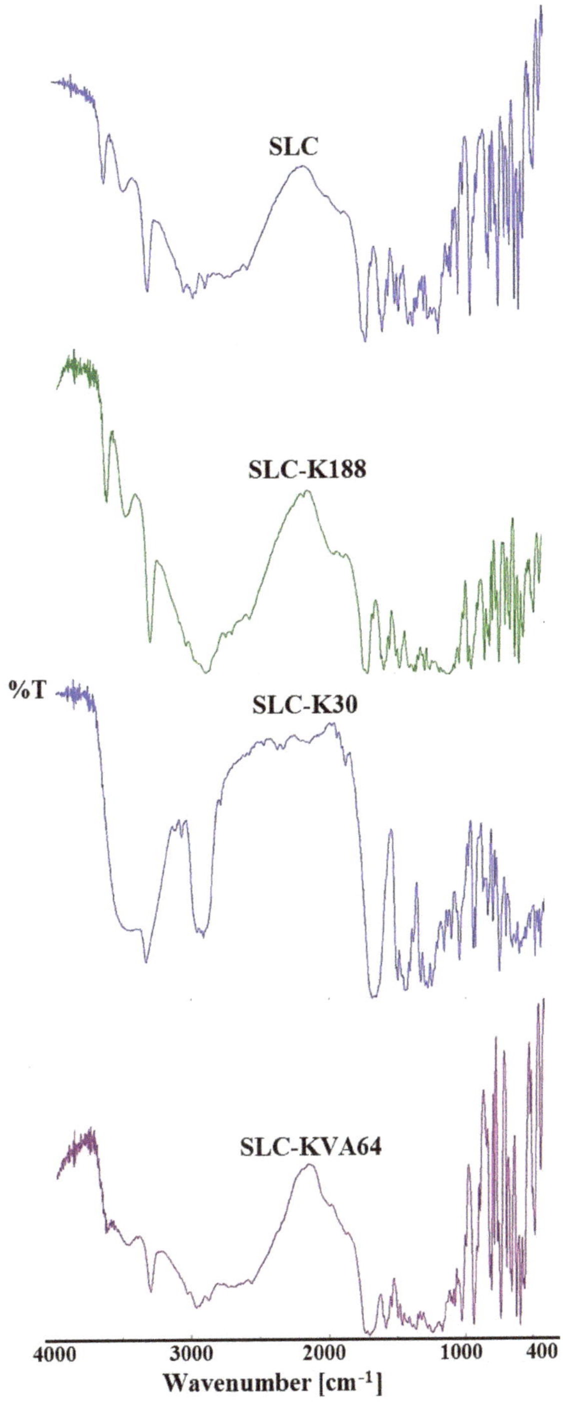

Figure 3. Comparative FTIR spectra of pure SLC, SLC-K188, SLC-K30, and SLC-KVA64.

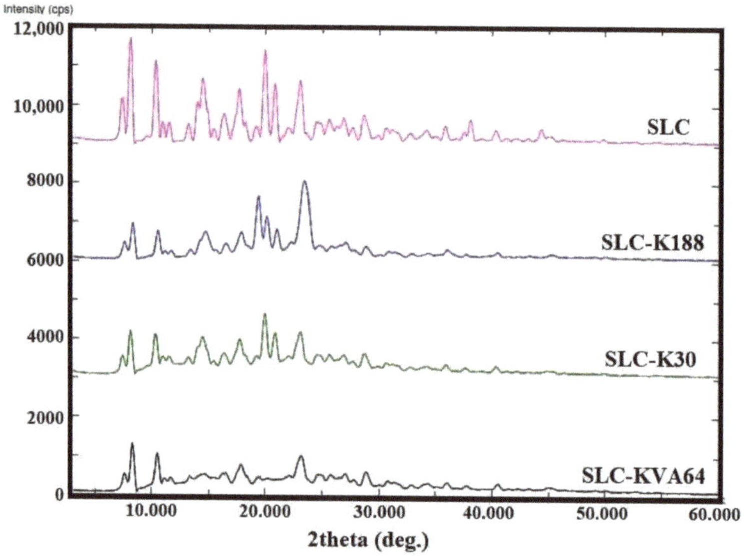

Figure 4. Comparative PXRD spectra of pure SLC, SLC-K188, SLC-K30, and SLC-KVA64.

3.6. SEM

Crystalline and amorphous solid dispersion can be differentiated visually using SEM images. According to the SEM images, the high crystallinity of SCL evidenced a large needle shaped powder. The SEM images of SDs (SLC-K188, SLC-K30, and SLC-KVA64) did not show any crystalline structure and the SDs appeared as aggregates of irregular shape. As can be seen, these polymers strongly disrupted the morphology of the SDs, due to transformation from a crystalline to amorphous state (Figure 5). The crystallinity of SLC was already confirmed by the XRD and DSC studies.

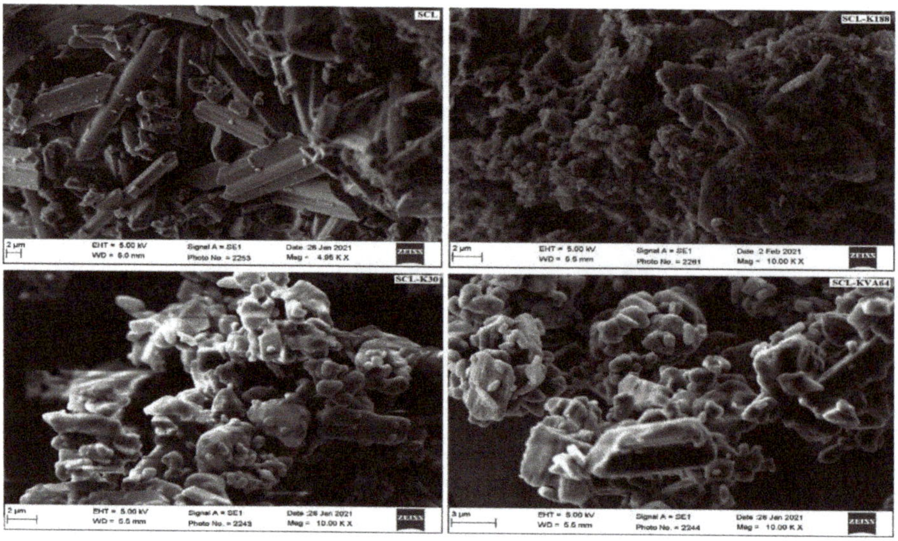

Figure 5. SEM images of pure SLC, SLC-K188, SLC-K30, and SLC-KVA64.

3.7. In Vitro Release Studies

Researchers have explored solid dispersion/inclusion complexation to improve the solubility, permeability, oral bioavailability, and therapeutic effectiveness of SLC through its encapsulation with polymers, with the aim of achieving an enhanced drug release [8]. In vitro release profiles of SLC from the prepared solid dispersions (SLC-K188, SLC-K30, and SLC-KVA64) are presented in Figure 6. Enhanced dissolution of SLC was noted in all solid dispersions. The pure SLC drug showed a much lower and incomplete release (29.08%) during the time-scale of the study (2 h). However, an almost complete release of SLC was observed from all the prepared solid dispersions. The maximum drug release was recorded by the SLC-KVA64 system (99.74%) after 2 h. The improvement in dissolution rate of SLC in a solid dispersion can be attributed to the dispersion of hydrophilic polymers, due to the wettability of the drug, which resulted in an enhancement in solubility [44]. These results suggest that the use of hydrophilic polymers as the carrier transformed the SLC crystals into an amorphous state, which improved the solubility of the drug. However, the highest release of SLC from SLC-KVA64 could possibly have been due to the maximum amorphization of SLC, wetting properties of SLC-KVA64 polymer with SLC.

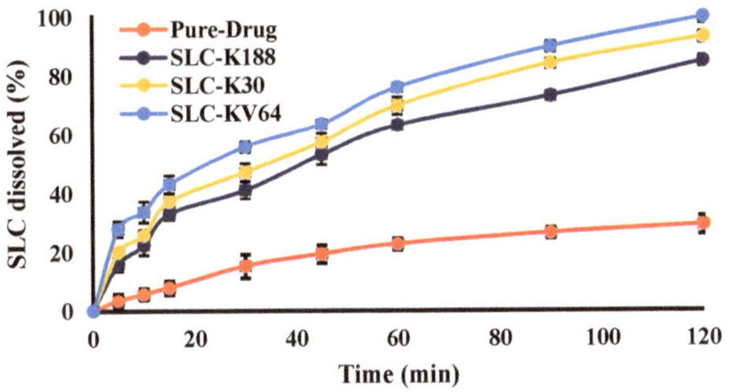

Figure 6. In vitro release profile of pure SLC, SLC-K188, SLC-K30, and SLC-KVA64.

The estimated values of similarity factor (f2) and MDT for all solid dispersions are summarized in Table 1. It was found that the f2 values for all formulae (SLC-K188, SLC-K30, and SLC-KVA64) were less than 50, suggesting all formulae are statistically not different ($p > 0.05$). The mean dissolution time is an indication of the dissolution process. The MDT for SLC-K188, SLC-K30, and SLC-KVA64 were estimated as 39.61, 37.65, and 35.52 min. The lowest MDT was measured for SLC-KVA64 among all the formulae, indicating a faster release of drug compared to the other formulae. However, the release kinetic models of fitted f and coefficient of correlation (R^2) values were obtained as first order ($R^2 = 0.999$), matrix diffusion ($R^2 = 0.996$), and matrix diffusion ($R^2 = 0.998$) for the formulae SLC-K188, SLC-K30, and SLC-KVA64, respectively.

3.8. In Vivo Sexual Behavior Studies

In this study, the SLC-KV64 formulation was tested and compared with pure SLC for its aphrodisiac effect on male rats. The male rats showed an improved sexual activity towards their female rat partner, shown by their eager and quick movement and visible signs of pre-copulatory action, such as anogenital exploration, body sniffing, and moving around, which finally resulted in mounting [35]. The data presented in Tables 2 and 3 show that SLC (5 mg/kg) significantly reduced the ML (67.27 ± 2.18 s) and IL (108.15 ± 3.52 s) and caused a significant increment in the MF (8.57 ± 0.37) and IF (5.36 ± 0.36) compared to the control. MF and IF are important indicators to measure the vigor, libido, and potency and that reflect the sexual motivation and efficiency of erection, respectively. The increase in MF and IF following administration of SLC-KV64 formulation was observed, suggesting the improved sexual behavior of rats [45]. As ML and IL values decreased following administration of SLC-KV64, this suggests stimulation of sexual motivation and arousal [46]. One formulation (SLC-KV64) also prolonged ejaculatory latency in the first and second series (EL-1 and EL-2) and caused a significant reduction in the post ejaculatory interval (PEI) compared to the control group. Due to EL and PEI events, the refractory period between the first and second series of mating, revealed that SLC in the formulation improved the sexual activity. The CE (%) and ICE (%) are presented in Figures 7 and 8. Improvements in CE (81.6%) and ICE (44.9%) were observed when rats were exposed to optimized SLC-KVA64 formulae in comparison to the pure SLC drug [37]. All observed sexual activity parameters of the optimized SLC-KVA64 were remarkably improved compared to the control and pure SLC drug; these parameters are statistically significant.

Table 1. Similarity factors and release kinetics of prepared SDs.

Solid Dispersions	MDT (min)	f2	Correlation Coefficient (R^2)				
			Zero Order	First Order	Matrix Diffusions	Korsmeyer-Peppas	N
SLC-K188	39.61	25.22	0.910	0.999	0.996	0.991	0.524
SLC-K30	37.65	21.68	0.907	0.989	0.996	0.992	0.495
SLC-KVA64	35.52	18.59	0.876	0.840	0.998	0.995	0.411

Table 2. Effect of SLC-KV64 on the mount latency (ML), mount frequency (MF), intromission latency (IL), intromission frequency (IF), and copulatory efficiency (CE) of male rats.

Groups	ML (s)	MF	IL (s)	IF
NC	122.71 ± 5.27	4.24 ± 0.48	245.15 ± 7.27	2.42 ± 0.15
SLC-STD	67.27 ± 2.18 *	8.57 ± 0.37 *	108.15 ± 3.52 *	5.36 ± 0.36 *
SKV64	59.16 ± 2.11 *‡	10.36 ± 0.42 *‡	97.78 ± 3.16 *‡	8.45 ± 0.50 *‡

Values are expressed as mean ± S.E.M., $n = 6$ rats/group. * indicates significance compared to NC group at $p < 0.05$. ‡ indicates significance compared to SLC-STD group at $p < 0.05$.

Table 3. Effect of SLC-KV64 on the ejaculation latency in the 1st series (EL-1), post ejaculatory interval (PEI) and ejaculation latency in the 2nd series (EL-2) of male rats.

Groups	EL-1 (s)	PEI (s)	EL-2 (s)
NC	376.62 ± 8.72	496.64 ± 14.73	395.20 ± 7.12
SDL-STD	423.16 ± 11.46 *	397.84 ± 8.46 *	421.22 ± 8.25 *
SKV64	458.38 ± 10.27 *‡	370.50 ± 8.50 *‡	449.20 ± 9.28 *‡

Values are expressed as mean ± S.E.M., $n = 6$ rats/group. * indicates significance compared to NC group at $p < 0.05$. ‡ indicates significance compared to SLC-STD group at $p < 0.05$.

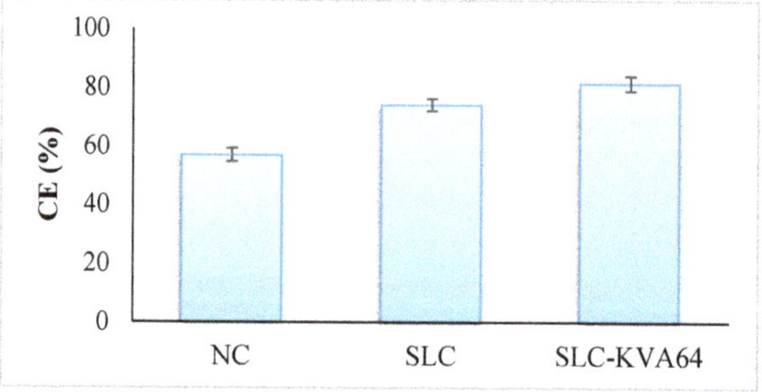

Figure 7. Effect of pure SLC and SLC-KV64 on CE.

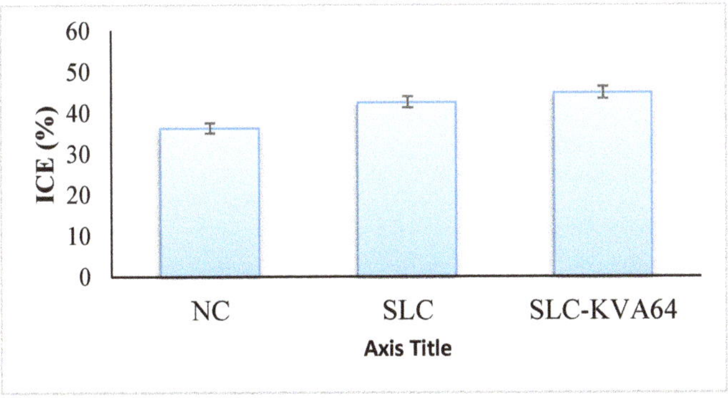

Figure 8. Effect of pure SLC and SLC-KV64 on ICE.

4. Conclusions

In this study, SDs of SLC were prepared by solvent evaporation method using three hydrophilic polymers as drug carriers, namely Kolliphor®P188, Kollidon®30, and Kollidon®-VA 64. The prepared SDs succeeded in improving the dissolution rate and sexual behavior in male rats. A marked influence of the polymers on SCL dissolution was noted. All SDs significantly improved the SCL dissolution compared to the pure drug. The optimized SD (SLC-KV64) system showed the maximum enhancement in dissolution rate compared to the pure SLC drug. The DSC and PXRD studies revealed the transformation of the crystalline state of SLC to an amorphous state, which required the lowest energy for drug solubilization. A significant improvement in sexual activity was observed in optimized SD (SLC-KV64) administered male rats, compared to pure SLC drug. Finally, we concluded from this research that the optimized SDs (SLC-KVA64) exhibited superior activity and are a promising strategy for improving solubility, dissolution rate, and aphrodisiac effects on male rats. Hence, the results suggest that the hydrophilic polymer Kollidon®-VA 64 could be an excellent carrier for enhancing the dissolution and therapeutic performance of SLC drug.

Author Contributions: Conceptualization, M.F.A. and M.K.A.; methodology, G.A.S.; M.M.A. and F.F.; formal analysis, M.M.A. and M.A.A.; investigation, M.F.A.; resources, M.K.A. and M.A.A.; writing—original draft preparation, M.K.A. and M.M.A.; writing—review and editing, A.Z., S.B. and M.A.A.; supervision, M.F.A., M.A.A. and M.K.A. All authors have read and agreed to the published version of the manuscript.

Funding: This research received no external funding.

Data Availability Statement: The data presented in this study are available on request from the corresponding author.

Acknowledgments: This publication was supported by the Deanship of Scientific Research at Prince Sattam Bin Abdulaziz University, Al-Kharj Saudi Arabia.

Conflicts of Interest: The authors declare no conflict of interest.

References

1. Yafi, F.A.; Jenkins, L.; Albersen, M.; Corona, G.; Isidori, A.M.; Goldfarb, S.; Maggi, M.; Nelson, C.J.; Parish, S.; Salonia, A.; et al. Erectile dysfunction. Nature reviews. *Dis. Primers* **2016**, *2*, 16003. [CrossRef]
2. Sikka, S.C.; Hellstrom, W.J.; Brock, G.; Morales, A.M. Standardization of vascular assessment of erectile dysfunction: Standard operating procedures for duplex ultrasound. *J. Sex Med.* **2013**, *10*, 120–129. [CrossRef] [PubMed]
3. Manecke, R.G.; Mulhall, J.P. Medical treatment of erectile dysfunction. *Ann. Med.* **1999**, *31*, 388–398. [CrossRef] [PubMed]
4. Jung, S.-Y.; Seo, Y.-G.; Kim, G.K.; Woo, J.S.; Yong, C.S.; Choi, H.-G. Comparison of the solubility and pharmacokinetics of sildenafil salts. *Arch. Pharm. Res.* **2011**, *34*, 451–454. [CrossRef] [PubMed]

5. Shin, H.S.; Bae, S.K.; Lee, M.G. Pharmacokinetics of sildenafil after intravenous and oral administration in rats: Hepatic and intestinal first-pass effects. *Int. J. Pharm.* **2006**, *320*, 64–70. [CrossRef]
6. Sawatdee, S.; Phetmung, H.; Srichana, T. Sildenafil citrate monohydrate-cyclodextrin nanosuspension complexes for use in metered-dose inhalers. *Int. J. Pharm.* **2013**, *455*, 248–258. [CrossRef]
7. Atipairin, A.; Sawatdee, S. Inclusion complexes between sildenafil citrate and cyclodextrins enhance drug solubility. *Asian J. Pharm. Sci.* **2016**, *11*, 104–105. [CrossRef]
8. Hosny, K.M.; El-Say, K.M.; Ahmed, O.A. Optimized sildenafil citrate fast orodissolvable film: A promising formula for overcoming the barriers hindering erectile dysfunction treatment. *Drug Deliv.* **2016**, *23*, 355–361. [CrossRef]
9. Sawatdee, S.; Atipairin, A.; Yoon, A.S.; Srichana, T.; Changsan, N.; Suwandecha, T.; Chanthorn, W.; Phoem, A. Oral bioavailability and pharmacokinetics of sildenafil citrate dry foam tablets in rats. *Cogent Med.* **2018**, *5*. [CrossRef]
10. Sanphui, P.; Tothadi, S.; Ganguly, S.; Desiraju, G.R. Salt and Cocrystals of Sildenafil with Dicarboxylic Acids: Solubility and Pharmacokinetic Advantage of the Glutarate Salt. *Mol. Pharm.* **2013**, *10*, 4687–4697. [CrossRef]
11. Hosny, K.M.; Alhakamy, N.A.; Almodhwahi, M.A.; Kurakula, M.; Almehmady, A.M.; Elgebaly, S.S. Self-Nanoemulsifying System Loaded with Sildenafil Citrate and Incorporated within Oral Lyophilized Flash Tablets: Preparation, Optimization, and In Vivo Evaluation. *Pharmaceutics* **2020**, *12*, 1124. [CrossRef]
12. Ahmed, M.M.; Fatima, F.; Anwer, M.K.; Aldawsari, M.F.; Soliman, G.A.; Fayed, M.H. Development and Characterization of Spray-dried Amorphous Solid Dispersion of Sildenafil: In vivo Evaluation. *Int. J. Pharmacol.* **2020**, *16*, 460–469. [CrossRef]
13. Vo, C.L.; Park, C.; Lee, B. Current trends and future perspectives of solid dispersions containing poorly water-soluble drugs. *Eur. J. Pharm. Biopharm.* **2013**, *85*, 799–813. [CrossRef]
14. Aldawsari, M.; Ahmed, M.; Fatima, F.; Anwer, M.K.; Katakam, P.; Khan, A. Development and Characterization of Calcium-Alginate Beads of Apigenin: In Vitro Antitumor, Antibacterial, and Antioxidant Activities. *Mar. Drugs* **2021**, *19*, 467. [CrossRef] [PubMed]
15. Zhang, X.; Xing, H.; Zhao, Y.; Ma, Z. Pharmaceutical Dispersion Techniques for Dissolution and Bioavailability Enhancement of Poorly Water-Soluble Drugs. *Pharmaceutics* **2018**, *10*, 74. [CrossRef] [PubMed]
16. Meng, F.; Gala, U.; Chauhan, H. Classification of solid dispersions: Correlation to (i) stability and solubility (ii) preparation and characterization techniques. *Drug Dev. Ind. Pharm.* **2015**, *41*, 1401–1415. [CrossRef]
17. Szafraniec-Szczęsny, J.; Antosik-Rogóz, A.; Kurek, M.; Gawlak, K.; Górska, A.; Peralta, S.; Knapik-Kowalczuk, J.; Kramarczyk, D.; Paluch, M.; Jachowicz, R. How Does the Addition of Kollidon®VA64 Inhibit the Recrystallization and Improve Ezetimibe Dissolution from Amorphous Solid Dispersions? *Pharmaceutics* **2021**, *13*, 147. [CrossRef]
18. Nair, A.R.; Lakshman, Y.D.; Anand, V.S.K.; Sree, K.S.N.; Bhat, K.; Dengale, S.J. Overview of Extensively Employed Polymeric Carriers in Solid Dispersion Technology. *AAPS PharmSciTech* **2020**, *21*, 309. [CrossRef] [PubMed]
19. Charalabidis, A.; Sfouni, M.; Bergström, C.; Macheras, P. The Biopharmaceutics Classification System (BCS) and the Biopharmaceutics Drug Disposition Classification System (BDDCS): Beyond guidelines. *Int. J. Pharm.* **2019**, *566*, 264–281. [CrossRef]
20. LaFountaine, J.S.; McGinity, J.W.; Williams, R.O., 3rd. Challenges and Strategies in Thermal Processing of Amorphous Solid Dispersions: A Review. *AAPS PharmSciTech* **2016**, *17*, 43–55. [CrossRef]
21. Li, B.; Konecke, S.; Harich, K.; Wegiel, L.; Taylor, L.S.; Edgar, K.J. Solid dispersion of quercetin in cellulose derivative matrices influences both solubility and stability. *Carbohydr. Polym.* **2013**, *92*, 2033–2040. [CrossRef]
22. Solanki, N.G.; Lam, K.; Tahsin, M.; Gumaste, S.G.; Shah, A.V.; Serajuddin, A.T.M. Effects of Surfactants on Itraconazole-HPMCAS Solid Dispersion Prepared by Hot-Melt Extrusion I: Miscibility and Drug Release. *J. Pharm. Sci.* **2019**, *108*, 1453–1465. [CrossRef] [PubMed]
23. Ali, W.; Williams, A.C.; Rawlinson, C.F. Stochiometrically governed molecular interactions in drug: Poloxamer solid dispersions. *Int. J. Pharm.* **2010**, *391*, 162–168. [CrossRef] [PubMed]
24. Ramadhani, N.; Shabir, M.; McConville, C. Preparation and characterisation of Kolliphor®P 188 and P 237 solid dispersion oral tablets containing the poorly water soluble drug disulfiram. *Int. J. Pharm.* **2014**, *475*, 514–522. [CrossRef]
25. Islam, N.; Irfan, M.; Khan, S.-U.-D.; Syed, H.K.; Iqbal, M.S.; Khan, I.U.; Mahdy, A.; Raafat, M.; Hossain, M.A.; Inam, S.; et al. Poloxamer-188 and D-α-Tocopheryl Polyethylene Glycol Succinate (TPGS-1000) Mixed Micelles Integrated Orodispersible Sublingual Films to Improve Oral Bioavailability of Ebastine; In Vitro and In Vivo Characterization. *Pharmaceutics* **2021**, *13*, 54. [CrossRef] [PubMed]
26. Yen, S.; Chen, C.; Lee, M.; Chen, L. Investigation of Dissolution Enhancement of Nifedipine by Deposition on Superdisintegrants. *Drug Dev. Ind. Pharm.* **2008**, *23*, 313–317. [CrossRef]
27. Alves, L.D.S.; Soares, M.F.D.L.R.; de Albuquerque, C.T.; da Silva, É.R.; Vieira, A.; Fontes, D.A.F.; Figueirêdo, C.B.M.; Sobrinho, J.L.S.; Neto, P.R. Solid dispersion of efavirenz in PVP K-30 by conventional solvent and kneading methods. *Carbohydr. Polym.* **2014**, *104*, 166–174. [CrossRef]
28. Kolter, K.; Flick, D. Structure and dry binding activity of different polymers, including Kollidon VA 64. *Drug Dev. Ind. Pharm.* **2000**, *26*, 1159–1165. [CrossRef]
29. Chaudhary, R.S.; Patel, C.; Sevak, V.; Chan, M. Effect of Kollidon VA®64 particle size and morphology as directly compressible excipient on tablet compression properties. *Drug Dev. Ind. Pharm.* **2018**, *44*, 19–29. [CrossRef] [PubMed]
30. Moseson, D.E.; Parker, A.S.; Beaudoin, S.P.; Taylor, L.S. Amorphous solid dispersions containing residual crystallinity: Influence of seed properties and polymer adsorption on dissolution performance. *Eur. J. Pharm. Sci.* **2020**, *146*, 105276. [CrossRef]

31. Chavan, R.B.; Lodagekar, A.; Yadav, B.; Shastri, N.R. Amorphous solid dispersion of nisoldipine by solvent evaporation technique: Preparation, characterization, in vitro, in vivo evaluation, and scale up feasibility study. *Drug Deliv. Transl. Res.* **2020**, *10*, 903–918. [CrossRef]
32. Anwer, M.K.; Mohammad, M.; Ezzeldin, E.; Fatima, F.; Alalaiwe, A.; Iqbal, M. Preparation of sustained release apremilast-loaded PLGA nanoparticles: In vitro characterization and in vivo pharmacokinetic study in rats. *Int. J. Nanomed.* **2019**, *14*, 1587–1595. [CrossRef]
33. Anwer, M.K.; Ahmed, M.M.; Alshetaili, A.; Almutairy, B.K.; Alalaiwe, A.; Fatima, F.; Ansari, M.N.; Iqbal, M. Preparation of spray dried amorphous solid dispersion of diosmin in soluplus with improved hepato-renoprotective activity: In vitro anti-oxidant and in-vivo safety studies. *J. Drug Deliv. Sci. Technol.* **2020**, *60*, 102101. [CrossRef]
34. Alshehri, S.M.; Shakeel, F.; Ibrahim, M.A.; Elzayat, E.M.; Altamimi, M.; Mohsin, K.; Almeanazel, O.T.; Alkholief, M.; Alshetaili, A.; Alsulays, B.; et al. Dissolution and bioavailability improvement of bioactive apigenin using solid dispersions prepared by different techniques. *Saudi Pharm. J.* **2019**, *27*, 264–273. [CrossRef] [PubMed]
35. Al-Shdefat, R.; Ali, B.E.; Anwer, M.K.; Fayed, M.H.; Alalaiwe, A.; Soliman, G.A. Sildenafil citrate-Glycyrrhizin/Eudragit binary spray dried microparticles: A sexual behavior studies on male rats. *J. Drug Deliv. Sci. Technol.* **2016**, *36*, 141–149. [CrossRef]
36. Fouad, S.A.; Malaak, F.A.; El-Nabarawi, M.A.; Abu Zeid, K.; Ghoneim, A.M. Preparation of solid dispersion systems for enhanced dissolution of poorly water soluble diacerein: In-vitro evaluation, optimization and physiologically based pharmacokinetic modeling. *PLoS ONE* **2021**, *16*, e0245482. [CrossRef] [PubMed]
37. Ahmed, M.M.; Fatima, F.; Abul Kalam, M.; Alshamsan, A.; Soliman, G.A.; Shaikh, A.A.; Alshahrani, S.M.; Aldawsari, M.F.; Bhatia, S.; Khalid Anwer, M. Development of spray-dried amorphous solid dispersions of tadalafil using glycyrrhizin for enhanced dissolution and aphrodisiac activity in male rats. *Saudi Pharm. J.* **2020**, *28*, 1817–1826. [CrossRef] [PubMed]
38. Besong, E.B.; Ateufack, G.; Babiaka, S.B.; Kamanyi, A. Leaf-Methanolic Extract of *Pseudopanax arboreus* (Araliaceae) (L. F. Phillipson) Reverses Amitriptyline-Induced Sexual Dysfunction in Male Rats. *Biochem. Res. Int.* **2018**, *15*, 2869727. [CrossRef] [PubMed]
39. Maria, J.; Noordin, M.I. Fast detection of sildenafil in adulterated commercial products using differential scanning calorimetry. *J. Therm. Anal. Calorim.* **2014**, *115*, 1907–1914. [CrossRef]
40. Melnikov, P.; Corbi, P.P.; Cuin, A.; Cavicchioli, M.; Guimarães, W.R. Physicochemical properties of sildenafil citrate (Viagra) and sildenafil base. *J. Pharm. Sci.* **2003**, *92*, 2140–2143. [CrossRef]
41. Almutairy, B.; Alshetaili, A.; Alali, A.; Ahmed, M.; Anwer, K.; Aboudzadeh, M. Design of Olmesartan Medoxomil-Loaded Nanosponges for Hypertension and Lung Cancer Treatments. *Polymers* **2021**, *13*, 2272. [CrossRef] [PubMed]
42. Kanaze, F.I.; Kokkalou, E.; Niopas, I.; Georgarakis, M.; Stergiou, A.; Bikiaris, D. Thermal analysis study of flavonoid solid dispersions having enhanced solubility. *J. Therm. Anal. Calorim.* **2006**, *83*, 283–290. [CrossRef]
43. Van Den Mooter, G. The use of amorphous solid dispersions: A formulation strategy to overcome poor solubility and dissolution rate. *Drug Discov. Today Technol.* **2012**, *9*, e79–e85. [CrossRef] [PubMed]
44. El-nawawy, T.; Swailem, A.M.; Ghorab, D.; Nour, S. Solubility enhancement of olmesartan by utilization of solid dispersion and complexation techniques. *Int. J. Novel Drug Deliv. Tech.* **2012**, *2*, 297–303.
45. Tajuddin Ahmad, S.; Latif, A.; Qasmi, I.A. Effect of 50% ethanolic extract of *Syzygium aromaticum* (L.) Merr. & Perry. (clove) on sexual behaviour of normal male rats. *BMC Complement. Altern. Med.* **2004**, *4*, 17. [CrossRef]
46. Yakubu, M.T.; Afolayan, A.J. Effect of aqueous extract of Bulbine natalensis (Baker) stem on the sexual behaviour of male rats. *Int. J. Androl.* **2009**, *32*, 629–636. [CrossRef]

polymers

MDPI

Article

Development and Evaluation of Polyvinylpyrrolidone K90 and Poloxamer 407 Self-Assembled Nanomicelles: Enhanced Topical Ocular Delivery of Artemisinin

Chandrasekar Ponnusamy [1], Abimanyu Sugumaran [2], Venkateshwaran Krishnaswami [1], Rajaguru Palanichamy [3], Ravichandiran Velayutham [4] and Subramanian Natesan [4,*]

1 Department of Pharmaceutical Technology, University College of Engineering, Bharathidasan Institute of Technology Campus, Anna University, Tiruchirappalli 620024, Tamil Nadu, India; chandruu0079@gmail.com (C.P.); venkpharm@gmail.com (V.K.)
2 Department of Pharmaceutics, SRM College of Pharmacy, SRM Institute of Science and Technology, Kattankulathur 603203, Tamil Nadu, India; abipharmastar@gmail.com
3 Department of Life Sciences, School of Life Sciences, Central University of Tamil Nadu, Tiruvarur 627007, Tamil Nadu, India; rajaguru62@gmail.com
4 Department of Pharmaceutics, National Institute of Pharmaceutical Education and Research (NIPER)—Kolkata, Chunilal Bhawan, 168, Maniktala Main Road, Kolkata 700054, West Bengal, India; directorniperkolkata@gmail.com
* Correspondence: natesansubbu1@gmail.com; Tel.: +91-996-563-0370

Citation: Ponnusamy, C.; Sugumaran, A.; Krishnaswami, V.; Palanichamy, R.; Velayutham, R.; Natesan, S. Development and Evaluation of Polyvinylpyrrolidone K90 and Poloxamer 407 Self-Assembled Nanomicelles: Enhanced Topical Ocular Delivery of Artemisinin. *Polymers* **2021**, *13*, 3038. https://doi.org/10.3390/polym13183038

Academic Editors: M. Ali Aboudzadeh and Shaghayegh Hamzehlou

Received: 3 June 2021
Accepted: 26 July 2021
Published: 8 September 2021

Abstract: Age-related macular degeneration is a multifactorial disease affecting the posterior segment of the eye and is characterized by aberrant nascent blood vessels that leak blood and fluid. It ends with vision loss. In the present study, artemisinin which is poorly water-soluble and has potent anti-angiogenic and anti-inflammatory properties was formulated into nanomicelles and characterized for its ocular application and anti-angiogenic activity using a CAM assay. Artemisinin-loaded nanomicelles were prepared by varying the concentrations of PVP k90 and poloxamer 407 at different ratios and showed spherical shape particles in the size range of 41–51 nm. The transparency and cloud point of the developed artemisinin-loaded nanomicelles was found to be 99–94% and 68–70 °C, respectively. The in vitro release of artemisinin from the nanomicelles was found to be 96.0–99.0% within 8 h. The trans-corneal permeation studies exhibited a 1.717–2.169 μg permeation of the artemisinin from nanomicelles through the excised rabbit eye cornea for 2 h. Drug-free nanomicelles did not exhibit noticeable DNA damage and showed an acceptable level of hemolytic potential. Artemisinin-loaded nanomicelles exhibited remarkable anti-angiogenic activity compared to artemisinin suspension. Hence, the formulated artemisinin-loaded nanomicelles might have the potential for the treatment of AMD.

Keywords: nanomicelles; artemisinin; cornea; toxicity

1. Introduction

Current treatment options adopted for posterior segment eye diseases such as age-related macular degeneration (AMD) are far from satisfactory due to the limited exposure of therapeutic drugs to the posterior segment of the eye with poor bioavailability [1,2]. Attempts to increase the dose of drugs may lead to posterior segment toxicity [3,4]. Site specific intravitreal injections/implants may improve the localized drug concentration which ultimately results in cataract development, endophthalmitis, haemorrhage and retinal detachment [5]. Topical drops, the most widely adopted method for ocular drug delivery, also suffer from precorneal clearance, lacrimation, tear dilution and tear turnover and result in a low bioavailability [6,7]. Topically administered drugs are penetrated and distributed into the posterior segment of the eye through diffusion across ocular tissues, trans-corneal permeation, direct entry through uvea and lateral diffusion across sclera and conjunctiva [8,9]. The topical application of particulate drug delivery systems such as nanosuspension, nanodispersion, nanocrystals, nanogels and nanomicelles has shown enhanced ocular bioavailability by minimizing the precorneal loss, increasing the corneal residence time, improving the corneal penetration and providing controlled or prolonged drug delivery to the disease site [10].

Nanomicelles are colloidal dispersions of surfactant (s) or polymeric surfactant (s) molecule aggregates with a fairly narrow size distribution of spherically shaped particles (10 to 100 nm). These systems have been reported to improve the solubility of poorly water-soluble drugs by encapsulating them into shells with improved stability [11,12]. Polymeric micelles are generally more stable compared to surfactant micelles and they are mainly used to treat the affected areas due to their bio adhesive properties [13].

Polyvinylpyrrolidone, a synthetic polymer, with the chemical term 1-Ethenyl-2-pyrrolidinone homopolymer, is widely used as a dissolution enhancer for poorly soluble drugs. Due to its strong hydrophilic characteristics, it may improve water penetration and the wettability of the hydrophobic drugs. Poloxamer 407, made up of nonionic polyoxy ethylene–polyoxy propylene copolymers, is used to enhance the solubilization of poorly water-soluble drugs and prolong the drug release [14,15]. The gel to sol transition has been reported for the poloxamers-based ocular delivery of (timolol maleate) upon topical administration [14,16].

Artemisinin is a poorly water-soluble cadinene-type sesquiterpene lactone isolated from Artemisia annua (Family-Compositae), chemically known as (3R, 5aS, 6R, 8aS, 9R, 12S, 12aR)-octahydro-3, 6, 9-trimethyl-3, 12-epoxy-12H-pyrano [4,3-j]-1,2-benzodioxepin-10 (3H)-one. Artemisinin possesses anti-inflammatory and anti-angiogenic activities along with its potent anti-malarial activity. The presence of the endoperoxide bridge (–C–O–O–C–) is responsible for its potent anti-cancer and anti-malarial activities with low toxicity. The endoperoxide bond in artemisinin forms an alkoxyl radical by accepting an electron from heme, thereby artemisinin becomes activated and exerts its anti-angiogenic effect [17–19]. Artemisinin has been reported to inhibit both NF-kB activation and the vascular endothelial growth factor (VEGF) which are the main factors in the development of AMD [20,21]. Hence, it is hypothesized that artemisinin can be useful for the treatment of AMD [20–22]. Further corneal permeation of artemisinin through the rabbit cornea has not been reported to best of our knowledge.

The aim of the present study was to enhance the aqueous solubility of artemisinin by using the combination of PVP K90 and poloxamer 407 nanomicelles and to evaluate its corneal permeability. Furthermore, the hemolytic potential, genotoxicity and critical micellar concentration of the developed nanomicelles were also evaluated. The anti-angiogenic potential of the developed nanomicelles was evaluated by chick embryo chorioallantoic membrane (CAM) assay.

2. Materials and Methods

2.1. Materials

Artemisinin was obtained from Herbochem, (Hyderabad, India). Polyvinylpyrrolidone (PVP K90) (Molecular weight—40,000 Da) and Poloxamer 407 (Molecular weight—95,000–110,000 Da) were procured from BASF Corporation, Mumbai, India. Potassium dihydrogen orthophosphate and sodium hydroxide were purchased from SD fine chemicals, Mumbai, India. All other solvents and reagents used were of HPLC grade. Milli Q water was used for HPLC analysis.

2.2. Compatibility Study

The drug and excipient compatibility were assessed by the Fourier Transform Infra-Red spectrometer (Spectrum Two, Perkinelmer, Waltham, MA, USA). The potassium bromide (KBr) pellets were prepared by grinding KBr with respective samples (artemisinin/PVP K90/poloxamer 407/physical mixture of artemisinin with excipients, processed mixture and/artemisinin loaded nanomicelles) and pressed into transparent pellets using a hydraulic press [22]. The IR spectra were recorded at the region of 4000–400 cm^{-1}.

2.3. Preliminary Screening for Nanomicelles Formation Capacity

Different mass ratios of PVP K90 and poloxamer 407 (1:1, 1:2, 1:3, 1:4, 1:5, 2:1, 2:2 etc., [1 = 5%]) and 100 mg of artemisinin were dissolved in ethanol, stirred to form homogeneous ethanolic solution and dried under vacuum to form artemisinin-loaded polymeric particles. The obtained mass was treated with sterile water, vortexed to form stable nanomicelles and examined for turbidity after 24 h.

2.4. Design of Experiment (DoE)

The central composite rotatable design–response surface methodology (CCRD–RSM) at two factor and three levels (i.e., −1, 0, +1) was used to optimize the artemisinin-loaded nanomicelles. The polymer (PVP K90) and polymeric surfactant (Poloxamer 407) concentrations were kept as input variables. The particle size and transparency of the developed formulations were kept as the response variables. The concentrations of the PVP K 90 (0–10% *w/v*) and poloxamer 407 (0–5% *w/v*) were selected based on the acceptable levels reported by USFDA for ophthalmic use [16,17]. The experimental design and statistical analysis of the obtained data were performed using the Design Expert Software (Version 6, Stat-Ease Inc., Minneapolis, MN, USA).

2.5. Preparation of Artemisinin Loaded Nanomicelles

A weighed amount of artemisinin was dissolved along with PVP K90 and Poloxamer 407 (2:1) in ethanol and was spray dried to form artemisinin-loaded polymeric particles. These polymeric particles were dissolved in a phosphate buffer with a pH of 7.4 and were vortexed to form stable artemisinin-loaded nanomicelles (Table 1).

Table 1. Composition of artemisinin nanomicelles (ANM) and blank nanomicelles (BNM).

Nanomicelles	Composition (%)		
	Polyvinyl Pyrolidone K90	**Poloxamer 407**	**Artemisinin**
ANM 1	5	2.5	0.05
ANM 2	8	4	0.05
ANM 3	10	5	0.05
BNM 1	5	2.5	-
BNM 2	8	4	-
BNM 3	10	5	-

2.6. Determination of the Critical Micellar Concentration (CMC)

The critical micellar concentration (CMC) of artemisinin-loaded nanomicelles was determined by the pyrene-based fluorescent probe method clearly described by Li et al. and Mohr et al. [23,24]. In this method, a solution of 10 mM pyrene was prepared using a phosphate buffer saline (PBS). Artemisinin nanomicelles and blank nanomicelles were prepared at a concentration of 30–1500 µg/mL using a saline phosphate buffer. Pyrene was added in order to achieve a 0.1 µM solution in concentration and was incubated for 30 min at room temperature in the dark. The sample was excited at 336 nm and fluorescence intensity was measured with dual emission at 375/384 nm using a microplate reader (Molecular Devices, San Jose, CA, USA). The ratio of intensity (I384/I375) was computed and compared to the nanomicelles' log concentrations.

2.7. Characterization of Artemisinin-Loaded Nanomicelles

2.7.1. Transparency

The transparency of the formulated artemisinin-loaded nanomicelles was evaluated at 400 nm by a UV spectrophotometer using distilled water as a blank [25].

2.7.2. Cloud Point

The cloud point of artemisinin-loaded nanomicelles was examined by placing a 100-fold diluted sample with water in a water bath and subjecting it to increases in temperature gradually. The temperature at which a decrease in transmittance occurred due to cloudiness formation was noted visually as the cloud point [26].

2.7.3. Particle Size and Zeta Potential

The average particle size and zeta potential of the developed artemisinin-loaded nanomicelles was determined by using a Zetasizer (Nano ZS, Malvern Instruments, Malvern, UK). The homogeneity of the particle size distribution was indicated by its polydispersity index (PDI).

2.7.4. Transmission Electron Microscopy (TEM)

Transmission electron microscopy (Philips EM-430; Philips Electronics, Eindhoven, The Netherlands) was employed to investigate the surface morphology of nanomicelles and it was operated at a driving voltage of 200 kV. A drop of 1.3% phosphotungstic acid was applied to the samples before they were placed over the carbon-coated copper grid. The grid was vacuum dried before being mounted on a grid holder and its morphology was examined.

2.7.5. Drug Content

The amount of artemisinin present in the artemisinin-loaded nanomicelles was checked by RP-HPLC (Shimadzu Corporation, Kyoto, Japan, LC- 20AD). Data acquisition was performed using spinchrome-1 software. A reverse phase phenomenex-C18 (5 μm, 4.6 mm × 250 mm) analytical column was used. The mobile phase consisting of the mixture of acetonitrile: water (65:35% *v/v*) was used at a flow rate of 1.0 mL/min and the UV detection was carried out at 219 nm [27].

2.8. In Vitro Hemolytic Potential

The in vitro hemolytic potential of the nanomicelles was evaluated by the method described by Amin et al. (2006) [28]. Briefly, 100 μL of blank nanomicelles were combined with 900 μL of fresh peripheral blood from healthy human volunteers and equilibrated at 25 °C for 2 min and centrifuged at 1900× *g* at 25 °C for 5 min. The precipitates were washed 4 times with 5 mL of normal saline after discarding the supernatant. The unlysed red blood cells were treated and vortexed with 4 mL of sterile water for supplementation with haemoglobin release and had a centrifugal effect for 5 min at 1900× *g*. The obtained supernatant was diluted with distilled water and absorbance was measured at 540 nm. Normal saline (0% lysis) and a 1% sodium carbonate solution (100% lysis) were used as negative and positive controls, respectively. The hemolytic capacity of the nanomicelles was estimated by the formula:

$$\%\ Hemolysis = \frac{Negative\ Control - Blank\ Nanomicelles}{Negative\ Control - Positive\ Control} \times 100 \tag{1}$$

2.9. Evaluation of the Genotoxicity of the Nanomicelles (Alkaline Comet Assay)

The toxic effect of the blank nanomicelles (BNM 2 and BNM 3) on the genetic materials of the cells was evaluated by the alkaline comet assay as previously described by Natesan et al. [29] using cells treated with H_2O_2 (100 μM) as the positive control.

2.10. In Vitro Drug Release

The in vitro drug release of the artemisinin-loaded nanomicelles was evaluated using a dialysis bag method (cut off 5000 Da, Himedia, Mumbai, India). The drug-loaded nanomicelles (1.0 mL) were placed in a dialysis bag which was hermetically sealed and suspended in 50 mL of a phosphate buffer solution at a pH of 7.4. The temperature was maintained at 34 ± 0.1 °C using a closed double jacketed thermostatic chamber and stirred at 600 RPM using a magnetic stirrer [30,31]. Aliquots (2.0 mL) were withdrawn at predetermined intervals and replaced by an equal volume of the fresh dialyzing medium. The samples were analyzed using RP-HPLC.

2.11. In Vitro Trans-Corneal Permeation Studies

The in vitro trans-corneal permeation effect of artemisinin-enriched nanomicelles were studied with a modified side-by-side of a Franz diffusion cell. Both chambers (donor and receptor) were built with an internal water jacket for water circulation and side braces to keep the system temperature constant. The protocol for animal research was authorized by the Institutional Animal Ethical Committee and animals were kept in accordance with approved guidance. In combination with the 2–4 mm surrounding scleral tissue, the male albino rabbit cornea was carefully removed and rinsed with cold saline. The excised cornea was positioned between the cells of the donor and the receivers in such a manner to allow the cornea's epidermal surface to face the donor compartment with a diffusion area of 0.78 cm^2. The phosphate buffer with a pH of 7.4 was deposited into the compartment and maintained at 34 ± 0.1 °C followed by a magnetic stirring of 150 rpm. The pre-heated (34 °C) drug-loaded nanomicelles were inserted into the donor cell at specified time intervals and the samples of the reservoir cell (0.3 mL) were removed and replaced by fresh media. The samples were evaluated using the RP-HPLC technique as previously described by Chandrasekar et al. [27].

2.12. Apparent Permeability Coefficient

The apparent permeability coefficient was calculated using the following formula:

$$Papp = \frac{\Delta Q}{\Delta t} \cdot \frac{1}{(A.Co.60)} \quad (2)$$

where, $\Delta Q/\Delta t$ is the flux across the cornea (ng/min), A is the diffusion area exposed, Co is the starting concentration of the artemisinin in a donor region, and 60 is regarded as the minute-to-second factor. The corneal flux was derived from the path of the regression line, which was drawn for the amount of drug (Q) penetrated by time (t) [32,33].

2.13. Histological Examination of the Drug-Permeated Cornea

The corneal permeate tissue obtained after the permeation study was washed with a phosphate buffer with a pH of 7.4 and was stored in a 10% formalin solution. The corneal permeate tissue was dehydrated with ethanol, embedded in paraffin, cut into vertical sections using microtome, stained with hematoxylin eosin and observed under a light microscope for any pathological change using untreated corneal tissue as a control [34].

2.14. Evaluation of the Anti-Angiogenic Effect of Artemisinin Nanomicelles Using a Chorioallantoic Membrane Assay (CAM Assay)

The anti-angiogenic effect of the nanomicelles formulation was evaluated using a Chorioallantoic Membrane Assay (CAM) as described by Ponnusamy et al. and Velpandian et al. [22,35]. Fertilized chicken eggs were incubated in an incubator at 37 °C for 3 days. The eggs were turned horizontally many times and then swabbed with 70% alcohol on the third day. By withdrawing the albumin (2.0 mL) from the fertilized eggs, the growing CAM was separated from the eggshell and an incision was made to create a window which was utilized as an entry point to the CAM. Sterile parafilm was used to close the window. The viable eggs were horizontally inserted and incubated for up to

5 days. On the 5th day the embryos were imaged by digital camera to reveal excising blood vessels on the window. A solution of artemisinin/blank nanomicelles/artemisinin-loaded nanomicelles on an impregnated filter paper disc (each 50 µg/disk) was put directly onto an exposed blood vessel using sterile surgical forceps on an increasing CAM and further incubated for 2 days. On day 7, the filter paper discs were carefully removed from the CAM and their anti-angiogenic impact was assessed and photographed at the sample applied area. A semi-quantitative score system was used to access the anti-angiogenic impact.

2.15. Stability Studies

The stability study of the artemisinin-loaded nanomicelles was carried out based on ICH guidelines by storing the samples at 40 ± 2 °C/75 ± 5% RH for 180 days in the stability chamber.

2.16. Statistical Analysis

Triplication of all experiments were completed and the data are given as a mean ± standard deviation. The student's "*t*" test was used to examine statistical data.

3. Results and Discussions

3.1. Compatibility Studies (FTIR)

The interactions between the drug and the excipients were analyzed by comparing the FTIR spectra of the pure drug, the individual excipients, the physical mixture, the processed mixture of the polymer and the polymeric surfactant and the artemisinin-loaded nanomicelles.

An artemisinin IR spectrum contains unique bands of IR regions such as, 1116 cm^{-1} for the C–O–O–C bending vibrations of the endoperoxide ring, 1736 cm^{-1} for the C=O lactone ring stretching and 1012 cm^{-1}, 1200 cm^{-1} and 3455 cm^{-1} stretching for the C–O, C–O–C and O–H stretching vibrations, respectively (Figure 1). The symmetric CH_3 stretch was marked by identifying the fermi resonance at 2952 cm^{-1} with overtones of methyl bending. The IR absorption bands at 883 cm^{-1} and 831 cm^{-1} were responsible for O–O–C and O–O stretching as the boat/twist form of 1, 2, 4-trioxane [36]. The characteristic IR absorption pattern of PVP K90 was shown in the bands of 3447 cm^{-1} with O–H stretching vibrations, C–H stretching vibration at 2886 cm^{-1}, C=O carbonyl stretching at 1655 cm^{-1}, C–H stretching at 1375 cm^{-1}, and C–N stretching vibration at 1281 cm^{-1} [18]. The properties of Poloxamer 407 include the absorption bands for aliphatic IR vibration C–H at 2921 cm^{-1}, the vibration C–O stretches at 1117 cm^{-1} and the vibration O–H stretches at 3452 cm^{-1} [37,38].

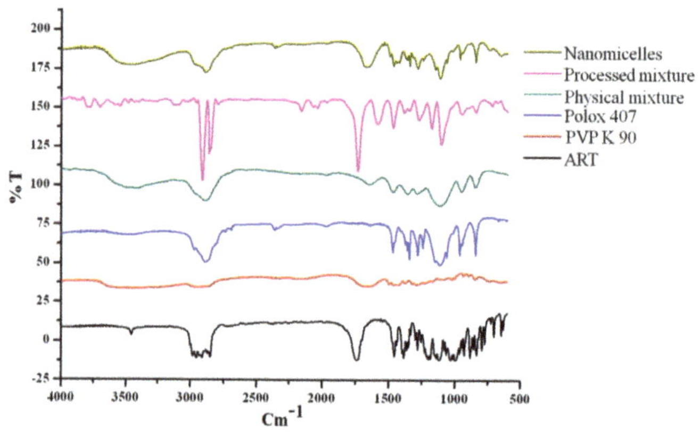

Figure 1. FTIR spectrum of artemisinin, PVP K 90, poloxamer 407, physical mixture, processed mixture and artemisinin nanomicelles.

The IR spectra of artemisinin nanomicelles and the processed mixture shows the presence of C=O stretching at 1644 and 1653 cm^{-1} (carbonyl bond) and 1117 cm^{-1} for C–O (ether bands) stretching vibrations which confirms the interaction between the poloxamer and PVP; this interaction was absent in the physical mixture [39]. The artemisinin-loaded nanomicelles exhibited a shift at the C=O stretching vibrations from 1736 cm^{-1} to 1644 cm^{-1} and 1653 cm^{-1}, respectively. The changes in the carbonyl (C=O) group of the lactone ring in artemisinin might contribute towards the artemisinin solubility.

The formation of a weaker hydrogen bond between the lactone ring carbonyl group (C=O) and the poloxamer hydroxy group could contribute to the enhancement of artemisinin solubility [40]. Moreover, the muster between the poloxamer hydroxy group and the PVP ketone group might increase the hydrophobic micelle corona, which might increase artemisinin solubility.

3.2. *Preliminary Screening for Nanomicelles Formation Capacity*

The stable nanomicelles were screened based on the transparency and the formation of a precipitate upon storage for up to 24 h. Poloxamar 407 and PVP k90 (1:2) produced clear nanomicelles, whereas other lower ratios produced unclear nanomicelles. The higher ratio of poloxamer 407 to PVP K 90 produced visually clear nanomicelles, but the amount required for the formation of nanomicelles was higher than the amount permitted to be used for ocular application as per USFDA [41,42]. The preliminary trial experiments indicated that the polymer and the polymeric surfactant concentration affects the transparency and particle size of the nanomicelles.

3.3. *Experimental Design*

The optimization to screen the smaller particle size with transparent artemisinin-loaded nanomicelles was obtained using central composite rotatable design–response surface methodology (CCRD–RSM). The optimization layout is shown in Table 2, and three-dimensional (3D) RSM graphs and contour graphs are shown in Figures 2 and 3, respectively. The mathematical relationships constructed for the studied response variables of particle size and transparency are expressed as given below in Equations (2) and (3). The best fitting was obtained with the quadratic model. The R-Squared, adjusted R-Squared, predicted R-Squared and adequate precision of the particle size were 0.5818, 0.2830, −1.9741 and 3.269, respectively. Similarly, the R-Squared, adjusted R-Squared, predicted R-Squared and adequate precision of the transparency of the systems were 0.7968, 0.6517, −0.4449 and 5.865, respectively.

Table 2. Experimental design variables and the observed responses in the response surface methodology for the formulation of nanomicelles.

Run	Factor 1 A: Poloxamer 407	Factor 2 B: PVP K 90	Response 1 Particle Size nm (PDI)	Response 2 Transparency %
1	0.00	0.00	44 (0.36)	96
2	−1.00	1.00	106 (0.54)	44
3	1.00	1.00	51 (0.42)	94
4	0.00	0.00	44 (0.36)	96
5	0.00	0.00	44 (0.36)	96
6	0.00	1.41	98 (0.43)	46
7	−1.00	−1.00	41 (0.35)	99
8	1.00	−1.00	78 (0.33)	44
9	−1.41	0.00	156 (0.41)	33
10	1.41	0.00	143 (0.56)	38
11	0.00	−1.41	139 (0.86)	37
12	0.00	0.00	44 (0.36)	96
13	0.00	0.00	44 (0.36)	96

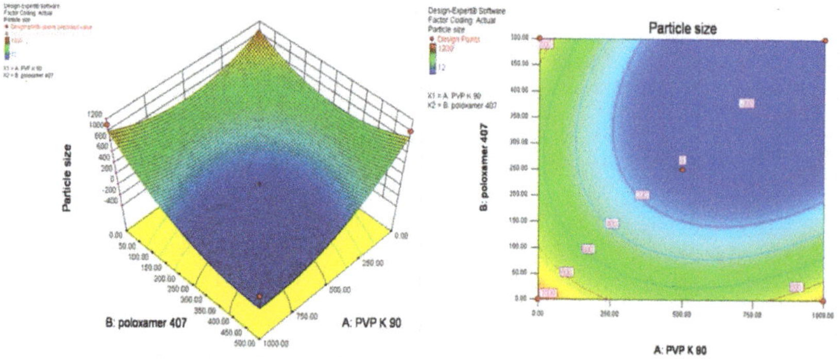

Figure 2. Three-dimensional (3D) and Contour plots of the response surface method graph for the effect of PVP K 90 and poloxamer 407 concentrations on the particle size of nanomicelles.

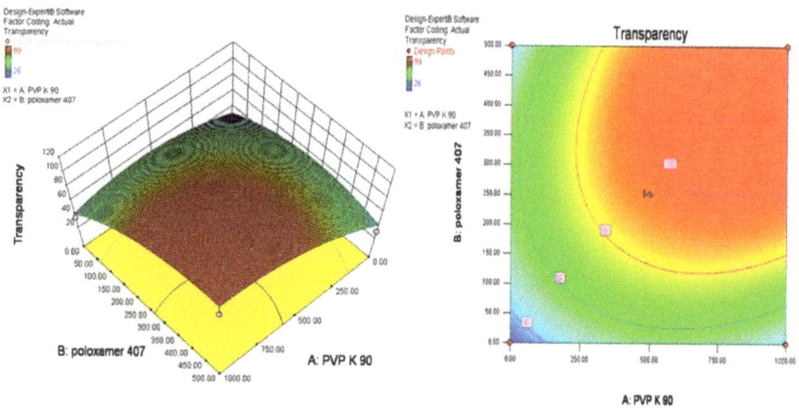

Figure 3. Three-dimensional (3D) and Contour plots of the response surface method graph for the effect of PVP K 90 and poloxamer 407 concentrations on the transparency of nanomicelles.

Final Equation in Terms of Coded Factors:

$$\text{Particle size} = +44.31 - 4.55 \times A - 2.50 \times B - 23.00 \times A \times B + 36.35 \times A^2 + 20.85 \times B^2 \quad (3)$$

$$\text{Transparency} = +96.00 + 0.26 \times A + 0.97 \times B + 26.25 \times A \times B - 22.31 \times A2 - 19.31 \times B2 \quad (4)$$

where A = Concentration of poloxamer, B = Concentration of PVP K 90.

The increasing concentration of poloxamer 407 contributed negatively towards the particle size and positively towards the transparency of nanomicelles, whereas PVP K90 exhibited opposite results. RSM diagrams showed that the higher the concentration of PVP K90 and poloxamer 407, the higher the particle size produced with a lower transparency. Transparent nanomicelles and lower particle size nanomicelles were obtained effectively within the limit of 5 to 10% of PVP K 90 and 2.5 to 5% of poloxamer 407.

3.4. Preparation of Artemisinin-Loaded Nanomicelles

Three different concentrations of PVP K90 and poloxamer 407 are permitted for ophthalmic applications at a 2:1 ratio and were used to formulate the artemisinin-loaded nanomicelles [30]. Previously, it has been reported that the addition of poloxamer in celecoxib microspheres showed a 5-fold enhanced solubility and dissolution rate (98.0% release in 30 min) of celecoxib. In addition, the viscosity of ocular formulations became enhanced upon the incorporation of the poloxamer and PVP K90 as thickening agents and this may retain the incorporated drugs for a prolonged period of time [43].

3.5. Determination of Critical Micellar Concentration

Poloxamers are non-ionic polymeric surfactants and PVP is a non-ionic homo copolymer. These two non-ionic polymers are grafted together with an increased solubility. The artemisinin carbonyl group may combine with the poloxamer hydroxyl group to produce self-assembled micelles in water [39].

The CMC results show that the addition of pyrene to the aqueous dispersion of blank nanomicelles and artemisinin-loaded nanomicelles causes a substantial quenching of pyrene fluorescence (Figure 4). An increase in the concentration of PVP K 90 and poloxamer 407 causes a red shift in the emission and the intensity ratio I384/I375 of pyrene is gradually increased. The CMC observed for the blank nanomicelles and artemisinin-loaded nanomicelles was around 263 ± 3 µg/mL and 295 ± 3 µg/mL, respectively.

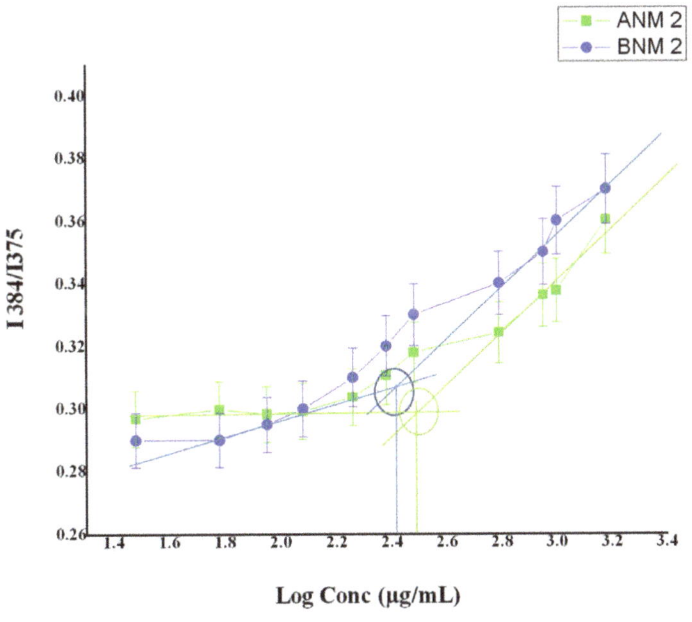

Figure 4. CMC of the blank nanomicelles and artemisinin nanomicelles by a pyrene-based fluorescent probe method.

3.6. Characterization of Artemisinin-Loaded Nanomicelles

3.6.1. Transparency

The artemisinin nanomicelles that were formed were clear and transparent in nature. The transparency of the formulations decreased from 99.0% to 94.0% upon the increase in the concentration of PVP K 90 and poloxamer 407 (Table 3).

Table 3. Evaluation of particle size, PDI, zeta potential, transparency and cloud point of the nanomicelles.

Formulation Code	Particle Size (nm)	PDI	Zeta Potential (mV)	Transparency (%)	Cloud Point (°C)
BNM 1	32 ± 0.7	0.31 ± 0.13	−4.0 ± 1.6	99 ± 1.1	69 ± 1
BNM 2	34 ± 0.3	0.30 ± 0.11	−7.0 ± 1.1	97 ± 1.4	68 ± 1
BNM 3	39 ± 0.6	0.34 ± 0.21	−10.0 ± 1.3	95 ± 1.8	69 ± 2
ANM 1	41 ±0.9	0.35 ± 0.11	−5.0 ± 1.2	99 ± 1.3	70 ± 1
ANM 2	44 ± 1.1	0.36 ± 0.12	−9.2 ± 1.4	96 ± 1.9	69 ± 2
ANM 3	51 ± 2.1	0.42± 0.13	−12.0 ± 2.8	94 ± 2.1	68 ± 1

Data are mean ± SD, n = 3.

3.6.2. Cloud Point Measurement

The cloud point of artemisinin-loaded nanomicelles was around 68–70 °C (Table 3). The cloudiness that appeared in the tested artemisinin-loaded nanomicelle disappeared upon the decrease in temperature just below the cloud point and this happened within a few minutes. These results indicate that the developed artemisinin-loaded nanomicelles might be stable at body temperature. The micellar enlargement and packing were formed due to the dehydration of the polymer block upon the increase in temperature. The non-ionic polymeric surfactant poloxamer undergoes phase separation at high temperatures due to the dehydration of the polyethylene oxide moiety [26].

3.6.3. Particle Size and Zeta Potential

The mean particle size and charge of the artemisinin-loaded nanomicelles was found to be 41 to 51 nm and –5 mV to –12 mV, respectively. The polydispersity index ranges from 0.353 to 0.422 (Table 3). The particle size of artemisinin-loaded nanomicelles increased upon the increase in the concentrations of the polymer and the polymeric surfactant [16,44].

3.6.4. Morphology and Drug Content

The TEM photomicrograph (Figure 5) shows that the dispersed particles had a smooth surface with a spherical shape and were distributed uniformly. The fragmentation pattern shows the hydration of the polymer blocks upon solubilizing in PBS. The surface of the dispersed particle photomicrograph displays the entrapped drug particles and the drug particle distribution within the nanomicelles. The amount of artemisinin present in the artemisinin-loaded nanomicelles ranged from 98 to 99% *w/v*.

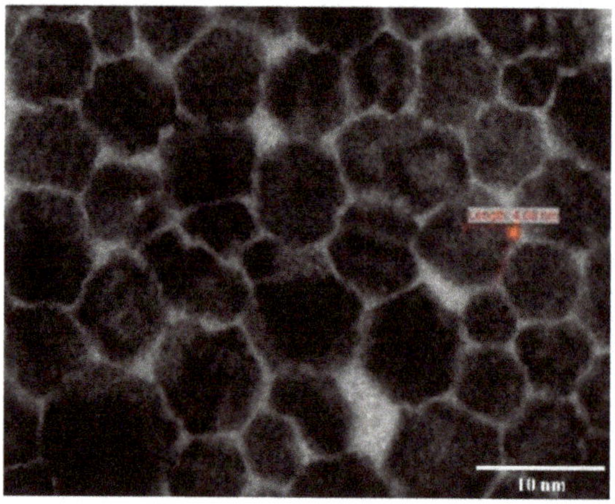

Figure 5. Transmission electron photomicrograph of artemisinin nanomicelles (ANM 2).

3.7. In Vitro Hemolytic Potential

The non-irritant potential of the blank nanomicelles with a concentration of PVP K90 and poloxamer 407 (8:4 and 10:5%) as evaluated by hemolytic studies is shown in Figure 6A. Topical ocular drops are intended for delivering the drugs through the cornea, aqueous humor and the vitreous humor to the retina. In wet AMD, a leakage of blood and fluid occurs in the macula region of the retina. The vehicle with the hemolytic properties will destroy the leaked blood cells that cause an enlargement of the macula. The blank nanomicelles BNM 2 and BNM 3 showed 20% and 22.5% hemolysis, respectively. It was observed that the percentage of hemolysis of the red blood cells produced by blank nanomicelles was within the acceptable range (25%). Hence, the formulated nanomicelles might be safe for topical administration.

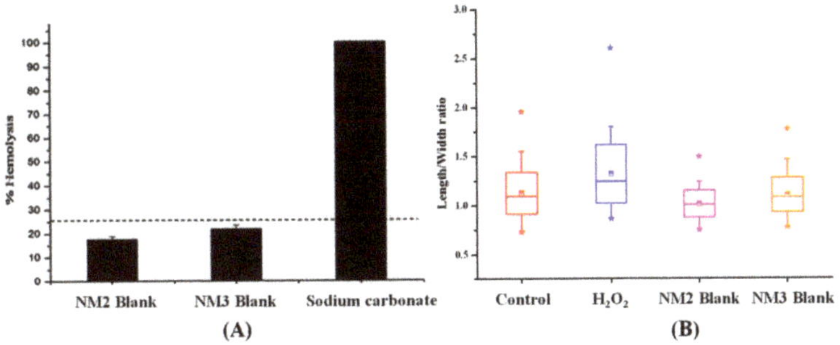

Figure 6. (**A**) In vitro hemolytic potential of blank nanomicelles in human peripheral blood. (**B**) In vitro genotoxicity of blank nanomicelles in human peripheral blood lymphocytes by an alkaline comet assay. * The minimum and maximum value (outliers) of the data.

3.8. Evaluation of the Genotoxicity of the Nanomicelles by the Alkaline Comet Assay

The genotoxic effect of the artemisinin-loaded nanomicelles was evaluated by an alkaline comet assay and the results are displayed in Figure 6B. The high concentration of the polymer (10%) and the polymeric surfactant (5%) containing blank nanomicelles (NM 3) produced no significant DNA damage (Figure 7) in comparison with hydrogen peroxide (positive control). The formulated artemisinin-loaded nanomicelles did not produce any remarkable damage in mammalian cells also. Hence, the developed artemisinin-loaded nanomicelles with higher concentrations of the polymer and the polymeric surfactant might be useful for ocular topical drug delivery.

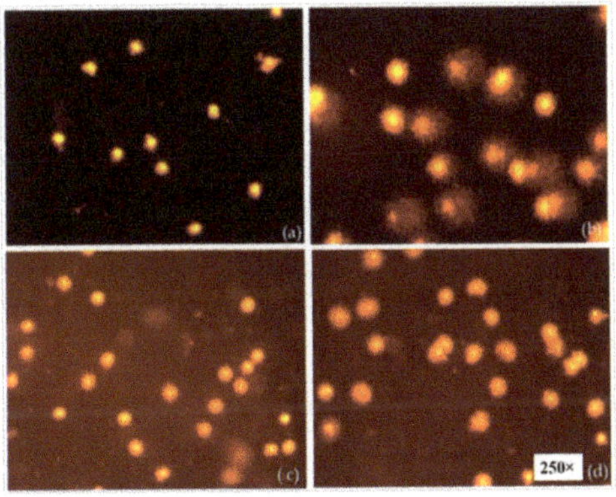

Figure 7. Photomicrographs of Ethidium/Bromide-stained white blood cells under fluorescent microscope. Control cells (**a**), hydrogen peroxide treated cells (**b**), blank NM 2 treated cells (**c**), blank NM 3 treated cells (**d**).

3.9. In Vitro Drug Release

The in vitro drug release studies indicate that around 96.0 to 99.0% of artemisinin was released from artemisinin nanomicelles ANM 1, ANM 2, ANM 3 and only 14% of artemisinin was released from pure drug suspension within 8 h (Figure 8A,B). The artemisinin-loaded nanomicelles (ANM 1 and ANM 2) showed a higher drug release than the nanomicelles formulated with higher concentrations of 10% PVP K 90 and 5% poloxamer 407 (ANM 3) and drug suspensions. In the present study, artemisinin release from the nanomicelles was indirectly proportional to the polymer and polymeric surfactant concentrations of the formulation. This might be due to the encapsulation of the drug particles inside the nano structure of the formulation. The slowest release of artemisinin suspensions was observed due to its hydrophobicity which leads to a floating of the drug powder in the dissolution medium. The increased dissolution rate of artemisinin-loaded nanomicelles might be due to the reduction in drug crystal size, the carrier solubilization effect, the reduction in drug aggregation and agglomeration, the conversion of the amorphous state and increased drug wettability [45].

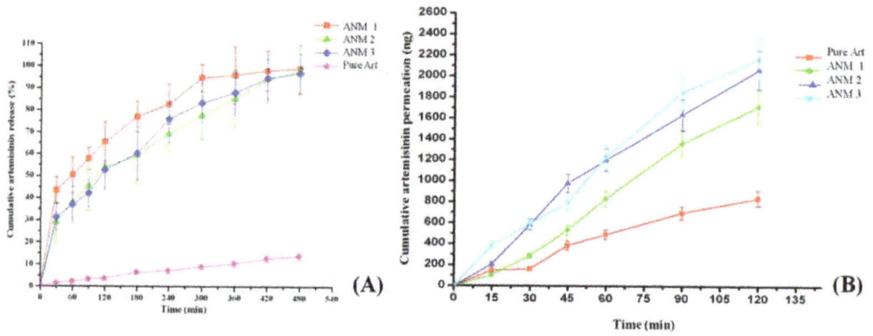

Figure 8. (**A**) In vitro drug release of artemisinin from artemisinin nanomicelles and pure drug suspension. (**B**) In vitro corneal permeation of artemisinin from artemisinin nanomicelles and pure drug suspension.

A similar result was reported for the in vitro release of timolol maleate from the ocular gel, where the release increased with a decreased concentration of pluronic due to the changes in the structural configuration of the polymeric gel. El-Kamel [43] emphasized that the increased concentration of Pluronic127 leads to a decrease in the amount of the drug released. It might be due to the reduction in the number and dimension of water channels through the polymeric gel.

3.10. In Vitro Trans-Corneal Permeation of All the Nanomicelles

The in vitro corneal permeation of artemisinin from the artemisinin-loaded nanomicelles was evaluated through the excised rabbit cornea. The permeation of artemisinin through the excised cornea followed the order of ANM 3 > ANM 2 > ANM 1 > drug dispersion (Figure 8B). The maximum amount of artemisinin (2169 ng) permeated from ANM 3 which is higher than that of drug suspension by 2.5-fold. Similarly, 2067 ng and 1717 ng of artemisinin permeated from ANM 2 and ANM 1 and 839 ng of artemisinin permeated from the artemisinin drug suspension. The permeation of artemisinin increased gradually upon the increased concentration of PVP and poloxamer 407. It has been reported that the ocular bioavailability of drugs is enhanced with an increased concentration of surfactants [33,46]. The ocular bioavailability of timolol maleate from ocular gel increased (2 to 2.5-folds) with an increased concentration of pluronics (15–25%). The increased corneal permeation observed with artemisinin-loaded nanomicelles might be due to the presence of poloxamer 407 which increases the solubility of the artemisinin, removes the phospholipids from the epithelial cell membrane without damage and relaxes the epithelium cell junction which leads to the influx of hydrophilic compounds through the cornea [33,47].

3.11. Apparent Permeability Coefficient (Papp)

The papp was calculated from the corneal permeation profiles of nanomicelles and the data are shown in Figure 9. The Papp of artemisinin from artemisinin-loaded nanomicelles ANM 3, ANM 2, ANM 1 and pure drug suspension was found to be 117.31 $\times$ 10^6 cm/s, 111.55 $\times$ 10^6 cm/s, 96.84 $\times$ 10^6 cm/s and 45.63 $\times$ 10^6 cm/s, respectively. The enhancement of Papp was 2.57, 2.44 and 2.12-fold for ANM3, ANM2 and ANM1 artemisinin-loaded nanomicelles when compared with artemisinin suspension. The corneal permeability of artemisinin significantly increased with an increased concentration of the polymeric surfactant [46]. It has been reported that the passive corneal transport of the drugs is improved in the presence of a surfactant [47]. It was observed that the apparent permeation coefficient of the artemisinin-loaded nanomicelles was directly proportional to the concentration of the surfactants.

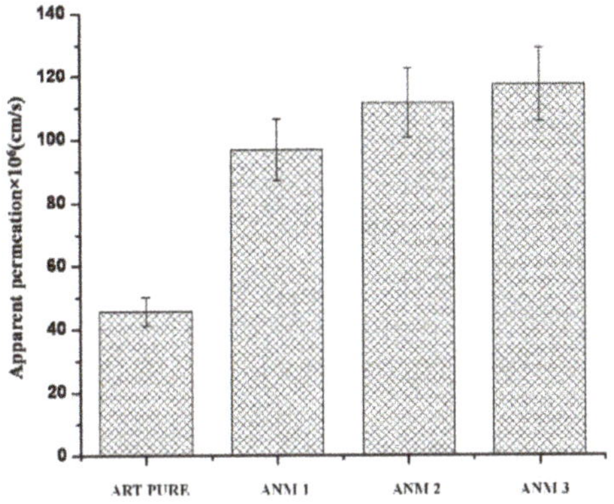

Figure 9. Apparent permeability coefficient of artemisinin from pure drug suspension and artemisinin nanomicelles.

3.12. Histological Examination of the Drug-Permeated Cornea

The realistic mechanism of drug permeation through the cornea was examined by histological evaluation (Figure 10). The photograph of the cornea after 2 h of corneal permeation shows that the surface of the epithelium and stroma were modified in comparison to the control. The pore sizes of the cornea and stroma expanded; hence, the drug permeation through cornea may be enhanced without damaging the membrane [48].

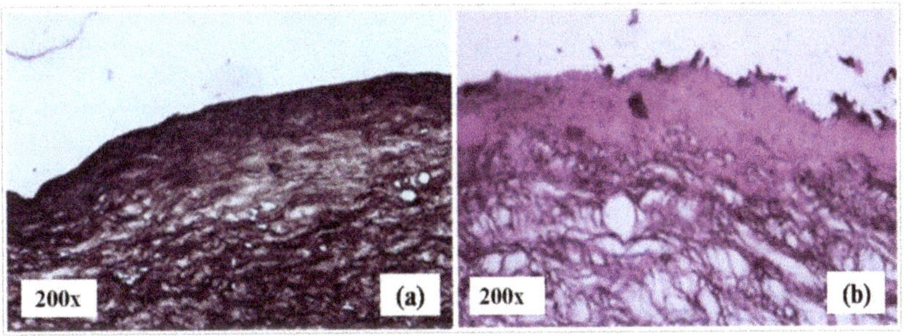

Figure 10. Histopathological image of a normal cornea (**a**) and an ANM 2 penetrated cornea (**b**).

The formulations consisting of Poloxamer 407 and PVP may become trapped in the corneal epithelium's lipid bilayer which may affect the physical characteristics of the epithelial membrane and cause membrane or corneal solubilization of artemisinin. The use of the formulation containing these combined surfactants also dissolves the phospholipids of epithelial cell membranes and establishes a trans-corneal pathway, allowing the drug to be entrapped in these areas more easily [48].

The use of the surfactant in ocular tissues was reported to show reversible changes in the histology and improve the penetration of encapsulated drugs; thereby, frequent administration and side-effects may be minimized [49].

3.13. *Anti-Angiogenic Effect of Artemisinin-Loaded Nanomicelles*

The artemisinin-loaded nanomicelles-treated embryo below the filter paper disc showed changes in blood vessels such as a reduced branching pattern and a capillary free area after 24 h of treatment (Figure 11). However, these changes were not noticeable in blank nanomicelles. The artemisinin suspension-treated embryo also showed very minor changes in the branching patterns of the blood vessels compared to artemisinin nanomicelles-treated CAM.

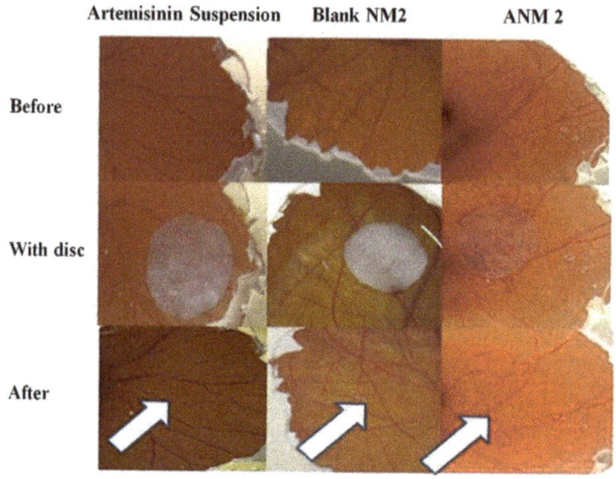

Figure 11. Anti-angiogenic effect of artemisinin drug suspension, drug free NM 2 and artemisinin nanomicelles (ANM 2).

The small capillaries below the filter paper disc in the developing CAM were absent, but the larger pre-existing vessels remain unaffected. The improved solubility of artemisinin in artemisinin-loaded nanomicelles might be responsible for the qualitative improvement in the anti-angiogenic effect. The anti-angiogenic agent was used to stop and eradicate the existing and newly formed blood vessels. Hence, the developed artemisinin-loaded nanomicelles are expected to have the potential to treat the wet AMD.

Blank nanomicelles did not produce any remarkable changes in the branching pattern of the blood vessels. It was also observed that small capillaries and larger preexisting vessels were unaffected below the disc. This suggests that the surfactant concentration used in the formulation does not produce any irritation and discoloration of the chorioallantoic membrane. Hence, the formulated artemisinin-loaded nanomicelles can be used for ophthalmic drug delivery.

3.14. Stability Studies

Artemisinin-loaded nanomicelles were examined at the end of every month to assess their instability parameters such as particle size, turbidity and drug content. The particle size and drug content of the formulations showed no significant changes during the study period.

4. Conclusions

Topical formulations of artemisinin-loaded nanomicelles were developed using PVP K90 and poloxamer 407 in the ratio of 2:1. An increase in the concentration of the polymers showed an enhanced particle size of the nanomicelles. The nanomicelles made with 8% of PVP K90 and 4% of poloxamer 407 (ANM 2) showed a moderate drug release and a higher corneal permeation of artemisinin when compared with artemisinin suspension. The artemisinin-loaded nanomicelles also showed a better anti-angiogenic potential than artemisinin suspension. Hence, developed artemisinin nanomicelles using 8% of PVP K 90 and 4% of Poloxamer 407 may have the potential for an effective treatment of AMD.

Author Contributions: C.P. wrote the manuscript and performed the main analysis and data collection. A.S. performed data interpretations. V.K. performed and interpreted the in vitro analysis. R.P., R.V. and S.N. supervised all the tasks, performed data interpretations and finalized the manuscript. All authors have read and agreed to the published version of the manuscript.

Funding: This research received no external funding.

Institutional Review Board Statement: The animal study protocol was approved by the Institutional Animal Ethical Committee and animals were maintained according to the approved guidelines of the Committee for the Purpose of Control and Supervision of Experiments on Animals (CPCSEA) in India. (Approval No: AUROT/IAEC/NOV2013—0017 Dt. 21.11.2013).

Informed Consent Statement: Informed consent was obtained from all subjects involved in the study.

Acknowledgments: The author, P. Chandrasekar, gratefully acknowledges the support received from the University Grant Commission, New Delhi for providing fellowships for Junior/Senior Research Fellows.

Conflicts of Interest: The authors declare no conflict of interest.

References

1. Stahl, A. The Diagnosis and Treatment of Age-Related Macular Degeneration. *Dtsch. Arztebl. Int.* **2020**, *117*, 513–520. [CrossRef]
2. Nayak, K.; Misra, M. A review on recent drug delivery systems for posterior segment of eye. *Biomed. Pharmacother.* **2018**, *107*, 1564–1582. [CrossRef]
3. Rodrigues, G.A.; Lutz, D.; Shen, J.; Yuan, X.; Shen, H.; Cunningham, J.; Rivers, H.M. Topical Drug Delivery to the Posterior Segment of the Eye: Addressing the Challenge of Preclinical to Clinical Translation. *Pharm. Res.* **2018**, *35*, 245. [CrossRef]
4. Gokulgandhi, M.R.; Vadlapudi, A.D.; Mitra, A.K. Ocular toxicity from systemically administered xenobiotics. *Expert Opin. Drug Metab. Toxicol.* **2012**, *8*, 1277–1291. [CrossRef]
5. Varela-Fernández, R.; Díaz-Tomé, V.; Luaces-Rodríguez, A.; Conde-Penedo, A.; García-Otero, X.; Luzardo-Álvarez, A.; Fernández-Ferreiro, A.; Otero-Espinar, F.J. Drug Delivery to the Posterior Segment of the Eye: Biopharmaceutic and Pharmacokinetic Considerations. *Pharmaceutics* **2020**, *12*, 269. [CrossRef]
6. Agrahari, V.; Mandal, A.; Agrahari, V.; Trinh, H.M.; Joseph, M.; Ray, A.; Hadji, H.; Mitra, R.; Pal, D.; Mitra, A.K. A comprehensive insight on ocular pharmacokinetics. *Drug Deliv. Transl. Res.* **2016**, *6*, 735–754. [CrossRef]
7. Sapino, S.; Chirio, D.; Peira, E.; Abellán Rubio, E.; Brunella, V.; Jadhav, S.A.; Chindamo, G.; Gallarate, M. Ocular Drug Delivery: A Special Focus on the Thermosensitive Approach. *Nanomaterials* **2019**, *9*, 884. [CrossRef]
8. Gote, V.; Sikder, S.; Sicotte, J.; Pal, D. Ocular Drug Delivery: Present Innovations and Future Challenges. *J. Pharmacol. Exp. Ther.* **2019**, *370*, 602–624. [CrossRef] [PubMed]
9. Swetledge, S.; Jung, J.P.; Carter, R.; Sabliov, C. Distribution of polymeric nanoparticles in the eye: Implications in ocular disease therapy. *J. Nanobiotech.* **2021**, *19*, 10. [CrossRef] [PubMed]
10. Mandal, A.; Bisht, R.; Rupenthal, I.D.; Mitra, A.K. Polymeric micelles for ocular drug delivery: From structural frameworks to recent preclinical studies. *J. Control. Release* **2017**, *248*, 96–116. [CrossRef] [PubMed]
11. Hanafy, N.A.N.; El-Kemary, M.; Leporatti, S. Micelles Structure Development as a Strategy to Improve Smart Cancer Therapy. *Cancers* **2018**, *10*, 238. [CrossRef]

12. Bose, A.; Roy Burman, D.; Sikdar, B.; Patra, P. Nanomicelles: Types, properties and applications in drug delivery. *IET Nanobiotech.* **2021**, *15*, 19–27. [CrossRef]
13. Vaishya, R.D.; Khurana, V.; Patel, S.; Mitra, A.K. Controlled ocular drug delivery with nanomicelles. *Wiley Interdiscip. Rev. Nanomed. Nanobiotechnol.* **2014**, *6*, 422–437. [CrossRef]
14. Giuliano, E.; Paolino, D.; Fresta, M.; Cosco, D. Mucosal Applications of Poloxamer 407-Based Hydrogels: An Overview. *Pharmaceutics* **2018**, *10*, 159. [CrossRef]
15. Sugumaran, A.; Ponnusamy, C.; Kandasamy, P.; Krishnaswami, V.; Palanichamy, R.; Kandasamy, R.; Lakshmanan, M.; Natesan, S. Development and evaluation of camptothecin loaded polymer stabilized nanoemulsion: Targeting potential in 4T1-breast tumour xenograft model. *Eur. J. Pharm. Sci.* **2018**, *116*, 15–25. [CrossRef] [PubMed]
16. Bodratti, A.M.; Alexandridis, P. Formulation of Poloxamers for Drug Delivery. *J. Funct. Biomater.* **2018**, *9*, 11. [CrossRef] [PubMed]
17. Khanal, P. Antimalarial and anticancer properties of artesunate and other artemisinins: Current development. *Mon. Chem. Chem. Mon.* **2021**, *152*, 387–400. [CrossRef]
18. Cheong, D.H.J.; Tan, D.W.S.; Wong, F.W.S.; Tran, T. Anti-malarial drug, artemisinin and its derivatives for the treatment of respiratory diseases. *Pharmacol. Res.* **2020**, *158*, 104901. [CrossRef] [PubMed]
19. Wang, J.; Zhang, J.; Shi, Y.; Xu, C.; Zhang, C.; Wong, Y.K.; Lee, Y.M.; Krishna, S.; He, Y.; Lim, T.K.; et al. Mechanistic Investigation of the Specific Anticancer Property of Artemisinin and Its Combination with Aminolevulinic Acid for Enhanced Anticolorectal Cancer Activity. *ACS Cent. Sci.* **2017**, *3*, 743–750. [CrossRef] [PubMed]
20. Lu, B.-W.; Xie, L.-K. Potential applications of artemisinins in ocular diseases. *Int. J. Ophthalmol.* **2019**, *12*, 1793–1800. [CrossRef]
21. Chong, C.-M.; Zheng, W. Artemisinin protects human retinal pigment epithelial cells from hydrogen peroxide-induced oxidative damage through activation of ERK/CREB signaling. *Redox Biol.* **2016**, *9*, 50–56. [CrossRef]
22. Ponnusamy, C.; Sugumaran, A.; Krishnaswami, V.; Kandasamy, R.; Natesan, S. Design and development of artemisinin and dexamethasone loaded topical nanodispersion for the effective treatment of age-related macular degeneration. *IET Nanobiotech.* **2019**, *13*, 868–874. [CrossRef]
23. Li, H.; Hu, D.; Liang, F.; Huang, X.; Zhu, Q. Influence factors on the critical micelle concentration determination using pyrene as a probe and a simple method of preparing samples. *R. Soc. Open Sci.* **2020**, *7*, 192092. [CrossRef] [PubMed]
24. Mohr, A.; Talbiersky, P.; Korth, H.-G.; Sustmann, R.; Boese, R.; Bläser, D.; Rehage, H. A new pyrene-based fluorescent probe for the determination of critical micelle concentrations. *J. Phys. Chem. B* **2007**, *111*, 12985–12992. [CrossRef] [PubMed]
25. Subramanian, N.; Ray, S.; Ghosal, S.K.; Bhadra, R.; Moulik, S.P. Formulation design of self-microemulsifying drug delivery systems for improved oral bioavailability of celecoxib. *Biol. Pharm. Bull.* **2004**, *27*, 1993–1999. [CrossRef]
26. Elnaggar, Y.S.R.; El-Massik, M.A.; Abdallah, O.Y. Self-nanoemulsifying drug delivery systems of tamoxifen citrate: Design and optimization. *Int. J. Pharm.* **2009**, *380*, 133–141. [CrossRef] [PubMed]
27. Ponnusamy, C.; Krishnaswami, V.; Sugumaran, A.; Natesan, S. Simultaneous estimation of artemisinin and dexamethasone in nanodispersions and assessment of Ex-vivo corneal transport study by RP-HPLC. *Curr. Pharm. Anal.* **2014**, *10*, 44–50. [CrossRef]
28. Amin, K.; Dannenfelser, R.-M. In vitro hemolysis: Guidance for the pharmaceutical scientist. *J. Pharm. Sci.* **2006**, *95*, 1173–1176. [CrossRef] [PubMed]
29. Natesan, S.; Sugumaran, A.; Ponnusamy, C.; Jeevanesan, V.; Girija, G.; Palanichamy, R. Development and evaluation of magnetic microemulsion: Tool for targeted delivery of camptothecin to BALB/c mice-bearing breast cancer. *J. Drug Target.* **2014**, *22*, 913–926. [CrossRef]
30. Subramanian, N.; Abimanyu, S.; Vinoth, J.; Sekar, P.C. Biodegradable Chitosan Magnetic Nanoparticle Carriers for Sub-Cellular Targeting Delivery of Artesunate for Efficient Treatment of Breast Cancer. *AIP Conf. Proc.* **2010**, *1311*, 416–424. [CrossRef]
31. Danafar, H.; Jaberizadeh, H.; Andalib, S. In vitro and in vivo delivery of gliclazide loaded mPEG-PCL micelles and its kinetic release and solubility study. *Artif. Cells Nanomed. Biotechnol.* **2018**, *46*, 1625–1636. [CrossRef]
32. Begum, G.; Leigh, T.; Courtie, E.; Moakes, R.; Butt, G.; Ahmed, Z.; Rauz, S.; Logan, A.; Blanch, R.J. Rapid assessment of ocular drug delivery in a novel ex vivo corneal model. *Sci. Rep.* **2020**, *10*, 11754. [CrossRef] [PubMed]
33. Li, X.; Pan, W.; Ju, C.; Liu, Z.; Pan, H.; Zhang, H.; Nie, S. Evaluation of Pharmasolve corneal permeability enhancement and its irritation on rabbit eyes. *Drug Deliv.* **2009**, *16*, 224–229. [CrossRef] [PubMed]
34. Pescina, S.; Govoni, P.; Potenza, A.; Padula, C.; Santi, P.; Nicoli, S. Development of a Convenient ex vivo Model for the Study of the Transcorneal Permeation of Drugs: Histological and Permeability Evaluation. *J. Pharm. Sci.* **2014**, *104*, 63–71. [CrossRef]
35. Velpandian, T.; Bankoti, R.; Humayun, S.; Ravi, A.K.; Kumari, S.S.; Biswas, N.R. Comparative evaluation of possible ocular photochemical toxicity of fluoroquinolones meant for ocular use in experimental models. *Indian J. Exp. Biol.* **2006**, *44*, 387–391.
36. Ansari, M.; Haneef, M.; Murtaza, G.; Dyspersja, S. Solid Dispersions of Artemisinin in Polyvinyl Pyrrolidone and Polyethylene Glycol. *Adv. Clin. Experimetal Med.* **2010**, *19*, 745–754.
37. Garg, A.K.; Sachdeva, R.K.; Kapoor, G. Comparison of crystalline and amorphous carriers to improve the dissolution profile of water insoluble drug itraconazole. *Int. J. Pharm. Bio Sci.* **2013**, *4*, 934–948.
38. Vyas, V.; Sancheti, P.; Karekar, P.; Shah, M.; Pore, Y. Physicochemical characterization of solid dispersion systems of tadalafil with poloxamer 407. *Acta Pharm.* **2009**, *59*, 453–461. [CrossRef]
39. Zhang, Y.; Lam, Y.M. Controlled synthesis and association behavior of graft Pluronic in aqueous solutions. *J. Colloid Interface Sci.* **2007**, *306*, 398–404. [CrossRef]

40. Saluja, H.; Mehanna, A.; Panicucci, R.; Atef, E. Hydrogen Bonding: Between Strengthening the Crystal Packing and Improving Solubility of Three Haloperidol Derivatives. *Molecules* **2016**, *21*, 719. [CrossRef]
41. Li, G.; Zhong, M.; Zhou, Z.; Zhong, Y.; Ding, P.; Huang, Y. Formulation optimization of chelerythrine loaded O-carboxymethylchitosan microspheres using response surface methodology. *Int. J. Biol. Macromol.* **2011**, *49*, 970–978. [CrossRef]
42. Singh, B.; Chakkal, S.K.; Ahuja, N. Formulation and optimization of controlled release mucoadhesive tablets of atenolol using response surface methodology. *AAPS Pharmscitech* **2006**, *7*, E19–E28. [CrossRef]
43. El-Kamel, A.H. In vitro and in vivo evaluation of Pluronic F127-based ocular delivery system for timolol maleate. *Int. J. Pharm.* **2002**, *241*, 47–55. [CrossRef]
44. Kim, H.; Csaky, K.G. Nanoparticle-integrin antagonist C16Y peptide treatment of choroidal neovascularization in rats. *J. Control. Release* **2010**, *142*, 286–293. [CrossRef]
45. Paradkar, A.; Ambike, A.; Mahadik, K. Characterization of curcumin-PVP solid dispersion obtained by spray drying. *Int. J. Pharm.* **2004**, *271*, 281–286. [CrossRef]
46. Ahuja, M.; Dhake, A.S.; Sharma, S.K.; Majumdar, D.K. Diclofenac-loaded Eudragit S100 nanosuspension for ophthalmic delivery. *J. Microencapsul.* **2011**, *28*, 37–45. [CrossRef] [PubMed]
47. Majumdar, S.; Srirangam, R. Solubility, stability, physicochemical characteristics and in vitro ocular tissue permeability of hesperidin: A natural bioflavonoid. *Pharm. Res.* **2009**, *26*, 1217–1225. [CrossRef]
48. Toropainen, E.; Ranta, V.P.; Talvitie, A.; Suhonen, P.; Urtti, A. Culture model of human corneal epithelium for prediction of ocular drug absorption. *Investig. Ophthalmol. Vis. Sci.* **2001**, *42*, 2942–2948. [PubMed]
49. Naguib, S.S.; Hathout, R.M.; Mansour, S. Optimizing novel penetration enhancing hybridized vesicles for augmenting the in-vivo effect of an anti-glaucoma drug. *Drug Deliv.* **2017**, *24*, 99–108. [CrossRef]

polymers

MDPI

Review

Development of a New Polymeric Nanocarrier Dedicated to Controlled Clozapine Delivery at the Dopamine D_2-Serotonin 5-HT_{1A} Heteromers

Sylwia Łukasiewicz

Department of Physical Biochemistry, Faculty of Biochemistry, Biophysics and Biotechnology, Jagiellonian University, 30-387 Krakow, Poland; sylwia.lukasiewicz@uj.edu.pl; Tel.: +48-012-664-61-34; Fax: +48-012-664-6902

Abstract: Clozapine, the second generation antipsychotic drug, is one of the prominent compounds used for treatment of schizophrenia. Unfortunately, use of this drug is still limited due to serious side effects connected to its unspecific and non-selective action. Nevertheless, clozapine still remains the first-choice drug for the situation of drug-resistance schizophrenia. Development of the new strategy of clozapine delivery into well-defined parts of the brain has been a great challenge for modern science. In the present paper we focus on the presentation of a new nanocarrier for clozapine and its use for targeted transport, enabling its interaction with the dopamine D_2 and serotonin 5-HT_{1A} heteromers (D_2-5-HT_{1A}) in the brain tissue. Clozapine polymeric nanocapsules (CLO-NCs) were prepared using anionic surfactant AOT (sodium docusate) as an emulsifier, and bio-compatible polyelectrolytes such as: poly-L-glutamic acid (PGA) and poly-L-lysine (PLL). Outer layer of the carrier was grafted by polyethylene glycol (PEG). Several variants of nanocarriers containing the antipsychotic varying in physicochemical parameters were tested. This kind of approach may enable the availability and safety of the drug, improve the selectivity of its action, and finally increase effectiveness of schizophrenia therapy. Moreover, the purpose of the manuscript is to cover a wide scope of the issues, which should be considered while designing a novel means for drug delivery. It is important to determine the interactions of a new nanocarrier with many cell components on various cellular levels in order to be sure that the new nanocarrier will be safe and won't cause undesired effects for a patient.

Keywords: encapsulation; clozapine; schizophrenia; polymeric nanocarriers; D_2-5-HT_{1A} receptor heterodimers; scFv antibodies

Citation: Łukasiewicz, S. Development of a New Polymeric Nanocarrier Dedicated to Controlled Clozapine Delivery at the Dopamine D_2-Serotonin 5-HT_{1A} Heteromers. *Polymers* **2021**, *13*, 1000. https://doi.org/10.3390/polym13071000

Academic Editors: M. Ali Aboudzadeh and Stefano Leporatti

Received: 18 February 2021
Accepted: 18 March 2021
Published: 24 March 2021

1. Introduction

Currently, the use of nanotechnology in molecular pharmacology has been attracting more and more attention. One of the leading trends in nanomedicine is the attempt to use drugs attached to nanoparticles in the therapy of many diseases. The nanoparticles delivery systems of active compounds create new possibilities, allowing, among other benefits, us to achieve a therapeutic effect only in a selected, well-defined target site, thus leading to the reduction or elimination of undesired side effects mainly related to non-selective action [1,2]. The main advantages of nanocarriers are their sub-cellular dimensions and tissue-cellular biocompatibility. The improvement of the compatibility of lipophilic, poorly water-soluble or even water-insoluble active compounds or increased drug permeability and absorption has been shown for various nanoformulations [3–6]. Drug encapsulation extends the duration of its action, protecting against rapid uptake and degradation, and by controlling the released dose it allows for maintaining concentration of the drug at the level of the required therapeutic concentration. Thus, such a strategy allows for reducing the size and frequency of doses [4–7]. In addition, the appropriate functionalization of the surface of the nanocarrier enables so-called "intelligent targeting", i.e., release of the drug at the appropriate destination [5,6,8,9]. Thanks to this, nanotherapeutics have the ability to "perform complex operations" in the right place at the right time in the patient's body, contrasting with previously developed preparations, whose greatest weakness has been non-selectivity. All of the above-mentioned features have consequently led to a reduction of the negative side effects of the therapy. Polymeric nanoparticles (PNp) are formed by biocompatible and biodegradable polymeric materials. This strategy enables a predictable decomposition process and complete metabolization of degradation products [10]. Polymeric nanocapsules (NCs) usually include an active compound immersed in a liquid core surrounded by a polymeric shell, which can be obtained from natural (chitosan, dextran, alginate, heparin, dextran sulfate, cellulose sulfate etc.) or synthetic polymeric (PLGA—poly(lactic-co-glycolic acid), PLA—poly(lactic acid), PGA—poly(glycolic acid), PCL—poly(caprolactone), PEI—poly(ethyleneimine), PLL—poly(L-lysine) etc.). To avoid serum protein adsorption, nonspecific binding to undesired cells and tissues, an outermost shell of the capsule is pegylated (grafted by polyethylene glycol–PEG). Selective delivery of active compounds requires the proper modification of a capsule shell, to which external elements (so-called targeting ligands) are embedded to obtain molecular recognition at the desired target location. One of the best targeting ligands are monoclonal scFv antibodies fragments (single chain variable fragments), which, due to their relatively small size, higher (as compared to other types of ligands) binding specificity as well as the lack of complement activating region and Fc domain (which directly translates into a reduction in the immunogenicity of the modified nanocapsules) [8–11], are increasingly used to functionalize nanocarriers. Moreover, the aptamer-functionalized nanocapsules show a good compatibility with the bloodstream and do not have a cytotoxic effect [5,6]. In conclusion, nanotechnology-based therapeutic methods provide unusual control over behaviour of the drug in the body, thus providing the possibility of targeted treatment.

Based on the current available literature, one may notice that in recent years there have been attempts at preparing a nanocarrier for antipsychotic drugs used to treat schizophrenia, a complex psychiatric disorder. Despite the better profile of atypical antipsychotics, they are not free of a negative impact on the patient's organism. Both clinical and experimental studies indicate that the reduction of side effects is not complete, which is probably related to the non-specific action of the drug [12]. Therefore, the studies focusing on finding new agents with greater therapeutic efficacy are still in progress. The reference for this search is clozapine, which belongs to the group of atypical antipsychotic drugs. This drug is used in the clinic, however, and is not free from serious side effects such as: myocarditis, arrhythmia, weight gain, metabolic disorders, and above all agranulocytosis [13,14]. Due to risks of the above-mentioned complications, therapy with clozapine is often limited. The possibility to direct this compound to the desired site of action would greatly enhance its specificity.

Wang et al. (2020) produced clozapine containing solid lipid nanoparticles (SLNs) by using ultrasonic technology. They indicate that the encapsulation improves drug stability in the carrier system, and also increases drug bioavailability in vivo [15]. Ishak et al. (2013) showed different pharmacokinetic profile and biodistribution behavior of clozapine (CZP)-loaded NPs which were coated with chitosan, pluronic F-68, PEG 4000 and polysorbate 80 [16]. Moreover, the factors affecting drug encapsulation efficiency, particle size, surface charge, and surface hydrophilicity have been studied [17]. Additionally one can find some research that provides a brief summary and discussion of the progress and development in the delivery of other antipsychotics (e.g., aripiprazol, olanzapine, paliperidone) with nanoparticle formulations [18–22].

The main purpose of the manuscript is to cover a wide scope of the issues, which should be considered while designing this novel means for drug delivery. Very often scientific literature brings very enthusiastic information concerning the possibilities of modern nanomedicine, nano pharmacology and drug delivery, but the presented data are usually limited to the description of interactions of synthesized nanoparticles at the site of their destination, without data concerning interactions with cells of an immunological system, or possibilities of new nano carriers to turn on or promote the inflammation processes, genotoxicity, or, last but not least, data on in-vivo interactions. The main topic of the present review focuses on efforts which need to be undertaken to obtain a reliable, well-defined nanocarrier dedicated to drug delivery, in order to be sure that it will be safe and not cause undesired effects for a patient.

The present review covers a set of papers concerning studies leading to encapsulation of clozapine into polymeric nanocarriers. It is important to determine the interactions of new nanocarriers with many cell components on various cellular levels in order to be sure that the new nanocarrier will be safe and not cause undesired effects for a patient.

Considering the above issue, below is the description of the research which was undertaken to obtain an encapsulated form of clozapine. Clozapine polymeric nanocapsules (CLO-NCs) were prepared using anionic surfactant AOT (sodium docusate) as an emulsifier, and biocompatible polyelectrolytes such as: PGA and PLL. The outer layer of the carrier was grafted by PEG. Several variants of nanocarriers containing the antipsychotic varying in physicochemical parameters were tested. It seems that increasing the efficacy and safety of the clinical use of clozapine can be achieved by designing an appropriate nanocarrier to deliver the above-mentioned therapeutic to the selected target which constitutes the areas of the brain rich in dopamine D_2 and serotonin $5\text{-}HT_{1A}$ ($D_2\text{-}5\text{-}HT_{1A}$) heteromers. The above strategy may contribute to increasing the availability and safety of the drug, as well as improve the selectivity of its action, resulting in increased effectiveness of schizophrenia therapy. However, the task is not easy and demands elaborate work, which is illustrated below.

1.1. *Preparation of Polymeric Nanocapsules (NCs)*

Polymeric nanocapsules (NCs) with a liquid core covered with a layer of biodegradable polyelectrolytes were prepared by the technique of sequential adsorption of the oppositely charged nanomaterials "layer-by-layer" LbL. The anionic surfactant AOT (sodium docusate, approved by FDA) as an emulsifier, and biocompatible polyelectrolytes such as: PGA polyanion, and PLL polycation were used. The pegylated outer layer of the capsule was prepared using PGA-g-PEG (PGA grafted by polyethylene glycol) (Figure 1). PEG is a neutral, hydrophilic polymer, and its high flexibility and mobility of the chain contributes to the stability of the NCs. Moreover, pegylated coatings are characterized by a reduced potential for protein adsorption, resulting in suppression of the opsonization process, and thereby a reduction of NCs uptake by cells of the immune system. The average size of the obtained NCs was 80–100 nm depending on the thickness of the outer layer. The obtained NCs were stable under physiological conditions at high ionic strength [23–25].

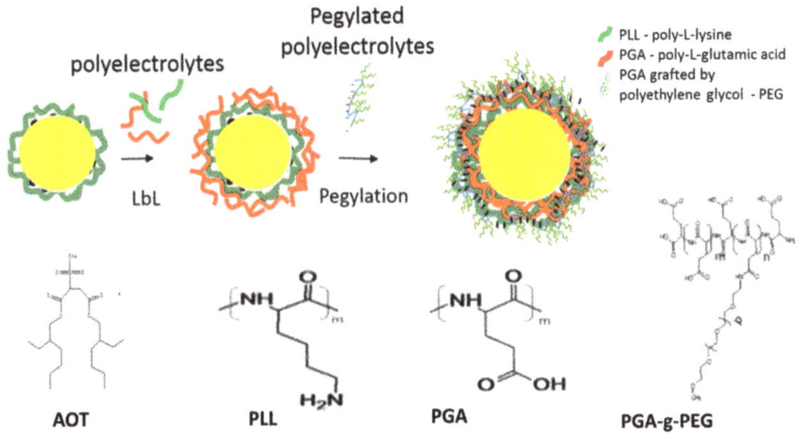

Figure 1. Structure of obtained polymeric NCs.

1.2. Interaction of the Nanocapsule/Target Cell—Cytotoxicity Studies of the Obtained Nanoformulations

Finding effective nanocarriers dedicated to the controlled delivery of active compounds requires systematic studies leading to the optimization of their interaction with target cells. This interaction depends on the type and physicochemical properties of the carrier and, above all, on the modification of its external layer. Quantification of cell viability allows to describe the toxicity of used nanomaterials. It is important to maintain a balance between the effective nanocarrier internalization and the induction of a toxic effect. The interaction between the NCs and the cell membrane is the main factor influencing this process and depends on the shape, size, flexibility, surface charge, modification and functionalization of a capsule [26–29]. In the case of NCs with a surface charge, their interaction with the cell membrane will be mainly determined by electrostatic interactions [26,27]. We have to take into account the fact that the sizes of capsules and their surface properties can significantly change in biological systems [30–33]. Due to varying ionic strength, as well as possible reactions with medium components (e.g., protein adsorption) spontaneous aggregation of NCs may occur [33–36]. Therefore, firstly, the biocompatibility and cytotoxicity of the obtained nanomaterials were determined, depending on the NCs dose, charge, size and modification of the outer layer. The experiments were carried out for various cell lines: HEK 293 (human embryonic kidney cell line), RAW 264.7 (mouse murine macrophages cell line), THP-1 (human leukemic monocyte cell line). The detailed results were presented in the publications [23,24]. Literature reports indicate a discrepancy in the obtained data using various tests dedicated to the estimation of cell viability [37,38]. Therefore, in order to obtain the most reliable results and avoid possible overinterpretation, several different tests were performed [23,24]. Generally, three main trends have been observed. Cytotoxicity depended on: (1) NCs concentration: for each type of tested NCs, the most toxic doses were defined, although it should be emphasized that for the most toxic types of NCs, the cell survival increased to 90% when the dose was reduced to 0.2×10^6 NCs per cell, which is much above the assumed theoretical amount of NCs sufficient to achieve a therapeutic effect; (2) the number of polyelectrolyte layers—the smaller the number of layer (when we compere layers with the same charge—even or odd number of layers), the greater the decrease in cell viability; (3) surface charge—the negative charge on the NCs surface was correlated with increased survivability. Moreover, the obtained results indicate a relationship between the surface charge of the NCs and the destabilization of the cell membrane. In conclusion, the more toxic ones turned out to be positively charged NCs. Below (Figure 2) an example of the distribution of cytotoxicity depending on number of layers forming the nanocarrier, measured in RAW 264.7 and THP cells after 24 h incubation with NCs. More detailed information concerning the issue can be found in [23,24].

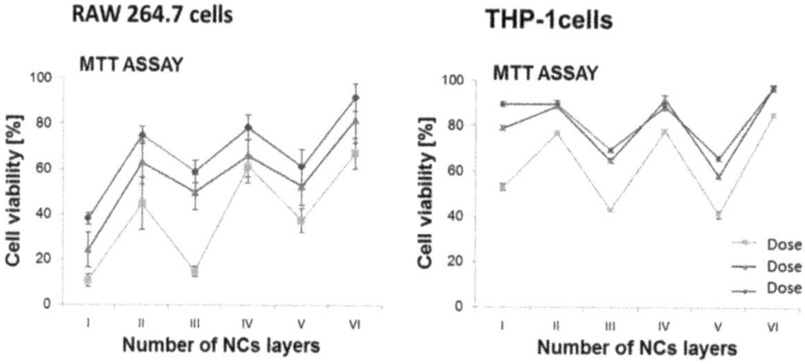

Figure 2. Cytotoxicity of obtained nanomaterials depending on their structure. Measurements for RAW 264.7 and THP-1 cells after a 24 h incubation with NCs. Detailed information [23,24].

The mechanism of the observed phenomenon is probably similar for all tested cell lines and may be associated with a more efficient adsorption of positively charged NCs on the cell surface, a tendency to reduce lipid density and eventually disruption of cell membrane function. Modifications of the external layer involving the PEG grafting have a positive effect on cell viability (no toxic effects were observed). Moreover, pegylation spatially stabilized the NCs and prevented their aggregation [23,24].

1.3. *Interactions of the Obtained NCs with Cells of the Immune System*

The use of nanotechnology in the development of new controlled drug delivery systems also requires extensive studies on the interaction of nanomaterials with cells of the immune system. Numerous reports point to the rapid elimination of nanocarriers from the blood stream [39]. Adsorption of plasma proteins on the surface of the nanocarriers allows macrophages of the mononuclear phagocytic system (MPS) to quickly recognize and remove NCs before they reach their destination [40]. This translates directly into reducing the half-life of the drug and thus limits the ability of nanomaterials to function as efficient nanocarriers. Therefore, it is extremely important to design a nanocarrier that is non-visible for phagocytic cells and at the same time has all the features allowing for performing the required function. As was mentioned previously, the interaction of nanocarriers with target cells mainly depends on the type and physicochemical properties of the nanocarrier. Therefore, in accomplishing the assumed goals, NCs were tested depending on their size and charge, as well as modification of the outer layer. Numerous studies indicate that the appropriate modification of the outer layer has the greatest impact on interaction with phagocytic cells. Decorating the particle surface with a neutral hydrophilic polymer such as PEG blocks the electrostatic and hydrophobic interactions, which leads to the reduction or complete elimination of protein adsorption, thereby minimizing the opsonization process, which in consequence increase the lifetime of the nanocarrier in the bloodstream. This effect is correlated with the PEG properties. The proper pegylation of the particle surface is a crucial step, because the PEG quality, chain size, number of chains, density and the way they are arranged have a huge impact on the interaction with the target cell and biodistribution of the nanocarrier in the body [41]. In summary, the formation of a hydrophilic shell around the NCs protects it against rapid phagocytic uptake. On the other hand, pegylation may also intensify the internalization process of nanomaterials by other cell types (e.g., tumor cells or blood-brain barrier cells) [39,42]. Reports indicate that both phagocytosis, endocytosis and micropetrocytosis may be involved in the internalization process [1]. Considering the above issues, the conducted experiments also focused on the study of the interaction between the obtained NCs (with different physicochemical parameters) and cells of the immune system. RAW 264.7 and THP-1 cell lines, as well as human monocyte-derived macrophages (HMDMs) cells that were differentiated from peripheral blood mononuclear (PMBC) cells from healthy donors were used in our studies [24,43]. It has been shown that all types of synthesized NCs are taken up by phagocytic cells; however, the uptake of pegylated NCs was substantially lower compared to unmodified NCs. The strongest inhibition of the process was observed in the case of blocking (in experiments specific inhibitors for specific endocytosis pathways were used) clathrin mediated endocytosis (RAW 264.7, THP-1). The presence of all types of obtained NCs in lysosomes was also visualized. In addition, it has been shown that unmodified NCs, in contrast to pegylated NCs, have an influence on the phagocytic potential. None of the obtained NCs variants also led to the differentiation of THP-1 cells. Based on the above observations, it was concluded that the obtained polymeric NCs can be successfully modified (by PEG grafting) in a way that allows them to be masked for phagocytic cells. This confirmed the earlier hypothesis that synthesized NCs are a promising candidate that can be used for controlled drug delivery (more detailed information concerning the issue one can find in [24,43]).

1.4. Obtaining and Characterization of the Encapsulated Form of Clozapine

Currently, encapsulation of active compounds is a promising strategy in modern molecular pharmacology. Therefore, in the light of the above-described issues, encapsulation of clozapine, allowing its controlled release and delivery, can lead to an improvement of the therapeutic potential of the drug, which may have a direct impact on the quality of schizophrenia therapy. Based on the data obtained in previous experiments, the type of nanocarrier used to encapsulate clozapine was defined [23,24]. These were six-layer polymer pegylated NCs. The polyelectrolyte layer (PLL/PGA) was formed by the sequential adsorption (LbL) method on the emulsion core, which contained dissolved clozapine. Several variants of capsules containing the above-mentioned drug, differing in physicochemical parameters (thickness of the outer layer, pegylation, charge) were obtained. Respectively, they were: positively charged five-layered NCs—CLO-NCs V-PLL, negatively charged six-layered NCs—CLO-NCs VI-PGA and neutral, pegylated six-layer NCs—CLO-NCs VI-PGA-g(x)-PEG (different PEG grafted). The synthesis, physicochemical properties stability, as well as he release profile of clozapine of the obtained NCs are well described in [25]. Schematic representation of prepared CLO-NCs is illustrated in the Figure 3.

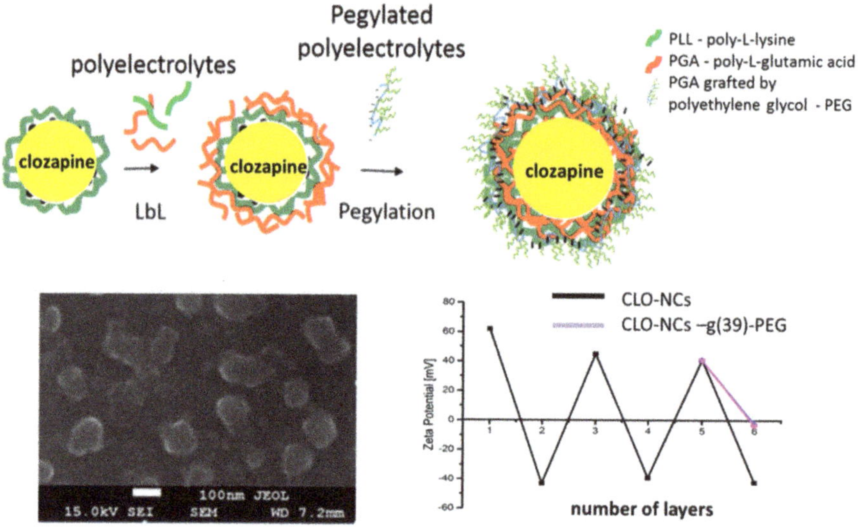

Figure 3. Upper panel—structure of CLO-NCs obtained using LbL technique. **Lower panel**—SEM micrograph of CLO-NCs VI PGA-g(39)-PEG and zeta potential measurements. Detailed information [25].

Based on the experience collected during the study of empty carriers, similar experiments were carried out to determine the behavior of carriers with encapsulated clozapine. Cytotoxicity and cell viability studies as well as interaction with phagocytic cells (Figure 4) showed similar results compared to those obtained with empty carriers [25]. CLO-NCs VI-PGA-g (39)-PEG was the formulation with the best parameters.

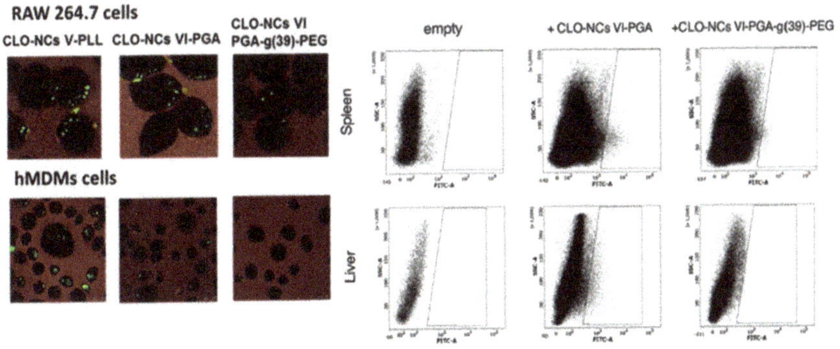

Figure 4. Left panel—In vitro CLO-NCs internalization studies performed in RAW 264.7 and hMDMs (human monocyte-derived macrophages) cells after a 2 h incubation with NCs. **Right panel**—In vivo-CLO-NCs biodistribution studies: the animals were injected with 150 mL suspension of CLO-NCs VI-PGA as well as CLO-NCs VI-PGA-g(39)-PEG. Flow cytometry studies were performed 4 h after injection. Detailed information [25].

While designing new drug carriers it is extremely important to estimate their biodistribution in an in-vivo system. Although the experiments with the use of mice were only qualitative, they clearly indicated dependence of the biodistribution profile on the modification of the outer layer of the capsule. Four hours after injection, the presence of CLO-NCs VI-PGA was confirmed mainly in the mouse liver and spleen (Figure 4), and on a smaller scale in the kidneys and lungs. Pegylation of the outer layer (CLO-NCs VI-PGA-g(39)-PEG) significantly reduced the accumulation level of capsules in the investigated organs [25].

Moreover, behavioral studies of the effectiveness of the encapsulated form of clozapine in experimental animals were performed [25]. The obtained results show that the encapsulated clozapine reduced the locomotor activity of mice in a manner characteristic of free clozapine; however, this effect was induced only by pegylated CLO-NCs (CLO-NCs VI-PGA-g(39)-PEG). Unpegylated NCs were not effective, probably due to their rapid elimination by macrophages. Although the obtained results are qualitative the effects of clozapine at significantly lower doses have been observed at this stage. In conclusion, the obtained results indicate the validity of clozapine encapsulation [25].

1.5. NCs Interactions with hCMEC/D3 Cells (Immortalized Human Cerebral Endocrine Cells, D3 Clone) Constituting the In Vitro Model of the Human Blood-Brain Barrier

Designing new drug delivery systems, especially those directed to the brain areas, it is first of all necessary to find an answer whether the new nanocarrier will be able to cross the blood brain barrier (BBB). Currently, several cell lines which have the characteristic of cells forming this natural barrier have been derived [44,45]. The best known in-vitro model of human BBB is the hCMEC/D3 cell line. This line was derived through the immortalization of human primary brain endothelial cells [45]. hCMEC/D3 cells show a morphology similar to primary cells, they form tight junction, exhibit trans-endothelial electrical resistance (TEER) and also maintain important and characteristic features of BBB, such as: expression of junctional proteins and efflux transporters [44,45]. This cell line was a convenient model dedicated to the study of the transcytosis process [46].

As mentioned earlier, the quantification of cell viability is a key element in understanding the interaction between NCs and the target cell. Therefore, when starting experiments using the human BBB model, these kinds of experiments were performed [43]. Also in this case, several different tests were carried out. Additionally, an attempt to answer the question whether cell death occurs through necrosis or apoptosis was made. As in the case of other tested cell lines, a decrease in viability, depending on NCs concentration, was shown, although in the case of hCMEC/D3 cells the scale of this phenomenon was much smaller, which probably could be related to a well-developed efflux transport and thus the rapid removal of excess capsules from the cell. Numerous reports [27,29,34], including our earlier studies [23–25], indicate a correlation between the positive charge on the NCs surface and the decrease in cell viability. This effect, associated with disruption of the cell membrane, is probably a common feature of all positively charged nanomaterials. The high level of LDH release due to the stimulation of the five-layered NCs supports this hypothesis. Unfortunately, based on the obtained data, it cannot be unambiguously determined whether apoptosis or necrosis has its contribution to cell death. In summary, the most promising results were obtained for six-layer capsules with a pegylated outer layer (CLO-NCs VI-PGA-g(39)-PEG), where the cell viability was almost 100%, after 24 and 48 h incubation with NCs.

BBB is an anatomical-functional system that regulates the exchange of substances between blood and the central nervous system (CNS). BBB maintains optimal homeostasis and protects of CNS against harmful substances, as well as enables selective transport of compounds circulating in the blood into the cerebrospinal fluid [47–49]. Due to the precise selectivity of the barrier, transport of therapeutic compounds to the brain is quite a challenge, because only uncharged, lipophilic and relatively small sizes substances can pass through the BBB without major obstacles. These are serious limitations that cannot be managed by currently available therapeutics [50]. Additional restrictions are precise transport mechanisms in endothelial cells, i.e., low level of pinocytic vesicles and selective transporters in the cell membrane [50]. Moreover, endothelial cells that have a polarized membrane to transport use transcytosis process (endocytosis on the apical side and exocytosis on the basolateral side) [47]. Therefore, the development of nanocarriers used to deliver drugs to defined areas of the brain requires detailed study of the transport process across BBB. Various endocytic pathways used for the internalization of exogenous substances by endothelial cells have been described in the literature [47,51]. Therefore, several experiments were carried out in order to find answers to three basic questions: (1) whether the mechanism of internalization is energy-dependent, (2) whether it occurs via endocytosis and (3) if yes, which endocytosis pathway is involved in the process. The obtained results [43] indicate the dependence of the process on NCs dose and time. The highest level of internalization has been described for positively charged, non-pegylated NCs, which is probably related to the facilitated interaction between the capsule and the cell membrane [34]. Moreover, the obtained results suggest an energy- and clathrine-dependent mechanism of internalization for all tested types of NCs and additional passive transport in the case of pegylated NCs. Confocal microscopy studies indicate the presence of synthesized NCs within clathrine vesicles as well as in the early endosomes and lysosomes. Considering the use of a new carrier for clozapine, it was necessary to carry out experiments showing not only the ability of model cells to internalize NCs, but also the ability of obtained NCs to cross the BBB.

In-vitro experiments involving hCMEC/D3 cell line revealed that the observed transcytosis process depended on NCs dose and time and the strongest effect was recorded for pegylated CLO-NCs (CLO-NCs VI-PGA-g(39)-PEG)—detailed information [43]. The most important data are presented in the Figure 5 where one can observe significant increase of transcytosis process (after 4 h incubation time) in case of pegylated CLO-NCs in comparison to PEG-unmodified CLO-NCs. Studies using a specific transcytosis inhibitor (filipin III) point to caveolae-dependent mechanism of the process. In conclusion, CLO-NCs VI-PGA-g(39)-PEG are able to cross the BBB and represent a promising model of the nanocarrier for clozapine.

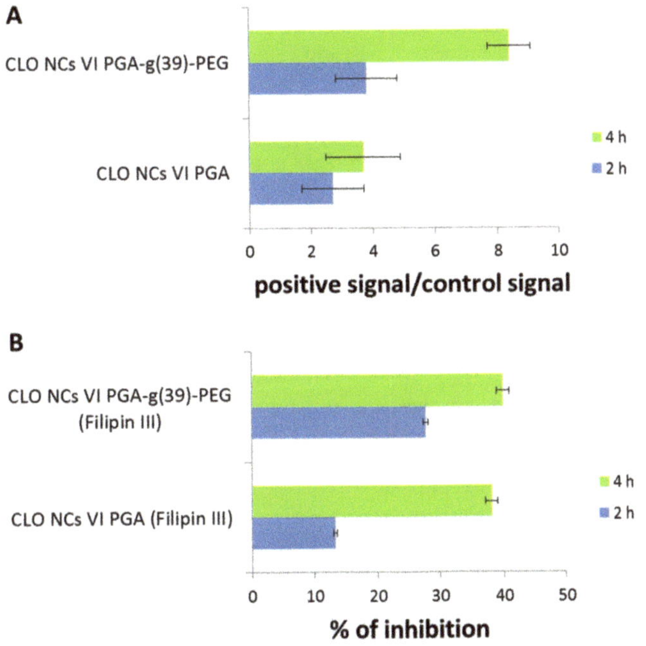

Figure 5. (**A**) Transcytosis experiment performed in hCMEC/D3 cells for CLO-NCs VI PGA and CLO-NCs VI PGA-g(39)-PEG (transwell pore-3mm). (**B**) Inhibition of the transcytosis process (incubation with filipin III—specific inhibitor of the process) for various types of CLO-NCs (detailed information [43]).

1.6. Heteromer of the D2-5-HT1A Receptors as an Important Target for Clozapine

The concept of oligomerization of G-protein coupled receptors (GPCRs) plays an important role in modern molecular pharmacology. The physical association of receptor proteins indicates a new level of signal complexity and the possibility of changing the pharmacological properties of the receptors included in the complex [52–54]. Research aimed at finding therapeutic substances operating via selective recognition of GPCRs heteromers is being undertaken more and more often. Such a strategy allows us to obtain a tissue-specific effect, since the interaction between receptors engaged in the complex formation can only take place when the receptors are simultaneously expressed on the same cell. Many recent reports indicate the existence of clinically relevant GPCRs heteromers, important in the treatment of, among others, pain, asthma or Parkinson's disease [54–58]. This evidence shows that GPCRs heteromers constitute extremely important targets in the design of modern treatment routes.

Both D_2R and 5-$HT_{1A}R$ are important sites of action of atypical antipsychotics [59], hence studies on the interaction between these receptors in the context of antipsychotic activity have been undertaken.

The obtained results are discussed in more detail in [60]. Constitutive dimerization of both investigated receptors in HEK 293 cells has been shown for the first time. To avoid the possibility of data misinterpretation, D_2-5-HT_{1A} receptor heteromers were confirmed by two independent techniques based on monitoring FRET (fluorescence resonance energy transfer) phenomenon (FLIM—fluorescence life time imagine microscopy and HTRF—homogenous time resolved FRET) (Figure 6). Moreover, the effect of various antipsychotics (inter alia: clozapine, aripiprazole, 8-OH-DPAT) on the heteromerization process has been determined. The highest effect was observed after incubation of cells with clozapine

(Figure 6), aripiprazole and simultaneous administration of clozapine and 8-OH-DPAT [60]. In the previous studies, [61–64], the opposite to the above-described effects of clozapine for dopamine D_1 (D_1R) and D_2R (D_1-D_2) heteromers, as well as serotonin 5-HT_{2A} (5-$HT_{2A}R$) and D_2R (D_2-5-HT_{2A}) heteromers, have been shown. These data indicated the specific effect of clozapine, depending on the type of receptors forming the complex. Furthermore, in vivo D_2R and $5HT_{1A}R$ co-localization in the mouse prefrontal cortex has been shown [60]. It suggests a potential presence of the above-mentioned heteromers in the brain. Additionally, to estimate the activation of intracellular signal transduction pathways as a result of antipsychotic action on D_2-5-HT_{1A} heteromers several functional tests have been performed. The experiments were carried out in HEK 293 cells expressing these receptors in various combinations. This approach enabled differentiation of the action of investigated compounds, depending on the presence of homo- or heteromeric complexes. Although the functional consequences of signal transmission via D_2-5-HT_{1A} heteromers are still not fully explained, the studies indicate the initiation of different signalling pathways depending on whether the receptors are co-expressed or produced individually in a cell. In summary, these results point to the possibility of antipsychotic action by specific targeting of active compounds on D_2-5-HT_{1A} heteromers (which have been shown to be present in cortical neurons [60] and they can also be an inspiration to improved pharmacotherapy of schizophrenia.

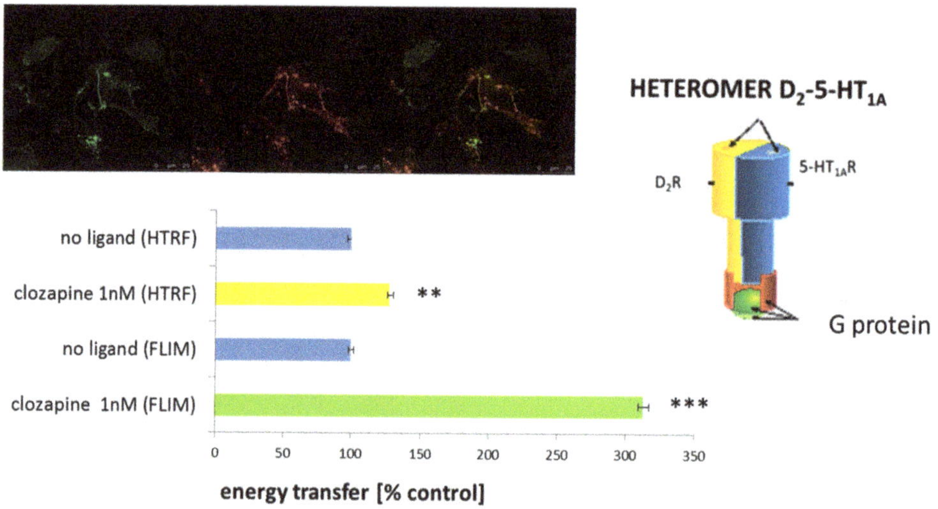

Figure 6. Constitutive dimerization of dopamine D_2 and serotonin 5-HT_{1A} receptors. **Upper panel**—HEK 293 cells expressing the dopamine D_2 (red) and serotonin 5-HT_{1A} (green) receptors; colocalization of both receptors (yellow). **Lower panel**—bar graph presenting FRET (fluorescence resonance energy transfer) measurements using HTRF (homogenous time resolved FRET) and FLIM (fluorescence life time imagine microscopy) techniques. The statistical significance was evaluated using student t-test and Mann–Whitney U-test, ** $p < 0.01$, *** $p < 0.001$. **Right panel**—schematic presentation of D_2-5-HT_{1A} heteromer. Detailed information [60].

1.7. Synthesis of a Targeting Ligand Specifically Recognizing the D2-5-HT1A Heteromer for Functionalization of the Obtained CLO-NCs VI-PGA-g(39)-PEG

The next step was the decorating of NC surfaces by attaching targeting ligands, which would allow selective delivery of drugs to defined target sites. Among others, human monoclonal antibody fragments—scFv (single-chain variable fragment) can be successfully used as targeting ligands. These antibodies consist of variable heavy (VH) and light (VL) regions of immunoglobulin chains linked by an elastic peptide linker designed to allow contact between the two chains and preserve the antigen binding site within a single linear molecule [57]. ScFvs fragments, compared to larger forms of monoclonal antibodies such as: Fab, $F(ab)_2$, IgG, are characterized by lower retention time in non-target tissues, better tissue penetration and reduced immunogenicity, which makes them attractive candidates for therapeutic applications [58].

Based on the data described above [60] we aimed to develop a targeting ligand in the form of a fragment of a human monoclonal scFv antibody specifically recognizing the D_2-5-HT_{1A} heteromer. To fulfil its role, such antibody must recognize the structural epitope formed within the heteromeric structure and, at the same time, not show specificity for monomeric or homomeric forms of the receptors. To accomplish this task the phage display technique—described for the first time by Smith [65]—Nobel laureate in 2018—was adapted [66]. The phagemid library of human scFv antibodies Tomlinson I + J (Geneservice) was used. This library allows for the preparation of approximately 3×10^8 different phages, the envelope of which the PIII-scFv fusion protein encoded by the pIT2 phagemid is embedded. Since both receptors included in the heteromer belong to the family of membrane proteins, it was extremely important to carry out the selection rounds, so-called bio-panning (Figure 7), under conditions most similar to those in which these receptors occur naturally in the cells, which allowed to preserve the native spatial conformation of the heteromer. For the isolation of phages specifically binding to the D_2-5-HT_{1A} heteromer, the immune-selection rounds were performed on CHO+ line cells (CHO-K1 stable line) overexpressing both receptors. Purified phages were incubated with antigen which constituted D_2-5-HT_{1A} heteromer presented on CHO+ cells. Then, by intensive rinsing, the unbound phages were removed. In the next stage the selected phages which possessed affinity to D_2-5-HT_{1A} heteromer were eluted, amplified, purified and used in the next round of the positive selection. Phages binding receptor monomers as well as other proteins present on the surface of CHO-K1 cells were eliminated by negative selection, using a CHO- cells. The CHO- cells constituted the mixture of stable CHO-K1 cells lines overexpressing only the single type of receptors forming the heteromer.

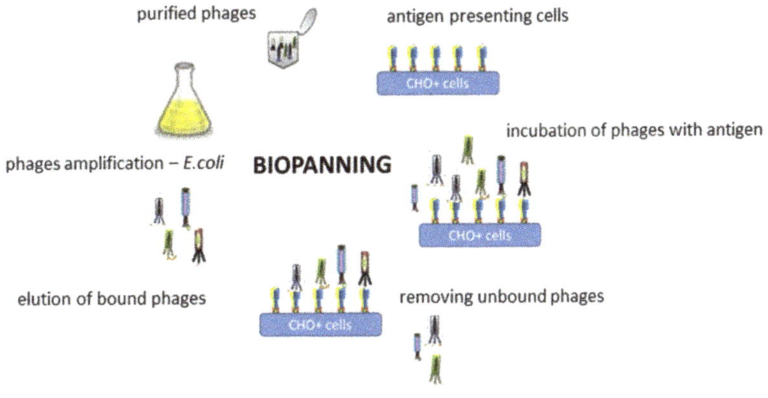

Figure 7. Phage display technique—biopanning process. For the isolation of phages specifically recognizing the D_2-5-HT_1A heteromer, the immune-selection rounds were performed. Purified phages were incubated with CHO+ line cells (CHO-K1 stable line) overexpressing both types of desired receptors. In the next step unbounded phages were removed in the process of intensive rinsing. Then the selected phages which point to affinity to D_2-5-HT_{1A} heteromer were eluted, amplified, purified and used in the next round of the positive selection. Detailed information [66].

To obtain a soluble form of scFv antibodies in the *E. coli* HB2151 expression system, the monoclonal phages, isolated in the selection process, that most strongly bind to the defined heteromer were used. The purification procedures based on affinity chromatography using Protein L-immobilized resin was performed. As a result of the experiments, the scFv monoclonal antibody specifically recognizing the D_2-5-HT_{1A} heteromers was isolated, and it has been used as a targeting ligand for functionalization of model NCs. The procedure is described in detail in [66].

2. Conclusions

The obtained NCs containing clozapine (CLO-NCs VI-PGA-g-PEG) represent a promising novel formulation of this compound. The encapsulated form of clozapine is safer since it does not influence the viability of diverse cells, does not cause activation of immunological system, and can cross BBB easily, not involving unsealing of the barrier. The experiments described here were carried out mainly using in-vitro models; however, preliminary studies showed that the CLO-NCs VI-PGA-g-PEG formulation allows clozapine effect in-vivo. In order to fully describe the behavior of nanocarriers, further detailed and extensive in-vivo studies are necessary. Moreover, the research points to the importance of modification of the most outer layer (surface) of the nanocarrier. NCs charge, pegylation process, as well as functionalization determining physico-chemical parameters of the nanocarriers enable its proper functioning. Table 1 includes summary of the results ob-tained for various variants of polymeric nanocapsules (NCs) constituting a new nanocar-rier for clozapine. As can be seen from the above review, each stage of the research towards obtaining the optimal nanocarrier is laborious and requires a lot of work. The full experimental paradigm is illustrated below (Figure 8). This kind of study engages specialists with wide knowledge in the field of chemistry, biochemistry, biophysics, biotechnology, molecular biology, pharmacology, molecular medicine etc. Designing of a nanocarrier from the chemical point of view, evaluation of its physico-chemical properties, description of its biochemical interaction in in vitro as well as in-vivo studies, designation of its pharmacological profile, and, last but not least, indication the desired site of action together with formulation of targeting ligand lead to professional development of new strategies of targeted delivery platform.

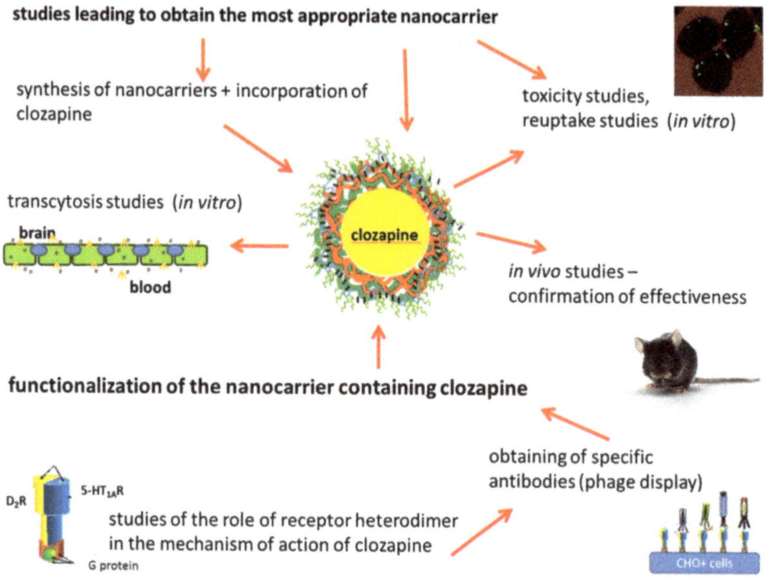

Figure 8. The full experimental paradigm.

Table 1. Summary of the results obtained for various variants of polymeric nanocapsules (NCs) constituting a new nanocarrier for clozapine.

	NCs I/III-PLL [b]	NCs II/IV-PGA [c]	NCs V-PLL [d]	NCs VI-PGA [e]	NCs VI-PGA-g(39)-PEG [f]	CLO-NCs V-PLL [g]	CLO-NCs VI-PGA [h]	CLO-NCs VI-PGA-g(39)-PEG [i]	Ref.
Charge of outer layer of nanocarrier	Positive	Negative	Positive	Negative	Neutral	Positive	Negative	Neutral	[23–25]
Toxicity [a]	High	Midium	Medium	Low	Low	Medium	Low	Very low	[23–25,43]
RAW and THP-1 cellls uptake	-	-	High +++ Clatrine path	High ++ Clatrine path	Low Clatrine path and passive transport	High +++ Clatrine path	High ++ Clatrine path	Low Clatrine path and passive transport	[24,25]
hMDMs cells Uptake	-	-	High +++	High ++	Low	High +++	High ++	Low	[43]
Phagocytic cells	Visible +++	Visible ++	Visible +++	Visible ++	Invisible	Visible +++	Visible ++	Invisible	[23–25,43]
hCMEC/D3 cells (in vitro BBB model) Uptake	-	-	High +++ Clatrine path	High ++ Clatrine path	High ++ Clatrine path and passive transport	High +++ Clatrine path	High ++ Clatrine path	High ++ Clatrine path and passive transport	[43]
Stimulation of phagocytic potential	-	-	-	-	-	High +++	High ++	Medium	[23,24,43]
Biodistribution In vivo	-	-	-	-	-	-	Lungs ++ Liver +++ Spleen +++ Kidney ++	Spleen + kidney +	[25]
Behavioral studies In vivo	-	-	-	-	-	-	Mice locomotor activity reductio, similarity to clozapine +	Mice locomotor activity reductio, similarity to clozapine ++	[25]
Transcytosis (hCMEC/D3 cells) In vitro BBB model)	-	-	High ++	High +	High +++	High ++	High + Caveole path	High +++ Caveole path	[43]

[a] Similar results have been recorded for various cells lines using different assays (see references). [b] **NCs I/III-PLL**—one or three layers polymeric nanocapsules (NCs) prepared using anionic surfactant AOT (sodium docusate) as an emulsifier, and biocompatible polyelectrolytes such as: PGA (poly(glycolic acid)) and PLL (poly(L-lysine)). Outer layer of the carrier constitutes PLL. [c] **NCs II/IV-PGA**—two or four layers polymeric nanocapsules (NCs) prepared using anionic surfactant AOT (sodium docusate) as an emulsifier, and biocompatible polyelectrolytes such as: PGA (poly(glycolic acid)) and PLL (poly(L-lysine)). Outer layer of the carrier constitutes PGA. [d] **NCs V-PLL**—five layers polymeric nanocapsules (NCs) prepared using anionic surfactant AOT (sodium docusate) as an emulsifier, and biocompatible polyelectrolytes such as: PGA (poly(glycolic acid)) and PLL (poly(L-lysine)). Outer layer of the carrier constitutes PLL. [e] **NCs VI-PGA**—six layers polymeric nanocapsules (NCs) prepared using anionic surfactant AOT (sodium docusate) as an emulsifier, and biocompatible polyelectrolytes such as: PGA (poly(glycolic acid)) and PLL (poly(L-lysine)). Outer layer of the carrier constitutes PGA. [f] **NCs VI-PGA-g(39)-PEG**—six layers polymeric nanocapsules (NCs) prepared using anionic surfactant AOT (sodium docusate) as an emulsifier, and biocompatible polyelectrolytes such as: PGA (poly(glycolic acid)) and PLL (poly(L-lysine)). Outer layer of the carrier constitutes PGA grafted by PEG (polyethylene glycol), grafting percentage was 39%. [g] **CLO-NCs V-PLL**—five layers polymeric nanocapsules (NCs) containing clozapine, prepared using anionic surfactant AOT (sodium docusate) as an emulsifier, and biocompatible polyelectrolytes such as: PGA (poly(glycolic acid)) and PLL (poly(L-lysine)). Outer layer of the carrier constitutes PLL. [h] **CLO-NCs VI-PGA**—six layers polymeric nanocapsules (NCs) containing clozapine, prepared using anionic surfactant AOT (sodium docusate) as an emulsifier, and biocompatible polyelectrolytes such as: PGA (poly(glycolic acid)) and PLL (poly(L-lysine)). Outer layer of the carrier constitutes PGA. [i] **CLO-NCs VI-PGA-g(39)-PEG**—six layers polymeric nanocapsules (NCs) containing clozapine, prepared using anionic surfactant AOT (sodium docusate) as an emulsifier, and biocompatible polyelectrolytes such as: PGA (poly(glycolic acid)) and PLL (poly(L-lysine)). Outer layer of the carrier constitutes PGA grafted by PEG (polyethylene glycol), grafting percentage was 39%. '+'—reflects intensity of the marked process.

Each stage of the presented research leading to full characterization of the functionalized nanocarrier requires elaborative work and is very demanding. However, the effort is worth it, because clozapine (or other antipsychotic drug) encapsulating and directing its activities in the defined areas enables enhancing its selectivity and specificity, as well as limits side effects which undoubtedly may contribute to increasing the safety of the schizophrenia therapy.

Funding: This research received no external funding.

Acknowledgments: The optimal nanocarrier for clozapine was found with the cooperation with Krzysztof Szczepanowicz from the Institute of Catalysis and Surface Chemistry, Polish Academy of Sciences (ICSC PAS). He was responsible for the synthesis and chemical characterization of the nanocapsules and encapsulation of clozapine. I am much obliged to Marta Dziedzicka-Wasylewska for her support during the project implementation. The open-access publication of this article was funded by the Priority Research Area BioS under the program "Excellence Initiative—Research University" at the Jagiellonian University in Krakow.

Conflicts of Interest: The author declare no conflict of interest.

References

1. Fond, G.; Macgregor, A.; Miot, S. Nanopsychiatry—The potential role of nanotechnologies in the future of psychiatry: A systematic review. *Eur. Neuropsychopharmacol.* **2013**, *23*, 1067–1071. [CrossRef] [PubMed]
2. Bhaskar, S.; Tian, F.; Stoeger, T.; Kreyling, W.; De La Fuente, J.M.; Grazú, V.; Borm, P.; Estrada, G.; Ntziachristos, V.; Razansky, D. Multifunctional Nanocarriers for diagnostics, drug delivery and targeted treatment across blood-brain barrier: Perspectives on tracking and neuroimaging. *Part. Fibre Toxicol.* **2010**, *7*, 3. [CrossRef]
3. Reinholz, J.; Landfester, K.; Mailänder, V. The challenges of oral drug delivery via nanocarriers. *Drug Deliv.* **2018**, *25*, 1694–1705. [CrossRef] [PubMed]
4. Jain, A.K.; Thareja, S. In vitro and in vivo characterization of pharmaceutical nanocarriers used for drug delivery. *Artif. Cells Nanomed. Biotechnol.* **2019**, *47*, 524–539. [CrossRef] [PubMed]
5. Rata, D.M.; Cadinoiu, A.N.; Atanase, L.I.; Popa, M.; Mihai, C.T.; Solcan, C.; Ochiuz, L.; Vochita, G. Topical formulations containing aptamer-functionalized nanocapsules loaded with 5-fluorouracil—An innovative concept for the skin cancer therapy. *Mater. Sci. Eng. C Mater. Biol. Appl.* **2021**, *119*, 111591. [CrossRef] [PubMed]
6. Rață, D.M.; Cadinoiu, A.N.; Atanase, L.I.; Bacaita, S.E.; Mihalache, C.; Daraba, O.-M.; Gherghel, D.; Popa, M. "In vitro" behaviour of aptamer-functionalized polymeric nanocapsules loaded with 5-fluorouracil for targeted therapy. *Mater. Sci. Eng. C Mater. Biol. Appl.* **2019**, *103*, 109828. [CrossRef]
7. Kadam, R.S.; Bourne, D.W.; Kompella, U.B. Nano-advantage in enhanced drug delivery with biodegradable nanoparticles: Contribution of reduced clearance. *Drug Metab. Dispos.* **2012**, *40*, 1380–1388. [CrossRef]
8. Bennewitz, S. Nanotechnology for delivery of drugs to the brain for epilepsy. *Neurotherapeutics* **2009**, *6*, 323–336. [CrossRef]
9. Mansor, N.I.; Nordin, N.; Mohamed, F.; Ling, K.H.; Rosli, R.; Hassan, Z. Crossing the Blood-Brain Barrier: A Review on Drug Delivery Strategies for Treatment of the Central Nervous System Diseases. *Curr. Drug Deliv.* **2019**, *16*, 698–711. [CrossRef]
10. Sánchez, A.; Mejía, S.P.; Orozco, J. Recent Advances in Polymeric Nanoparticle-Encapsulated Drugs against Intracellular Infections. *Molecules* **2020**, *25*, 3760. [CrossRef]
11. Pulgar, V.M. Transcytosis to Cross the Blood Brain Barrier, New Advancements and Challenges. *Front. Neurosci.* **2019**, *12*, 1019. [CrossRef] [PubMed]
12. Masri, B.; Salahpour, A.; Didriksen, M.; Ghisi, V.; Beaulieu, J.M.; Gainetdinov, R.R.; Caron, M.G. Antagonism of dopamine D2 receptor/beta-arrestin 2 interaction is a common property of clinically effective antipsychotics. *Proc. Natl. Acad. Sci. USA* **2008**, *105*, 13656–13661. [CrossRef]
13. De Berardis, D.; Rapini, G.; Olivieri, L.; Di Nicola, D.; Tomasetti, C.; Valchera, A.; Fornaro, M.; Di Fabio, F.; Perna, G.; Di Nicola, M.; et al. Safety of antipsychotics for the treatment of schizophrenia: A focus on the adverse effects of clozapine. *Ther. Adv. Drug Saf.* **2018**, *9*, 237–256. [CrossRef] [PubMed]
14. Berardis, D.; Serroni, N.; Campanella, D.; Olivieri, L.; Ferri, F.; Carano, A.; Cavuto, M.; Martinotti, G.; Cicconetti, A.; Piersanti, M.; et al. Update on the Adverse Effects of Clozapine: Focus on Myocarditis. *Curr. Drug Saf.* **2012**, *7*, 55–62. [CrossRef] [PubMed]
15. Wang, N.; Gao, X.; Li, M.; Li, Y.; Sun, M. Use of Solid Lipid Nanoparticles for the Treatment of Acute Acoustic Stress-Induced Cochlea Damage. *J. Nanosci. Nanotechnol.* **2020**, *20*, 7412–7418. [CrossRef] [PubMed]
16. Ishak, R.A.; Awad, G.A.; Zaki, N.M.; El-Shamy, A.E.-H.A.; Mortada, N.D. A comparative study of chitosan shielding effect on nano-carriers hydrophilicity and biodistribution. *Carbohydr. Polym.* **2013**, *94*, 669–676. [CrossRef]
17. Pinkerton, N.M.; Grandeury, A.; Fisch, A.; Brozio, J.; Riebesehl, B.U.; Prud'Homme, R.K. Formation of Stable Nanocarriers by in Situ Ion Pairing during Block-Copolymer-Directed Rapid Precipitation. *Mol. Pharm.* **2013**, *10*, 319–328. [CrossRef] [PubMed]
18. Sun, Y.; Kang, C.; Liu, F.; Song, L. Delivery of Antipsychotics with Nanoparticles. *Drug Dev. Res.* **2016**, *77*, 393–399. [CrossRef]

19. Masoumi, H.; Basri, M.; Samiun, S.; Izadiyan, Z.; Lim, C.J. Enhancement of encapsulation efficiency of nanoemulsion-containing aripiprazole for the treatment of schizophrenia using mixture experimental design. *Int. J. Nanomed.* **2015**, *13*, 6469–6476. [CrossRef] [PubMed]
20. Joseph, E.; Reddi, S.; Rinwa, V.; Balwani, G.; Saha, R. DoE based Olanzapine loaded poly-caprolactone nanoparticles decreases extrapyramidal effects in rodent model. *Int. J. Pharm.* **2018**, *25*, 198–205. [CrossRef] [PubMed]
21. Sherje, A.P.; Surve, A.; Shende, P. CDI cross-linked β-cyclodextrin nanosponges of paliperidone: Synthesis and physicochemical characterization. *J. Mater. Sci. Mater. Med.* **2019**, *13*, 74. [CrossRef] [PubMed]
22. Łukasiewicz, S.; Mikołajczyk, A.; Szczęch, M.; Szczepanowicz, K.; Warszyński, P.; Dziedzicka-Wasylewska, M. Encapsulation of clozapine into polycaprolactone nanoparticles as a promising strategy of the novel nanoformulation of the active compound. *J. Nanoparticle Res.* **2019**, *21*, 149. [CrossRef]
23. Łukasiewicz, S.; Szczepanowicz, K. In vitro interaction of polyelectrolyte nanocapsules with model cells. *Langmuir* **2014**, *30*, 1100–1107. [CrossRef] [PubMed]
24. Łukasiewicz, S.; Szczepanowicz, K.; Błasiak, E.; Dziedzicka-Wasylewska, M. Biocompatible Polymeric Nanoparticles as Promising Candidates for Drug Delivery. *Langmuir* **2015**, *31*, 6415–6425. [CrossRef]
25. Łukasiewicz, S.; Szczepanowicz, K.; Podgórna, K.; Błasiak, E.; Majeed, N.; Ogren, S.O.Ö.; Nowak, W.; Warszyński, P.; Dziedzicka-Wasylewska, M. Encapsulation of clozapine in polymeric nanocapsules and its biological effects. *Colloids Surf. B Biointerfaces* **2016**, *140*, 342–352. [CrossRef]
26. Nel, A.; Xia, T.; Mädler, L.; Li, N. Toxic Potential of Materials at the Nanolevel. *Science* **2006**, *311*, 622–627. [CrossRef] [PubMed]
27. Hong, S.; Leroueil, P.R.; Janus, E.K.; Peters, J.L.; Kober, M.-M.; Islam, M.T.; Orr, B.G.; Baker, J.J.R.; Holl, M.M.B. Interaction of Polycationic Polymers with Supported Lipid Bilayers and Cells: Nanoscale Hole Formation and Enhanced Membrane Permeability. *Bioconjugate Chem.* **2006**, *17*, 728–734. [CrossRef] [PubMed]
28. Lee, H.; Larson, R.G. Lipid Bilayer Curvature and Pore Formation Induced by Charged Linear Polymers and Dendrimers: The Effect of Molecular Shape. *J. Phys. Chem. B* **2008**, *112*, 12279–12285. [CrossRef]
29. Leroueil, P.R.; Berry, S.A.; Duthie, K.; Han, G.; Rotello, V.M.; McNerny, D.Q.; Baker, J.R., Jr.; Orr, B.G.; Holl, M.M.B. Wide Varieties of Cationic Nanoparticles Induce Defects in Supported Lipid Bilayers. *Nano Lett.* **2008**, *8*, 420–424. [CrossRef]
30. Verma, A.; Stellacci, F. Effect of Surface Properties on Nanoparticle? *Cell Interact. Small* **2010**, *6*, 12–21.
31. Bailly, A.L.; Correard, F.; Popov, A.; Tselikov, G.; Chaspoul, F.; Appay, R.; Al-Kattan, A.; Kabashin, A.; Braguer, D.; Esteve, M.A. In vivo evaluation of safety, biodistribution and pharmacokinetics of laser-synthesized gold nanoparticles. *Sci. Rep.* **2019**, *9*, 12890. [CrossRef]
32. Mailander, V.; Landfester, K. Interaction of Nanoparticles with Cells. *Biomacromolecules* **2009**, *10*, 2379–2400. [CrossRef]
33. Stark, W.J. Nanoparticles in Biological Systems. *Angew. Chem. Int. Ed.* **2011**, *50*, 1242–1258. [CrossRef]
34. Xiao, K.; Li, Y.; Luo, J.; Lee, J.S.; Xiao, W.; Gonik, A.M.; Agarwal, R.G.; Lam, K.S. The effect of surface charge on in vivo biodistribution of PEG-oligocholic acid based micellar nanoparticles. *Biomaterials* **2011**, *32*, 3435–3446. [CrossRef]
35. Donahue, N.D.; Acar, H.; Wilhelm, S. Concepts of nanoparticle cellular uptake, intracellular trafficking, and kinetics in nanomedicine. *Adv. Drug Deliv. Rev.* **2019**, *143*, 68–96. [CrossRef]
36. Gaucher, G.; Asahina, K.; Wang, J.; Leroux, J. Effect of Poly(NVinyl-Pyrrolidone)-Block-Poly(d,l-Lactide) as Coating Agent on the Opsonization, Phagocytosis, and Pharmacokinetics of Biodegradable Nanoparticles. *Biomacromolecules* **2009**, *10*, 408–416. [CrossRef]
37. Tadros, T.F.; Warszyński, P.; Zembala, M. The Influence of Polymer Adsorption on Deposition Kinetics of Colloid Particles II. Experimental Studies. *Colloids Surf.* **1989**, *39*, 93–105. [CrossRef]
38. Weyermann, J.; Lochmann, D.; Zimmer, A. A practical note on the use of cytotoxicity assays. *Int. J. Pharm.* **2005**, *288*, 369–376. [CrossRef]
39. Zhang, Y.; Kohler, N.; Zhang, M. Surface Modification of Superparamagnetic Magnetite Nanoparticles and Their Intracellular Uptake. *Biomaterials* **2002**, *23*, 1553–1561. [CrossRef]
40. Chaudhari, K.R.; Ukawala, M.; Manjappa, A.S.; Kumar, A.; Mundada, P.K.; Mishra, A.K.; Mathur, R.; Mönkkönen, J.; Murthy, R.S. Opsonization, biodistribution, cellular uptake and apoptosis study of PEGylated PBCA nanoparticle as potential drug delivery carrier. *Pharm. Res.* **2012**, *29*, 53–68. [CrossRef]
41. Poon, Z.; Lee, J.B.; Morton, S.W.; Hammond, P.T. Controlling in vivo stability and biodistribution in electrostatically assembled nanoparticles for systemic delivery. *Nano Lett.* **2011**, *11*, 2096–2103. [CrossRef]
42. Neves, A.R.; Queiroz, J.F.; Weksler, B.; Romero, I.A.; Couraud, P.-O.; Reis, S. Solid lipid nanoparticles as a vehicle for brain-targeted drug delivery: Two new strategies of functionalization with apolipoprotein E. *Nanotechnology* **2015**, *26*, 495103. [CrossRef]
43. Łukasiewicz, S.; Błasiak, E.; Szczepanowicz, K.; Guzik, K.; Bzowska, M.; Warszyński, P.; Dziedzicka-Wasylewska, M. The interaction of clozapine loaded nanocapsules with the hCMEC/D3 cells—In vitro model of blood brain barrier. *Colloids Surf. B Biointerfaces* **2017**, *159*, 200–210. [CrossRef] [PubMed]
44. Eigenmann, D.E.; Xue, G.; Kim, K.S.; Moses, A.V.; Hamburger, M.; Oufir, M. Comparative study of four immortalized human brain capillary endothelial cell lines, hCMEC/D3, hBMEC, TY10, and BB19, and optimization of culture conditions, for an in vitro blood-brain barrier model for drug permeability studies. *Fluids Barriers CNS* **2013**, *10*, 1–17. [CrossRef] [PubMed]
45. Weksler, B.; Romero, I.A.; Couraud, P.-O. The hCMEC/D3 cell line as a model of the human blood brain barrier. *Fluids Barriers CNS* **2013**, *10*, 16. [CrossRef] [PubMed]

46. Markoutsa, E.; Papadia, K.; Clemente, C.; Flores, O.; Antimisiaris, S.G. Anti-Abeta-MAb and dually decorated nanoliposomes: Effect of Abeta1-42 peptides on interaction with hCMEC/D3 cells. *Eur. J. Pharm. Biopharm.* **2012**, *81*, 49–56. [CrossRef] [PubMed]
47. Georgieva, J.H.; Kalicharan, D.; Couraud, P.O.; Romero, I.A.; Weksler, B.; Hoekstra, D.; Zuhorn, I.S. Surface characteristics of nanoparticles determine their intracellular fate in and processing by human blood-brain barrier endothelial cells in vitro. *Mol. Ther.* **2011**, *19*, 318–325. [CrossRef]
48. Tosi, G.; Duskey, J.T.; Kreuter, J. Nanoparticles as carriers for drug delivery of macromolecules across the blood-brain barrier. *Expert Opin. Drug Deliv.* **2019**, *17*, 23–32. [CrossRef]
49. Xiao, G.; Gan, L.-S. Receptor-Mediated Endocytosis and Brain Delivery of Therapeutic Biologics. *Int. J. Cell Biol.* **2013**, *2013*, 703545. [CrossRef]
50. Brzezińska, K.; Ziaja, M. Struktura i funkcje bariery krew-mózg. *Postępy Biologii Komórki* **2012**, *1*, 84–99.
51. Smith, M.W.; Gumbleton, M. Endocytosis at the blood-brain barrier: From basic understanding to drug delivery strategies. *J. Drug Target.* **2006**, *14*, 191–214. [CrossRef]
52. Gomes, I.; Ayoub, M.A.; Fujita, W.; Jaeger, W.C.; Pfleger, K.D.; Devi, L.A. G Protein–Coupled Receptor Heteromers. *Annu. Rev. Pharmacol. Toxicol.* **2016**, *56*, 403–425. [CrossRef]
53. Rozenfeld, R.; Devi, L.A. Exploring a role for heteromerization in GPCR signalling specificity. *Biochem. J.* **2010**, *433*, 11–18. [CrossRef] [PubMed]
54. Albizu, L.; Moreno, J.L.; González-Maeso, J.; Sealfon, S.C. Heteromerization of G protein-coupled receptors: Relevance to neurological disorders and neurotherapeutics. *CNS Neurol. Disord. Drug Targets* **2010**, *9*, 636–650. [CrossRef]
55. Fujita, W.; Gomes, I.; Devi, L.A. Revolution in GPCR signalling: Opioid receptor heteromers as novel therapeutic targets: IUPHAR Review 10. *Br. J. Pharmacol.* **2014**, *171*, 4155–4176. [CrossRef] [PubMed]
56. Derouiche, L.; Massotte, D. G protein-coupled receptor heteromers are key players in substance use disorder. *Neurosci. Biobehav. Rev.* **2019**, *106*, 73–90. [CrossRef] [PubMed]
57. Kamal, M.; Jockers, R. Biological Significance of GPCR Heteromerization in the Neuro-Endocrine System. *Front. Endocrinol. (Lausanne)* **2011**, *1*, 2. [CrossRef]
58. Carriba, P.; Ortiz, O.; Patkar, K.; Justinova, Z.; Stroik, J.; Themann, A.; Muller, C.; Woods, A.S.; Hope, B.T.; Ciruela, F.; et al. Striatal Adenosine A2A and Cannabinoid CB1 Receptors Form Functional Heteromeric Complexes that Mediate the Motor Effects of Cannabinoids. *Neuropsychopharmacology* **2007**, *32*, 2249–2259. [CrossRef]
59. Newman-Tancredi, A.; Kleven, M.S. Comparative pharmacology of antipsychotics possessing combined dopamine D2 and serotonin 5-HT1A receptor properties. *Psychopharmacology* **2011**, *216*, 451–473. [CrossRef]
60. Łukasiewicz, S.; Błasiak, E.; Szafran-Pilch, K.; Dziedzicka-Wasylewska, M. Dopamine D2 and serotonin 5-HT1A receptor interaction in the context of the effects of antipsychotics-in vitro studies. *J. Neurochem.* **2016**, *137*, 549–560. [CrossRef]
61. Faron-Górecka, A.; Górecki, A.; Kuśmider, M.; Wasylewski, Z.; Dziedzicka-Wasylewska, M. The role of D1-D2 receptor hetero-dimerization in the mechanism of action of clozapine. *Eur. Neuropsychopharmacol.* **2008**, *18*, 682–691. [CrossRef]
62. Łukasiewicz, S.; Faron-Górecka, A.; Kędracka-Krok, S.; Dziedzicka-Wasylewska, M. Effect of clozapine on the dimerization of serotonin 5-HT(2A) receptor and its genetic variant 5-HT(2A)H425Y with dopamine D(2) receptor. *Eur. J. Pharmacol.* **2011**, *659*, 114–123. [CrossRef]
63. Bird, R.E.; Hardman, K.D.; Jacobson, J.W.; Johnson, S.; Kaufman, B.M.; Lee, S.M.; Pope, S.H.; Riordan, G.S.; Whitlow, M. Single-chain antigen-binding proteins. *Science* **1988**, *242*, 423–426. [CrossRef]
64. Frenzel, A.; Hust, M.; Schirrmann, T. Expression of Recombinant Antibodies. *Front. Immunol.* **2013**, *4*, 217. [CrossRef]
65. Smith, G.P. Filamentous fusion phage: Novel expression vectors that display cloned antigens on the virion surface. *Science* **1985**, *228*, 1315–1317. [CrossRef]
66. Łukasiewicz, S.; Fic, E.; Bzowska, M.; Dziedzicka-Wasylewska, M. Isolation of Human Monoclonal scfv Antibody Specifically Recognizing the D2-5-HT1A Heteromer. *J. New Dev. Chem.* **2019**, *2*, 18–25. [CrossRef]

polymers

MDPI

Review

Designing Natural Polymer-Based Capsules and Spheres for Biomedical Applications—A Review

Kusha Sharma [1], Ze'ev Porat [2,3,*] and Aharon Gedanken [1,*]

[1] Department of Chemistry, Bar-Ilan Institute for Nanotechnology and Advanced Materials, Bar-Ilan University, Ramat-Gan 52900, Israel; kusha28793@gmail.com
[2] Department of Civil and Environmental Engineering, Ben-Gurion University of the Negev, Be'er Sheva 84105, Israel
[3] Department of Chemistry, Nuclear Research Center-Negev, Be'er Sheva 84190, Israel
* Correspondence: poratze@post.bgu.ac.il (Z.P.); Aharon.Gedanken@biu.ac.il (A.G.)

Abstract: Natural polymers, such as polysaccharides and polypeptides, are potential candidates to serve as carriers of biomedical cargo. Natural polymer-based carriers, having a core–shell structural configuration, offer ample scope for introducing multifunctional capabilities and enable the simultaneous encapsulation of cargo materials of different physical and chemical properties for their targeted delivery and sustained and stimuli-responsive release. On the other hand, carriers with a porous matrix structure offer larger surface area and lower density, in order to serve as potential platforms for cell culture and tissue regeneration. This review explores the designing of micro- and nano-metric core–shell capsules and porous spheres, based on various functions. Synthesis approaches, mechanisms of formation, general- and function-specific characteristics, challenges, and future perspectives are discussed. Recent advances in protein-based carriers with a porous matrix structure and different core–shell configurations are also presented in detail.

Keywords: natural polymers; polymeric capsules; porous polymeric spheres; active pharmaceutical carriers; drug delivery; stimuli-responsive release; cell culture platforms

Citation: Sharma, K.; Porat, Z.; Gedanken, A. Designing Natural Polymer-Based Capsules and Spheres for Biomedical Applications—A Review. *Polymers* **2021**, *13*, 4307. https://doi.org/10.3390/polym13244307

Academic Editors: M. Ali Aboudzadeh and Shaghayegh Hamzehlou

Received: 21 November 2021
Accepted: 6 December 2021
Published: 9 December 2021

1. Introduction

Conventional drug therapy involves administering the drug or pharmaceutical agent directly into the body, through oral, pulmonary, or parenteral routes. However, several demerits to this approach are the rapid release of the drug into the body at the site of administration, loss of drug dose on the way from the site of administration to the target site (due to biological degradation), the requirement for administering higher doses of the drugs to compensate for this loss, higher chances of over- or under-medication, side effects due to the interaction of the drugs with untargeted sites, the requirement of frequent dosing, lower drug bioavailability, lower per-unit cost (but higher overall healthcare cost), and higher total dosage requirement into the body. These demerits have led to the need for a different approach, which involves transporting the active pharmaceutical cargo (APC) and releasing it to the targeted (or affected) site in the body for therapeutic effect via drug delivery agents. Such a therapeutical approach has enabled the site-specific, slow, sustained, and controlled release of drugs, thus improving their bioavailability, pharmacokinetics, and increased efficacy, as well as minimizing the side effects to the untargeted sites and overall risk to the patient, thereby reducing the overall medication cost, due to the decreased frequency of drug administration and increasing patient compliance.

The development of drug delivery systems (DDS) began in the 1950s, when Jatzkewitz et al. (1955) reported that the conjugation of the psychedelic drug Mescaline, with co-polymer of N-vinylpyrrolidone and acrylic acid, prolonged its in-vivo residence time [1]. The first generation of drug delivery (1950–1980) involved the study of controlled-release mechanisms and development of oral and transdermal sustained-release systems [2]. Eventually, the first controlled delivery device, based on silicone rubber for delivering the drug isoproterenol, was reported in 1964 for its potential application as implants to treat heart block [3]. This was followed by several studies on developing a variety of polymeric and liposomal systems for the controlled release of various drugs and their underlaying release mechanisms [4–7]. The second-generation drug delivery (1981–2010) was basically focused on the study and development of constant-release, self-regulated drug delivery systems, and nanoparticle-based drug delivery systems. During this era, many sustained-release drug formulations (drugs-DDS), based on polymeric nanoparticles (Adagen, Gliadel, Copaxone), polymeric implants (Zoladex), liposomal carriers (Doxil, Abelcet), dendrimer-conjugates, and protein-based nanoparticles (Abraxane), were clinically tested and approved by the FDA. The past decade has been focused on designing smart, stimuli-responsive systems for targeted drug delivery. These systems have been shown to actively deliver the drug to the target site and enable controlled drug release by undergoing physical and/or chemical changes, in response to biological or external triggers.

In the past few decades, a wide variety of novel drug delivery approaches, in the form of micro- and nanoparticles (core–shell capsules, as well as matrix-type spheres), transdermal patches, gels, dendrimers, micelles, microneedles, and microfluidics-based devices have been developed (Figure 1). These were usually made of synthetic polymers (such as poly-lactic glycolic acid), natural polymers (such as polysaccharides, polypeptides, and polynucleotides) (see Table 1), liposomes, metallic formulations, metal oxides, carbon nanotubes, etc., aimed at a variety of functions, including site-selective, active, or passive targeted delivery of a wide variety of drugs for treating diseases, such as cancer and diabetes. Several parameters, such as the material of fabrication, size, shape, structural configuration, and surface characteristics of these APC carrier systems (ACSs), play a major role in their interaction with the in-vivo chemical environment, while passing, from the site of administration to the site of action, their function and in-vivo biodistribution. As such, these parameters are considered vital to designing better and smarter ACSs.

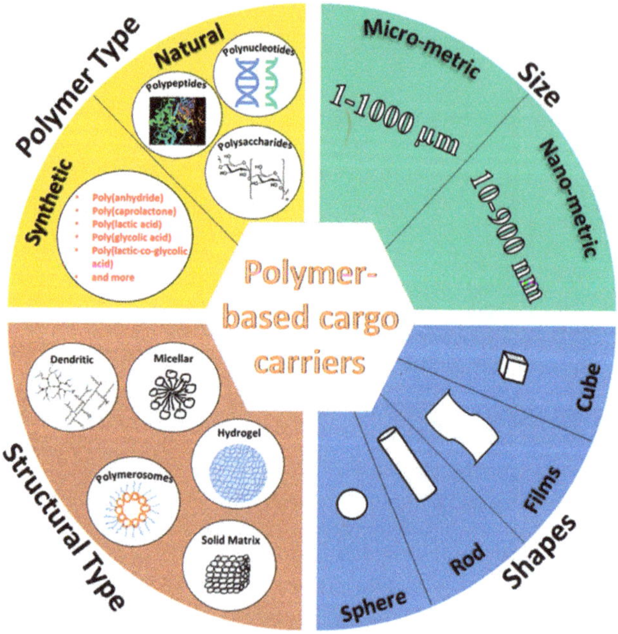

Figure 1. Classification of polymer-based carriers of biomedical cargo.

The following review focuses on function-specific aspects of designing micro- and nano-metric spherical APC carrier systems (SACS) made of natural polymers, such as polysaccharides and polypeptides, having structural configurations of core–shell and porous matrix. Chemical aspects involved in designing SACS, their synthesis approaches, formation mechanisms, and general- and application-specific characteristics are discussed. Finally, recent advances in the protein-based SACS, with a porous matrix structure, as well as different core–shell configurations, are presented in detail.

Table 1. List of natural polymers utilized to develop biomedical carriers.

Polymer Class	Polymer
Polysaccharides	Cellulose
	Cellulose derivatives
	Alginate
	Gellan gum
	Pectin
	Gum Arabica
	Gaur gum
	Locust bean gum
	Starch
	Carrageenan
	Chitin
	Chitosan
	Xanthan gum
	Shellac
	Dextran
	Cashew gum
	Pullulan
Polypeptides	Gelatin
	Bovine serum albumin
	Human serum albumin
	Egg albumin
	Casein
	Collagen
	Keratin
	Elastin
	Resilin
	Soy protein
	Gliadin
Hyaluronic acid	Hyaluronic acid
Phospholipids	Liposomes
Polynucleotides	Ribonucleic acid
	Deoxyribonucleic acid

2. Chemical Aspects of Designing Natural Polymer-Based Spherical Capsules and Spheres

Designing nano- and micro-capsules calls for the foremost consideration of the requirements laid down for their utilization in various biomedical functions. Such functions may involve sustained-release of cargo at the affected site in the body [8], the stimuli-responsive release of the cargo [9], its targeted delivery to the site of action [10], its protection from the hostile bodily environment [11], its better bioavailability in the body [12], better integration of the cargo into the body (such as the integration of the progenitor cells at tissue lesions) [13], blood vessel embolization [14] by the capsules, etc. The structural configurations of the SACS and mode of encapsulation of the cargo are chosen based on these biomedical requirements, upon which the eventual path/technique of capsule synthesis depends. Here, we discuss these requirements in detail, their decisive role in structural configuration choice and parameters, and the synthesis approaches that have been employed for years in the development of natural polymer-based SACS.

2.1. Function-Specific Carrier Design

2.1.1. Structural Configurations and Carrier Materials

Through the years of evolution in micro- and nano-capsule design, several core–shell configurations have been developed. These include core–shell capsules with solid, liquid, or hollow cores, encapsulated within a single or multiwalled shell and made of natural polymers, such as carbohydrates and proteins. Porous spherical matrices have also been prepared as cargo carriers. It is important to note that the polymer-based, micro- and nanometric core–shell capsules and porous spheres fall under a broader category of micro- and nano-particles, with sizes ranging from 1–1000 μm and 10–900 nm, respectively. A general schematic diagram of the different structural configurations is presented in Figure 2.

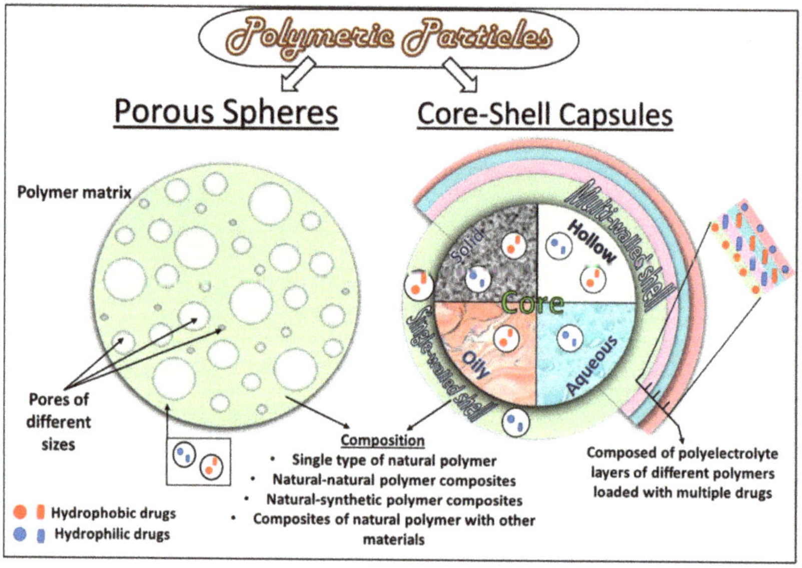

Figure 2. Structural configurations of core–shell and porous natural polymeric/protein particles.

Polymeric capsule shells, with a variety of compositions, have been prepared. These shell compositions may involve (a) a single type of natural polymer, such as chitosan [15] or albumin [16]; (b) composites of different types of natural polymers, such as BSA-alginate [17]; (c) composites of different types of natural and synthetic polymers, such as collagen-PLGA [18]; (d) natural polymers functionalized by other materials, including inorganic nanoparticles [19], functionalizing polymers [10], antibodies [20], and a variety of other materials. Diverse core materials have been encapsulated, in solid [21] or liquid form [22,23], within these shells, either as carriers of different types of active pharmaceutical cargo (APC) or made directly of the solid or liquid APCs (see Figure 2). A solid core of various natural [11] as well as synthetic polymers [24], metallic particles [23], and composites, have been prepared to make up a hydrophilic or a hydrophobic core, depending upon the type of moieties present in the precursor materials and application-based requirements. Liquid cores of organic solvent [22], oils [16], and a variety of aqueous media [25] have also been prepared to disperse either hydrophobic or hydrophilic APCs. In addition, hollow/porous capsules, made of natural polymers, have also been developed [18,26]. Depending upon the biomedical applications, the cargo may be dispersed or dissolved in the liquid/solid core as a reservoir/matrix or/and embedded in the shell of the capsules with a liquid/solid/hollow core. Depending upon the desired applications, the encapsulated cargo can be medicinal drugs [10], growth factors [12], stem cells, progenitor cells [27], probiotic bacterial strains [11], nutritional molecules (such as vitamins [28]), hormones (such as insulin [29]), and several more.

Desired Functions

Sustained-release. The choice of the core–shell materials depends primarily upon the physical and chemical properties of the natural polymer, type of applications, and mode of action required. Sustained-release formulations are prepared from the natural polymeric shell and core materials that facilitate the prolonged-release of APC via a combination of processes, such as diffusion, erosion, osmosis, and swelling. These processes are discussed in detail in the next section. Purely diffusion-controlled release from a capsule primarily involves the mass transfer of the cargo from the capsule to the release media, driven solely by their concentration gradient [30]. However, generally, release capsules made of natural polymers undergo a combination of dissolution, swelling, and erosion processes to release the cargo at the target site. Silk fibroin-based microcapsules have shown swelling-controlled release of doxorubicin (Dox) [14], wherein the microcapsules experienced enormous water uptake, leading to the enhanced initial release of Dox, and eventually swelled-up, due to which the Dox release rate slowed down. Collagen microcapsules have shown erosion-controlled release of human vascular endothelial growth factor (rhVEGF) and basic fibroblast growth factor (bFGF) over time [28,31].

Stimuli-responsive release. In addition, the cargo release processes may also be triggered in response to certain stimuli, such as a change in pH and temperature or the presence of digestive enzymes. Ionic polysaccharide-based capsules of chitosan, alginate, agar, carrageenan, cellulose, gaur, and xanthan gum have shown pH- and temperature-responsive release, due to the sensitivity of certain groups (such as amine group in chitosan) towards certain pH and higher temperature. Similarly, pectin and chondroitin sulfate show pH sensitivity and enzymatic degradation [32]. Various proteins, such as albumins, show pH, temperature, and enzyme responsive release of cargo. The pH-responsive action has also been shown in polymer–polymer composite microcapsules of BSA-alginate [17]. Additionally, the enzyme-catalyzed release of 3,4,9,10-tetra-(hectoxy-carbonyl)-perylene (THCP) was observed from BSA/polyphenol microcapsules, due to their degradation by α-chymotrypsin [33]. Cargo release, in the cases above, may involve both the release from the capsule core, as well as the capsule shell, depending upon the location and the state of the APC in the capsule.

Targeting. Targeted delivery of cargo refers to delivering an APC to the target site, selectively and independently of the route (site and method) of administration, through a delivery agent. Targeting can be organ-specific, tissue-specific, specific to pathogens (such as parasites), receptor-specific, or specific at the organelle-level for targeting mitochondria, cytoplasm, DNA, etc. A higher concentration of the drug at the desired site can be ensured through targeted delivery by preventing undesired drug loss and adverse effects at the untargeted sites. For targeted delivery of APC, the capsule shell is usually functionalized with various ligands, such as peptides [34], polymers [35], antibodies [20], nucleic acids, and vitamins [22]. Physically stimulated targeted delivery formulations have also been developed, wherein superparamagnetic particles have been functionalized on the microcapsule shell for the magnetically stimulated delivery of capsules at the target site [19].

Protection of the cargo. Another function of the capsules involves the protection of the cargo from the hostile bodily environment. Cargo, such as hydrophilic drugs and probiotic bacterial strains, have been shown to directly degrade when introduced into the body. Their protection from biodegradation, before their release at the target site, can be ensured by their encapsulation inside hydrophobic shells made of polymers, such as zein protein [11]. Hydrophilic and hydrophobic drugs have also been protected within composite capsule shells and their organic cores, respectively [36]. Similarly, many such strategies have been employed for the protection of the cargo inside the polymeric capsules.

Increasing cargo bioavailability. Encapsulation in polymeric capsules has also been applied to increase the bioavailability and dissolution rate of the cargo. Such cargo materials are usually hydrophobic, and their better absorption in the body requires structural modifications and changes in their degree of crystallinity. These modifications can be introduced by making biphasic, amorphous solid dispersions (ASDs) of the hydrophobic, crystalline cargo with a natural polymeric material [37], or by changing the microenvironment of the encapsulated cargo. Such a strategy involves the entrapment of the cargo in a polymer matrix or an acidic compound, such as citric acid. ASDs have been made to serve as solid cores encapsulated within protein microcapsules [12,21]. ASDs of various drugs have also been made in composition with a variety of natural polymers, such as gaur gum, xanthan gum, and acacia [38]. The encapsulation of ASDs in hydrophilic capsules also ensures the enhanced bioavailability of the cargo.

Carriers as cell-culture platforms. Hollow core capsules and porous spheres have served as 3D culture platforms/scaffolds for various types of cells for their better integration into the body at the tissue lesion-affected area and tissue regeneration. Depending upon the site of the lesion and type of tissue, various natural polymers can be selected for the synthesis of capsules and spheres that may serve as platforms for cell and tissue culture. Porous microspheres have been shown to provide a larger surface area to serve as effective cell culture platforms. It has been shown that spheres with pore diameter $\geq$20 μm are suitable for cell culture inside the sphere pores [26]. Microcapsules and porous spheres made of various natural polymers, such as collagen [13], gelatin [23], silk fibroin [39], pectin [40], chitosan/gellan gum [41], chondroitin sulfate, alginate, etc., have been shown to serve as excellent scaffold materials for cell culture, especially in bone tissue regeneration strategies. In addition, these capsules and spheres have been supplied with bioactive strategies that assist in cell attachment, proliferation, and differentiation. Thus, cell carriers can not only be made to act as 3D cell culture platforms but also induce cell differentiation to assist in easy, fast, and better integration of cultured cells at lesion sites.

Blood-vessel embolization. Microcapsules of natural polymers have also been made to enable blood vessel embolization, a strategy concerned with the deliberate blockage of blood flow in the vessels and arteries to cut off nutrition and oxygen supply of tumor [14,42]. Biocompatible, biodegradation, and non-toxic properties of natural polymers are advantageous for this strategy. An ideal embolizing agent must possess good mechanical strength and be of appropriate size that can adapt to the target blood vessel diameter. Moreover, it should be visible under X-rays and potentially impair angiogenesis [14]. Controllable degradation, good biocompatibility, and blood compatibility are other essential properties of an embolizing agent. Microcapsules and spheres of chitosan [43], gelatin [44], starch [45], alginate [46], etc. [47], have been used as embolizing agents in the treatment of various cancer therapies. In addition to their embolizing effect, these agents can also act as carriers for anti-cancer drugs (to act on cancer cells synergistically).

Various micro- and nano-capsule core–shell configurations have enabled the introduction of multi-functionalities in the capsules, thus employing one or more of the aforementioned strategies. For instance, microcapsules have been developed to enable simultaneous functions of sustained-release of drug and blood vessel embolization [14,43], targeted delivery and sustained-release [10,35], sustained-release, and tissue regeneration by cell delivery [26] and the like.

2.1.2. Modes of Encapsulation

A cargo is encapsulated into the micro- or nano-capsules either during (in-process encapsulation) or after the capsule synthesis (post-synthesis encapsulation) [39], depending upon the type of application or design convenience. In-process encapsulation involves the introduction of cargo in the appropriate precursor solutions before applying one of the capsule synthesis techniques described in the following sub-section. Post-synthesis encapsulation is mainly achieved by incubating the capsules in the cargo solutions, leading to their absorption by the capsules. The cargo can be introduced at the desired location in the capsules using both ways.

The APC can be dissolved or dispersed in either the core or the shell matrix (Figure 2). In the case of liquid-core capsules, the APC can either be dissolved in an oily carrier [12] or exist as an aqueous core [25]. Alternatively, it can also be encapsulated by the polymeric shell in its free form. In both cases, the APC exists as a reservoir inside the capsule core. In a solid core capsule, the APC can be entrapped in the solid core as a matrix system [11].

2.2. Synthesis Approaches and Mechanisms of Carrier Formation

Over the decades, many techniques for synthesizing natural polymeric micro- and nano-capsules and spheres, involving various chemical, physical, or physiochemical processes, have been developed and reviewed in detail by many authors [48–53]. This section, therefore, refrains from discussing general procedural technicalities in detail. Instead, it takes a closer look at the structural configuration-specific synthesis approaches, processes involved, and modifications introduced in the preparation of natural polymer-based spherical capsules with liquid/hollow/solid cores and porous microspheres. Mechanisms and interactions involved in capsule formation have also been discussed wherever necessary and possible. Generally, polymeric micro- and nano-capsule synthesis techniques follow the approaches and processes discussed herewith.

2.2.1. Solid Templating

This route involves the deposition of layer/s of polymer over solid micro- or nanoparticles of oxides, carbonates ($CaCO_3$, $MnCO_3$, or $CdCO_3$), metallic particles, or natural [8] or synthetic polymers to yield core–shell capsules. Polyelectrolytes with opposite charges can be easily alternatively deposited to form multiwalled capsules. The deposition is usually carried out by dipping the core template alternately in different polymeric solutions to achieve the desired number of polymeric shell layers and is facilitated by non-covalent and covalent interactions between the core and first polymer layer, as well as the consecutive polymer–polymer layers. The method, thus, enables the formation of layers of different polymers, capable of carrying a variety of drugs possessing different physical and chemical properties. The drugs can be introduced into the polymeric layers during layer assembly and into the core via co-precipitation during the core formation (or after the synthesis of the system, through absorption). Figure 3 presents a schema of the general procedure involved in the solid templating technique. Solid core multiwalled, as well as hollow core multiwalled micro- and nano-capsules, can be made using this process (refer to Table 2 for examples). Solid templating is a promising approach that provides more refined control over the capsule size, thickness, functionalities, encapsulation mode, type of solid core, and morphologies.

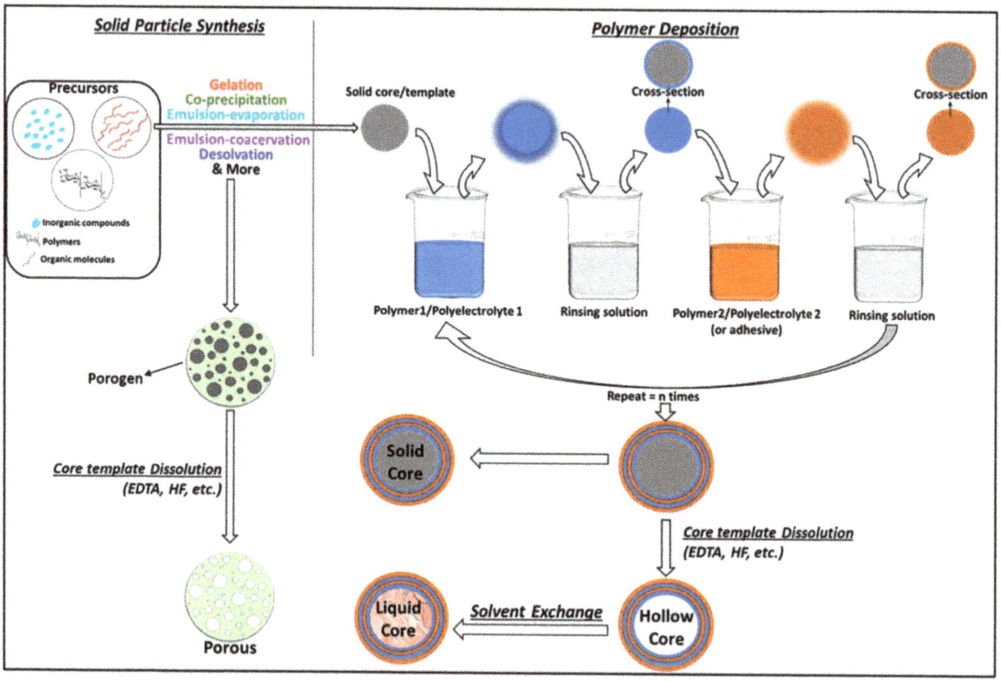

Figure 3. Schematic diagram of the solid templating approach.

Preparation of hollow core capsules is straightforward by solid templating. Typically, natural polymers are deposited over a sacrificial template core to create single, double, or multilayer shells solid core microcapsules. After the deposition of the shell layer/s, the template is dissolved to give a hollow core. Many types of materials have been used as sacrificial template cores, amongst which silica and calcium carbonate [51] nano- or microparticles are the most common. The core template dissolution is carried out by immersing the capsules in a chelating solvent, such as 8% hydrofluoric acid [54] and ethylenediaminetetraacetic acid (EDTA) [55]. During this process, the solvent molecules diffuse into the capsules to dissolve the solid core. It is significant to note that the core template must be completely dissolved. To ensure that, the core removal step is repeated multiple times. However, it has been shown that the core chelating solvents are not thoroughly removed during the capsule purification step, which may pose toxicity-related issues for biological applications. Yitayew et al. gave a proof-of-concept, using endotoxin-free cell lines as sacrificial template cores to mitigate these issues. They used live *E. coli* DH5 cells as a sacrificial template for synthesizing hollow core chitosan–alginate multiwalled capsules [56]. The microcapsules were dispersed in lysis buffer (0.1% Triton X-100, 2 mM EDTA in 10 mM Tris-pH8) overnight and washed with acetic acid buffer to remove the template cells. It is also worth noting that the core-removal step can cause shell deformities and engage in undesired reactions with the APC [57]. As mentioned earlier, the core materials in solid core micro- and nano-capsules may carry functionalities such as drug entrapment by the co-precipitation of the APC into a solid polymeric core [14], magnetically guided systems involving paramagnetic nano- or microparticles as solid cores [58], as implantable capsules with titanium microparticle core [59], and several more. Oily core polymeric shell capsules have also been indirectly prepared by using solid templating to prepare hollow core capsules and filling the capsule core with an organic solvent by solvent exchange [60]. Solid templating can also be employed to prepare porous spheres. To do so, solid template particles (also known as porogen) are dispersed in the aqueous or oil phase containing the dissolved polymer [61]. The obtained phase with dispersed solid porogen particles is then emulsified with the water–oil phase to obtain porogen-containing microspheres [62]. The template moieties are then dissolved to give porous spheres [63,64]. Different types of solid porogens can be employed, including polymer particles, such as polystyrene [62] and gelatin [64]. Various examples of porous spheres prepared by solid templating are presented in Table 3.

The surface charge of the cores (solid cores or sacrificial templates) is modified to facilitate attractive forces and interactions for the deposition of polymeric layers [65]. Shell-forming polymeric materials are more often oppositely charged polyelectrolytes. The most commonly used oppositely-charged (positive-negative) natural polyelectrolyte pairs for LbL deposition are chitosan–alginate [65], chitosan–hyaluronic acid [55,59], gelatin–epigallocatechin gallate [36], and BSA polycation–alginate [17]. Traditionally driven by electrostatic attractions between the opposite charges, these oppositely-charged polyelectrolytes sequentially self-assemble around the core during the dipping process to form micro- or nano-capsules [17,66]. The self-assembly can also be facilitated by hydrogen bonding between neutral polymers, as well as charged polyelectrolytes [67,68], generally by introducing modifications. Manna et al. used adenine modified neutral chitosan (CS) and thymine modified negative hyaluronic acid (HA) polyelectrolyte to mimic DNA base-pairing between adenine and thymine, enabling the self-assembly of these polymers into thin layers [68]. In another study, silk fibroin multilayers were deposited on a silica template using tannic acid (TA) as an adhesive between the silk layers aided by hydrogen bonding between the protonated hydroxyl group of TA and carbonyl groups present in silk fibroin [54]. These hydrogen-bonded capsule layers are often exploited to enable pH-stimulated cargo release from the capsules.

However, non-covalent interactions, such as hydrogen bonding and electrostatic attractions, are not robust enough to sustain the drastic pH differences and variations in ionic strengths across different biological environments, which may result in premature disassembly of capsules, unintentional release of cargo, aggregation, and fusion of multilayers, resulting in the loss of their multi-functionalities [69]. To remedy this, covalent interactions have been induced between polyelectrolyte layers, before or after their assembly over a solid core. Post-assembly covalent interactions between the polymeric layers are usually established by incubating the capsules in a solution containing cross-linkers, such as genipin and glutaraldehyde [65,70]. Glutaraldehyde crosslinks hydroxyl groups and amino groups in natural polymeric layers of the capsules [71]. To avoid the use of external crosslinkers, modified polyelectrolytes have been used to facilitate crosslinking. Oxidized sodium alginate (OSA) and CS were covalently assembled by crosslinking between the aldehyde groups of OSA and the amino groups of CS [72]. In another study, chitosan and hyaluronic acid were thiolated before their assembly into alternative layers on the $CaCO_3$ template. Disulfide cross-linking between thiolated polyelectrolyte layers was induced post-assembly, mediated by horseradish peroxidase and tyramine hydrochloride [55]. Liu reviewed and classified several methods employed to stabilize LbL assembled core–shell capsules of various synthetic and natural polymers [69]. Apart from covalent cross-linking, they described surface concealing as one of the methods to protect the capsules from adhesion and collapse, while retaining their ionic-responsive properties. It is noteworthy that, although covalently assembled and stabilized hollow and solid core shell capsules can sustain drastic pH and ionic strength changes, non-covalent interactions can facilitate the stimuli-responsive release of the cargo at the target with characteristic pH. Hence, a tradeoff must be achieved between the two. We believe that surface concealing of the capsule shell layer/s may prove worthy on such occasions.

2.2.2. Emulsion Templating

These methods utilize micro- or nano-emulsions between two or more types of mutually immiscible solvents as templates for polymeric capsule growth. Depending upon their solubility, the type of capsule configuration required and mode of APC encapsulation, the natural polymer, the APC, stabilizers, cross-linkers, and/or surfactants are dissolved in the appropriate solvents. Two types of emulsions can be achieved, i.e., (a) single emulsion: water in oil (w/o), oil in water (o/w); and (b) double emulsion: w/o emulsion, dispersed in water to give w/o/w or vice versa to give o/w/o. The resulting emulsion is then subjected to different types of chemical, physical, or physiochemical processes, such as diffusion-evaporation, coacervation, ultrasonication, crosslinking, interfacial deposition, solidification, spray-drying, freeze-drying, etc., to achieve stable biopolymeric capsules having liquid/solid/hollow cores. The steps involving the emulsion formation and diffusion-evaporation/coacervation/interfacial deposition are modified as needed. A procedural schema of the emulsion templating technique is represented in Figure 4. It is important to remember that emulsion templating can be used to achieve both spheres and capsule configurations by introducing variations during the synthesis procedure. However, we will mainly focus on the synthesis of core–shell capsule configurations and porous spheres. Various recent examples of capsules prepared using the emulsion templating technique are listed in Table 4. Porous spheres synthesis, using emulsion templating, involves the addition of porogen such as effervescent salts like ammonium bicarbonate or other inorganic salts (like sodium chloride) [73], in the appropriate phase, prior to the emulsion formation [64]. Ice crystals have also been employed as porogens. During the procedure, the polymer-containing emulsion is rapidly cooled to freezing temperatures to form ice crystals before initiating crosslinking or polymer precipitation. Ice crystals are then removed by sublimation or vacuum drying to produce highly porous polymeric spheres [26,74,75]. Table 3 lists recent examples of porous spheres prepared using the emulsion templating technique.

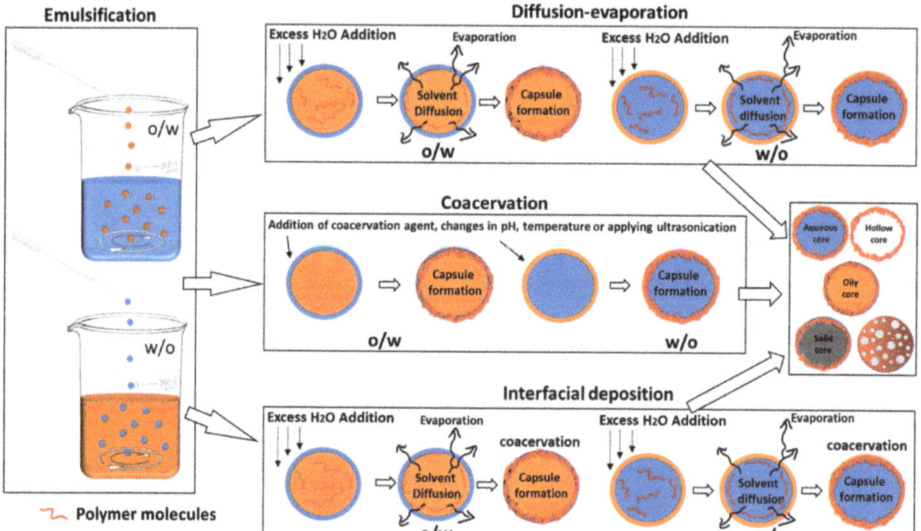

Figure 4. Schematic diagram of emulsion templating approach.

Emulsion–Diffusion–Evaporation

This method involves a mixture of partially water-miscible solvent (such as ethanol, acetone, or ethyl acetate), water, and an immiscible solvent (oils such as soyabean oil, Miglyol, or oleic oil). The process usually requires the preparation of mutually saturated organic solvent and aqueous solution [76]. For an oily core capsule formation, the organic phase includes preparing the APC for encapsulation, an optional hydrophobic stabilizer, and a water-immiscible oil/organic phase, dissolved in the water-saturated partially water-miscible organic solvent. The external aqueous phase consists of the polymeric shell material and one or several hydrophilic stabilizers, dissolved in the solvent-saturated water. To prepare a hollow core capsule, similar steps to those mentioned above are used, except for adding a water-immiscible oil/organic solvent in the organic phase. An o/w emulsion is made by introducing the water-saturated organic phase into the solvent-saturated aqueous phase, under constant stirring. The emulsion is then subjected to diffusion and/or evaporation as follows:

(a) Fast diffusion, by dilution with water: an excess of water is added to the emulsion, such that the partially-water miscible organic solvent from the organic droplets of the emulsion diffuses out, leaving behind the polymer-stabilized capsules. The amount of water required should be enough to diffuse out and dissolve the inner, partially water-miscible organic phase.

(b) Further, the solvent-dissolved diluted water is removed by evaporation under reduced pressure. Sometimes, the undiluted emulsion is directly subjected to the rapid displacement of the organic solvent from the internal to the external phase by evaporation under reduced pressure.

Similarly, aqueous core capsules can be prepared by making w/o emulsions. The unwanted solvent, the excess of untrapped APC, and the stabilizers are then eliminated by reduced pressure, dialysis, ultracentrifugation, or crossflow filtration. Preparation of porous spheres through this approach involves the formation of double emulsions and depends upon the rate of solvent diffusion from the inner phase to the outer phase and its evaporation [64]. While diffusing from the inner to the outer phase, the volatile solvent evaporates, leaving a porous polymer matrix behind.

Emulsion-Coacervation

The process involves three main steps: (a) preparation of an emulsion (o/w or w/o) (b) coacervation, which involves the separation of the liquid phases in a colloidal solution brought about either chemically (by changing the pH, temperature, or ionic environment) or physically (by ultrasonication, to encapsulate the dispersed core material), and (c) stabilization of the polymer as shells by physically- or chemically-induced crosslinking. Both single and double emulsions can be coacervated using this technique. Usually, a double emulsion is used for the entrapment of hydrophilic drugs in the capsule's core to ensure their efficient encapsulation. Additionally, it is easier to prepare aqueous core capsules using double emulsions. During the chemical coacervation step, the core and shell materials (polymers and APCs), dissolved in different/same solvents, are precipitated around the emulsion droplets by changing the pH of the system (with the addition of an acid or a base), lowering/increasing the temperature, or salting out, a process wherein the addition of appropriate ionic salts brings about a decrease in the solubility of the non-electrolytic biomolecules in the system. Simple coacervation is usually carried out for a system with one type of non-electrolytic biopolymeric precursor solution, which is usually coacervated using the salting out process [77]. Complex coacervation involves oppositely-charged polyelectrolytes of two or more biopolymers, and changing the pH usually brings about coacervation between them [50]. Stabilization of the polymeric shell, after its precipitation around the core material, during the chemical coacervation, is carried out by adding external crosslinkers, such as divinyl sulfone, 2,3-dibromopropanol, glutaraldehyde, etc. The concentration of the cross-linkers determines the thickness of the polymeric shell of the capsules and can be tweaked to achieve the desired thickness [52]. Physical coacervation can be induced by initiating electrostatic interactions between polyelectrolytes of different polymeric materials or polymeric sidechains of the same type of polymer or oxidative cross-linking. Electrostatic attractive forces between various polyelectrolytes of opposite surface charge have been utilized to enable physical coacervation to form capsules of two or more types of polymers [52].

Ultrasonication-assisted emulsification-coacervation. Physically-induced coacervation generally involves utilizing physical forces, such as ultrasonication, without the need for external cross-linkers. In a typical process, ultrasonication is applied at the oil–water interface when an aqueous solution of a polymer is overlayered with an organic phase, such as oil or other water-immiscible organic solvents (see Figure 5). Depending upon its solubility and desired location, the APC can be dissolved either in the aqueous or the organic phase. After a few minutes of sonication (~3 min), an o/w emulsion is formed and coacervated, due to various physical phenomena, induced by sonication. The first liquid-filled protein microspheres, prepared by Suslick, were composed of bovine serum albumin (BSA) and were filled with air [78] or organic phases [79], such as n-dodecane, n-decane, n-hexane, cyclohexane, or toluene. Ultrasonic irradiation of human serum albumin (HSA) or hemoglobin (Hb) formed similar microspheres to those of BSA. Since then, ultrasonication has been increasingly utilized to synthesize capsules of both synthetic [80], as well as other natural polymers [81,82]. It has proven to be a facile, cost- and time-effective technique that enables highly efficient encapsulation of a variety of drugs/cargo in the shell, as well as the core of the capsules. Our group has utilized this technique to synthesize protein microcapsules of BSA [83], HSA, and egg albumin [84], encapsulating a variety of hydrophilic and hydrophobic cargo, including gemcitabine [20], ribonucleic acid (RNA) [16,35], rhodamine B [85], MSQ (12i) 1-methyl-4-(substituted) styryl-quinolinium derivative [85,86], etc.

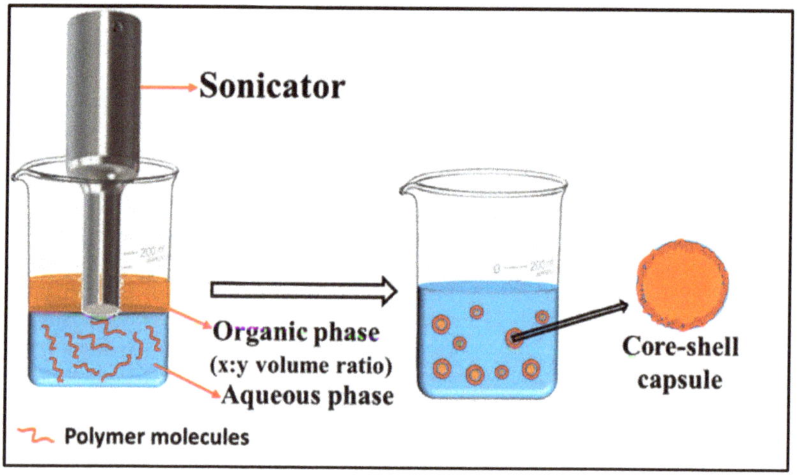

Figure 5. Schematic diagram of ultrasonication-assisted emulsification-coacervation.

During ultrasonication-assisted emulsification, the size of the protein microspheres depends on the nature of the oil–water interface, viscosity, surface tension, and hydrophobicity of the organic phase. The hydrophobicity of the material inside the protein microspheres determines the stability of the microspheres [87–91]. High viscosity leads to the formation of larger structures, which, in turn, results in a decrease in the stability and fraction of active material incorporated inside the microspheres. The ratio between the hydrophobic content and water phase also affects the stability and size of the capsules. A smaller ratio between the two leads to the formation of smaller capsules, which are less stable, as opposed to the ones formed at a larger ratio between the two. It has also been shown that an oil–water ratio of (>0.5) can cause phase inversion, giving a w/o emulsion [92]. The chemical and physical nature of the encapsulated material also affects the size of the microspheres as well as the hydrophobicity of the core of the capsule. In the case of proteins and polypeptides, the stability of the oil–water emulsions depends upon the protein sequence and the molecule size. The amphiphilic nature of proteins is also responsible for their self-assembly at the oil–water interface, thus stabilizing the emulsion [93]. Suslick found that protein microspheres are created only in the presence of oxygen or air [91]. He explained that the sonochemical process, which follows an implosive collapse of gas bubbles, produces OH· and H· radicals. These radicals form H_2, H_2O_2, and, in the presence of O_2, the superoxide radical HO_2. Hydroxyl, superoxide, and peroxide radicals are all potential protein cross-linking agents. Suslick and co-workers proposed that cysteine, a sulfide-containing amino acid present in these three proteins, is oxidized by the superoxide radical. The microcapsules are held together by protein cross-linking through disulfide linkages. Silva et al. alternatively proposed that amphiphilic polymers, such as proteins, can form stable microcapsules, due to the presence of hydrophilic and hydrophobic moieties that align themselves at the oil–water emulsion interface, due to the high shear forces generated by ultrasonication, and are entirely independent of cysteine content in the protein [88]. This alignment can also induce changes in the secondary structure of proteins, such as that of silk fibroin, which experiences an increase in its β-sheet content. Additionally, the cavitation produced during ultrasonication induces thermal denaturation of proteins, which in turn assists in the formation of the microcapsules [87].

Emulsion-Interfacial Deposition

This technique involves a combination of diffusion-evaporation and coacervation after the formation of an emulsion. In a typical process, an organic solvent (with oil and/or partially water-miscible solvent), containing the dissolved APC and/or the dissolved polymer, is introduced drop-by-drop into an aqueous solution, under constant stirring. Subsequently, large volumes of water are added, such as in the emulsion–diffusion–evaporation method. This is done to draw out the partially miscible organic solvents from the emulsion droplets, thus driving the polymeric molecules inside the organic emulsion droplets to precipitate at the droplet interface under the suitable pH, temperature, or ionic conditions (similar to the chemical coacervation method) [51]. The particles are recovered and cleaned using centrifugation and filtration. Narrow size distribution is obtained. The technique does not require the usage of external high-energy sources. However, it is limited by drug solubility, given that hydrophilic drugs cannot be encapsulated using this technique. In addition, the removal of residual solvent is challenging. Other disadvantages include the requirement of extensive optimization of parameters, such as the salt type (and its concentration), intensive purification of the obtained particles, and possible incompatibility of the salts with the bioactive drugs. Aqueous core capsules have been prepared using this technique, wherein an aqueous phase containing acetone (lower boiling point than water) and the dissolved polymer is added to the oil phase to give w/o emulsion, followed by subsequent evaporation of acetone at reduced pressure and ambient temperature [94]. During the evaporation process, the dissolved polymer precipitates at the water–oil interface of the water droplet, due to the decrease in acetone concentration, and forms the polymeric capsules with an aqueous core. The rationale here is to utilize a polymer that is soluble in the water–acetone solution (but insoluble in pure water or oil). Hence, the choice of polymer is crucial. This may be why the aqueous-core capsules that are made of natural polymeric shell and prepared using this technique are hard to find.

Emulsion-Spray Drying

In a typical spray-drying method for capsule formation, the polymer/s and the APC are dissolved in appropriate solvents to form shell and core materials. The core material is introduced into the shell material, and the resulting emulsion is dispersed as ultrafine droplets through a nozzle in a hot air flow [12]. The solvent evaporates instantaneously, and the dried capsules are collected under low pressure in a dry airflow. Solid core capsules are easily synthesized using this technique. Porous spheres can also be synthesized by spraying polymer and porogen, followed by removing the porogen templates. Spray-drying is easy to perform, yields consistent capsule sizes, is scaled up effortlessly, and fully automated. However, the adhesion of material on the walls of the instrument, agglomeration, and nozzle clogging hinder the yield, leading to high maintenance costs. Additionally, it is hard to get capsules under the 100 microns size range [77].

2.2.3. Other Techniques

Coextrusion–Coacervation

Precursor solutions of the core and shell material are fed into the concentric nozzles (of preset diameters) and extruded into a non-solvent (solidification liquid) at a specific rate to form core–shell droplets, which undergo coacervation to form core–shell capsules. Sometimes external crosslinkers are added to obtain stable capsules [95]. Porous microspheres have also been prepared using a similar approach of extrusion/injection of the polymeric solution into liquid nitrogen to form ice-crystals that act as porogens to give porous polymeric spherical matrices [39,96,97]. The size and shape of the capsules depend on the feeding rate, temperature, and type of precursor core–shell solvents, as well as the distance between the nozzle and the solidification liquid, its concentration, and surface tension [50]. Similar to spray drying, this technique is also limited by blockage of the nozzle and is high maintenance.

Microfluidics

The method involves the formation of emulsions (o/w or w/o) in various microfluidic devices. A microfluidic device is set with pre-requisite conditions, such as size, shape, and reproducibility. This allows the formation of carefully controlled polymeric capsules with entrapped drug molecules. However, it is not suitable for the synthesis of nano-sized capsules because of the inherent micron-length scale of the device. Different microfluidic systems, including T-mixer and co-flowing junction, hydrodynamic flow flowing, multi-inlet vortex mixers, staggered herringbone, and toroidal mixers, are used for achieving polymer particles or capsules of various sizes and shapes [97]. Using a microfluidics T-junction mixer, Mendes et al. produced hollow core polypeptide–polysaccharide (xanthan gum) microcapsules [98]. Porous microspheres have also been prepared with the assistance of microfluidics [99].

Table 2. Examples of capsules prepared by solid templating approach.

Polymer 1/Poly-electrolyte 1	Polymer 2/Poly-electrolyte 2	Solid Template/Core	Template Dissolving Agent	Template/Core Synthesis Method	Shell-Type and Deposition	APC and Location	EE (%)	Capsule Surface Charge (mV)	Template/Core Size and Capsule Size	Core-Polymer and Polymer–Polymer Interactions	Crosslinking between Core and Layers	Ref.
BSA polycation (+5.05 mV)	Alginate polyanion (−24.6 mV)	Template: amine modified-SiO_2 (+11.8 mV)	NH_4F/HF	Stöber process	Multiwalled (seven alternate layers of BSA and Alginate)	Betamethasone disodium phosphate (BSP); shell; post-synthesis introduction	56%	+5.05 mV	~128 nm; ~170 to188 nm	Non-covalent (hydrogen bonding, electrostatic, van der Waals, and hydrophobic interaction)	-	[17]
BSA	Tannic Acid	Template: $CaCO_3$. Core: BSA	Ethyl-enediaminetetra acetic acid trisodium salt (EDTA)	Co-precipitation	Multiwalled (six bilayers of BSA/Tannic Acid)	Tetramethylrhodamine-isothiocyanate labeled BSA; core; co-precipitated with the solid template during synthesis	-	(−30 ± 1.9) mV	-	Hydrogen bonding	-	[33]
Silk fibroin (anionic)	Aminopropyl triethoxysilane (APTES) (cationic)	Template: polystyrene	N,N-dimethyl formamide (DMF)	-	Multiwalled (nine layers of Silk fibroin)	chlorin e6 (Ce6) and doxorubicin (DOX); shell; post-synthesis introduction	DOX = 80% Ce6 = 90%	–	~150 to 250 nm; ~230 nm	Electrostatic interactions	-	[8]
Silk fibroin	-	Solid core: poly(lactic-co-glycolic acid)	-	Single emulsion-solvent evaporation method	Single layer of silk fibroin	Simvastatin; Core; in-synthesis encapsulation	59.4% to 70.3%	-	~15.3 µm	Covalent bonding	Chemical crosslinking by Glutaralde-hyde	[24]
calcium cross-linked k-carrageenan	k-carrageenan and chitosan polyelectrolyte complex	Template: $CaCO_3$. Core: BSA	EDTA	Co-precipitation	Multiwalled	Curcumin; after core synthesis, before layer assembly	6.25 to 8%	-	-	Electrostatic interactions	-	[100]
Gelatin A	(−)-epigallocatechin gallate (EGCG)	Template: $MnCO_3$	EDTA	-	Multiwalled (four layers)	-	-	−25 mV	~4.0 µm; ~4–5 µm	Non-covalent (hydrophobic and electrostatic interactions)	-	[36]
Chitosan polycation	Alginate Polyanion	Template: *E. coli* cells (−32.70 ± 3.2 mV)	Lysis buffer (0.1% Triton X-100, 2 mM EDTA in 10 mM Tris-pH8)	Cultured	Multiwalled (four bilayers of chitosan–alginate)	-	-	(−36.08 ± 8.8) mV	-	Electrostatic interactions	-	[56]

Table 2. *Cont.*

Polymer 1/Poly-electrolyte 1	Polymer 2/Poly-electrolyte 2	Solid Template/Core	Template Dissolving Agent	Template/Core Synthesis Method	Shell-Type and Deposition	APC and Location	EE (%)	Capsule Surface Charge (mV)	Template/Core Size and Capsule Size	Core-Polymer and Polymer–Polymer Interactions	Crosslinking between Core and Layers	Ref.
Thiolated-chitosan polycation	Thiolated-hyaluronic acid polyanion	Template: $CaCO_3$ −15.8 mV	EDTA	Co-precipitation	Multiwalled (four bilayers of chitosan/hyaluronic acid)	BSA and Dextran; Core; Co-precipitated with the solid template during synthesis	20.2%	−11 to −25 mV	3.0 µm; 4 to 6 µm	Covalent interactions by disulfide bonding	Enzymatic crosslinking using horseradish peroxidase and tyramine hydrochloride	[55]
Chitosan	-	Solid; Ca-alginate	-	Extrusion	A single layer of chitosan	Insulin and probiotic cells; post-synthesis	-	-	–	-	Electrostatic interactions	[101]

Table 3. Examples of porous spheres prepared by solid & emulsion templating approach.

Polymer Matrix	Porogen	Preparation Method	Porogen Removal Process	Crosslinkers; Precipitants	APC	Pore Size	Sphere Size	Ref.
Silk fibroin	Ice crystals ~(−195 °C)	Microinjection into liquid nitrogen and freeze-drying	Sublimation	-	Basic fibroblast growth factor (bFGF)	1.5–7.0 µm	95 µm to 260 µm	[39]
	Ice crystals (−20 °C)	w/o emulsion, rapid cooling, and freeze-drying	Sublimation	-	Strontium	(20 ± 5) to (34.8 ± 6.5) µm	-	[26]
		Microinjection into liquid nitrogen and freeze-drying	Sublimation	Ethanol-assisted precipitation	-	0.3–10.7 µm	208.4–727.3 µm	[102]
Chitosan	Ice crystals (−20 °C)	w/o emulsion, low temperature, thermally-induced phase separation, and pH-assisted coacervation	Drying under vacuum	-	-	20–50 µm	ca. 150 µm	[74]
	Ice crystals ~(−195 °C)	Microinjection into liquid nitrogen and freeze-drying	Sublimation	Saturated sodium tripolyphosphate (STPP) crosslinker	-	<30 µm	<400 µm	[96]
Chitosan/poly(L-glutamic acid) (PLGA) polyelectrolyte complex	Ice crystals (−20 °C)	w/o emulsion, low temperature, thermally-induced phase separation	Drying	-	-	(47.5 ± 5.4) µm	250 µm	[75]

Table 3. *Cont.*

Polymer Matrix	Porogen	Preparation Method	Porogen Removal Process	Crosslinkers; Precipitants	APC	Pore Size	Sphere Size	Ref.
Collagen/cellulose	Solid polystyrene	w/o emulsion	Washing with acetone	n-butyl al-cohol as precipitant	BSA	~198.5 nm	8–12 µm	[62]
Alginate	NaCl	w/o emulsion, freeze drying	-	Calcium chloride as crosslinker	-	200–300 nm	~158 µm	[73]
Soy protein	$CaCO_3$	Solid templating over porogen by incubation	Dissolution by EDTA	Transglutaminase as crosslinker	-	-	3–12 µm	[61]
Silk sericin and hydroxylapatite	Silk sericin	Nucleation and growth of hydroxyapatite, induced by the sericin template in simulated body fluid	-	-	Doxorubicin	-	1–3 µm	[103]

Table 4. Examples of capsules prepared by emulsion templating approach.

Polymer Shell	Core & Type	Template & Organic Solvent	Emulsion Type	Method	APC & Location	Interactions	Crosslinkers; Stabilizers; & Surfactants	Surface Charge	Size	Encapsulation Efficiency	Ref.
Human serum albumin (HSA)	Lauroglycol 90; oily	Lauroglycol 90; Acetone	o/w single emulsion	Diffusion-evaporation	Exemestane and hesperetin; core	Electrostatic interactions	None; 1:1w/w poloxamer/Tween 80 mixture; benzalkonium chloride	20.7 ± 1.26 mV	172.4 ± 8.6 nm	95–98%	[10]
Folic acid-functionalized HSA	Oily; dodecane	Dodecane	o/w single emulsion	Ultrasonic emulsification	-	Oxidative crosslinking	-	−20 mV	~440 nm	-	[22]
Wheat germ agglutinin-functionalized HSA	Biocompatible plant oils; oily	Almond oil, rapeseed oil, olive oil, and linseed oil	o/w single emulsion	Ultrasonic emulsification	-	Oxidative crosslinking	-	−12.4 ± 9.4 mV	(662.1 ± 7.6) nm to (862.2 ± 59.5 nm)	-	[104]
Fluorescently tagged bovine serum albumin (BSA) shell; Shell filled with PLGA and unsaturated fatty linoleic acid	Lecithin; aqueous	Dichloromethane and ethanol	w/o/w double emulsion	Double emulsion–evaporation	lipophilic paclitaxel in the oily shell and hydrophilic transcription factor p53 in the aqueous core	-	Pluronic F-68 & Lecithin	−36.4 mV	~180 nm	-	[25]

Table 4. *Cont.*

Polymer Shell	Core & Type	Template & Organic Solvent	Emulsion Type	Method	APC & Location	Interactions	Crosslinkers; Stabilizers; & Surfactants	Surface Charge	Size	Encapsulation Efficiency	Ref.
BSA	Soya bean oil; oily	Soya bean oil	o/w single emulsion	Ultrasonic emulsification	Ribonucleic acid (RNA); shell	Oxidative crosslinking	-	−40 meV	(0.5 µm to 2.5 µm)	~60%	[16]
Polyvinyl alcohol (PVA) functionalized-BSA								0 meV			
Polyethyleneimine (PEI) functionalized-BSA								+20 meV			
Silk fibroin	Sodium alginate; solid	Paraffin oil	w/o single emulsion	Emulsion-coacervation	-	Chemical crosslinking using glu-taraldehyde	Span 80	-	Avg. 141.839 µm.	-	[14]
Collagen and PLGA layers	Hollow	Dichloromethane	o/w single emulsion	Emulsion–evaporation	MnO_2 nanoparticles; shell	Carbodiimide initiated covalent crosslinking	Crosslinking facilitated by N-(3-Dimethylamino propyl)-N′-ethyl carbodiimide hydrochloride (EDC), N-Hydroxysuccinimide (NHS); stabilizer: polyvinyl alcohol (PVA)	-	-	-	[18]
Anti-epidermal growth factor receptor (EGFR) modified-BSA	Dodecane; oily	dodecane	o/w single emulsion	Ultrasonic emulsification	Gemcitabine; shell	Oxidative crosslinking	-	-	~1.1 µm	30%	[20]
Whey protein isolate (WPI)	Sunflower oil; solid	Sunflower oil	o/w single emulsion	Spray- and freeze-drying	Vitamin E; core	-	-	-	~145.3 µm	89.3%	[12]
Gelatin	Citric acid; solid	Dichloromethane and ethanol	o/w single emulsion	Spray drying	Itraconazole; core	Physical crosslinking	-	-	-	-	[21]

Table 4. *Cont.*

Polymer Shell	Core & Type	Template & Organic Solvent	Emulsion Type	Method	APC & Location	Interactions	Crosslinkers; Stabilizers; & Surfactants	Surface Charge	Size	Encapsulation Efficiency	Ref.
Tetramethylrhodamine-isothiocyanate labeled-BSA, tannic acid, and BSA layers	Sunflower oil; oily	Sunflower oil	o/w single emulsion	Emulsion-coacervation	3,4,9,10-tetra-(hectoxy-carbonyl)-perylene (THCP); core	Hydrogen bonding between the shell layers	-	(−30 ± 1.9) mV	-	-	[33]
Chitosan	Soybean oil, oily	Soybean oil; benzyl benzoate	o/w single emulsion	Emulsion-microfluidic	Tea tree oil; core	Covalent interactions by chemical crosslinking	Terephthalal dehyde (TPA)	-	~106 µm	19.5–49.3%	[105]
Gelatin and gum arabica	Soybean oil; aqueous	Soybean oil	w/o/w double emulsion	Emulsion-complex coacervation	Sucralose; core	Covalent interactions	Lecithin		81 to 113 µm	43.04 to 89.44%	[106]
Folic acid-modified hyaluronic acid	Ethyl acetate; oily	Ethyl acetate	o/w single emulsion	Ultrasonication	Curcumin; core	Oxidative crosslinking	-	-	400 to 600 nm	91.3 to 93.2%	[107]
Soy protein and gum arabica	(80 vol% NEOBEE M5 + 20 vol% limonene); oily	80 vol% NEOBEE M5 + 20 vol% limo-nene	o/w single emulsion	Complex coacervation	-	Heat-induced gelation crosslinking	-	-	-	-	[108]
Pea protein isolate and sugar beat pectin	Hemp seed oil; oily	Hemp seed oil	o/w single emulsion	Complex coacervation, followed by spray-drying	Hempseed oil	pH-induced crosslinking	-	-	(12.80 ± 2.17) to (23.70 ± 1.23) µm	(79.65 ± 5.99) to (94.42 ± 6.63)%	[109]

3. Natural Polymer-Based Capsule Characterization

3.1. General Characteristics

3.1.1. Size

One of the primary characteristics of any biomedical formulation is its operating size. It is a critical parameter that determines the suitability of the capsules to penetrate the target biological site, as well as its applicability arising from in-vivo pharmacokinetics [110]. In addition, the capsule size influences the drug-loading capacity, drug release rate and profile, and capsule stability [111]. Smaller capsules may provide a larger surface area for the entrapment of a surface-bound drug, leading to potentially higher loading capacity. However, a smaller core, achieved due to smaller size capsules, may not ensure sufficient loading capacity for a drug-loaded in the capsule core as a reservoir. Alternatively, a larger capsule, with a thicker shell (or multiwalled shell), can have a higher loading capacity for a shell-bound drug but may or may not have a higher loading capacity for a core-bound drug, in which case the core size is of paramount importance. A shell-bound drug releases at an accelerated rate from smaller size capsules, due to the increased surface area [111,112]. Larger polymeric capsules have been shown to degrade/dissolve faster than smaller capsules, due to bulk erosion [113]. However, it has also been previously shown that the particle size had a minimal effect on the polymer degradation rate [114]. Hence, it is safe to draw that the dependence of capsule degradation on its size may be system- and parameter-specific.

Capsule size can be affected by the type of precursor polymer and its concentration [115], emulsion homogenization speed, agitation rate [116], type and concentration of the emulsifying agent [117], volume of the aqueous and the oil phase, size of the solid template/core, type and concentration of the surfactant, storage conditions, thickness of the polymeric shell, and synthesis technique employed. Valot et al. studied process the influence of process parameters on the size distribution of ethyl cellulose microcapsules synthesized, using the emulsion–evaporation technique [118]. They found that the mean capsule size decreases with the increase in the volume ratio of the dispersed organic phase to the continuous aqueous phase and an increase in the stirring rate. They also concluded that a decrease in the surfactant concentration leads to increased mean capsule size.

Size distribution measurements are usually performed using the dynamic light scattering (DLS) method, wherein the micro- or nano-capsules are dispersed in a solvent media during measurements. Size and morphological studies are also conducted using scanning electron microscopy (SEM) and transmission electron microscopy (TEM). However, care must be taken during sample preparation. We have observed that the liquid core microcapsules are prone to bursting during air drying and vacuum conditions in the SEM instrument. Lyophilization of the sample for ESEM measurements can be an option to avoid such a scenario.

3.1.2. Stability

The stability of micro- and nano-capsule concerns their storage, as well as operating in-vivo stability. After synthesis and purification, microspheres are either stored as colloidal solutions at lower temperatures, solid freeze-dried samples, lyophilized into powders, or in the form of spray-dried or vacuum-dried powders. Proper capsule storage ensures a better shelf-life of capsule formulations and their subsequent usage. Sonochemically prepared liquid-core human serum albumin capsules have shown to be stable for long-terms in suspension, as well as in freeze-dried conditions [104].

In-vivo stability of a capsule can be increased to avoid the initial burst release of the drug [31], which is usually an undesirable feature of a drug delivery formulation, and to extend the drug release rate. Moreover, the capsules can be stabilized and programmed to release drugs that target particular conditions, as in the case of stimuli-responsive release systems.

3.1.3. Moisture Content

Moisture content is an important physical property for the dried micro- and nano-capsules and spheres that influences the stability of the core after drying and affects the processibility, shelf-life, usability, and quality of the pharmaceutical product [119]. Furthermore, the maximum permissible moisture content in certain products depends on the guidelines established by regulatory bodies, such as the FDA. In general, products with moisture content between 3–10 g/100 g possess good storage stability [12].

Moisture content is determined using a thermogravimetric approach by measuring weight loss upon drying. Many moisture content measuring instruments are available. During a typical measurement procedure, the sample is heated, and the weight loss, due to moisture evaporation, is recorded [12].

3.1.4. Surface Charge

Another important property of any micro- or nano-capsule is its surface charge, which is usually determined by zeta potential measurements. The surface charge establishes the in-vivo capsule distribution and affects the drug release rate from the capsules. The surface charge can be modified using functionalizing polymers to enable targeted delivery of micro- and nano- capsules, for instance, to the cell nucleus [35].

3.1.5. Encapsulation Efficiency

The efficiency of the drug encapsulation is calculated using the expression:

$$Encapsulation\ Efficiency\ (\%) = \frac{C_t - C_{un}}{C_t} \times 100\%$$

C_t is the total concentration of the drug initially present in the precursor solution before capsule or sphere formation, and C_{un} is the drug concentration measured in the residual precursor solution after the capsule or sphere formation.

3.1.6. Drug-Loading Capacity

The drug-loading capacity is defined as the amount (weight) of drug-loaded per unit weight of micro- or nano-capsules and is calculated by the expression:

$$Loading\ capacity = \frac{W_d}{W_c}$$

in which W_d is the total entrapped drug and W_c is the total weight of the capsules.

3.1.7. Cytotoxicity

To determine the suitability and biocompatibility of capsule formulations, in-vitro cytotoxicity analysis is done in-vitro on tissue cells using cell viability and cytotoxicity assays [120]. These assays measure the cellular or metabolic changes associated with viable or nonviable cells and detect structural changes, such as loss of membrane integrity upon cell death or physiological and biochemical activities, indicative of living cells. Various types of cytotoxicity assays are available on the market, including MTT (methyl thiazolyl tetrazolium) and CCK-8 (Cell Counting Kit-8). The testing protocol for each is different and is explicitly defined by the assay manufacturers. In a typical procedure, the cells are incubated in 96-well plates at 37 °C, until adherent to the culture plates, followed by the addition of sterilized capsule suspensions. To these capsule-containing culture wells, prescribed volumes of cytotoxicity assay are added each day, incubated for 2 h, and scanned for absorbance at a particular wavelength to measure the optical density for counting the number of surviving cells and analyze their metabolic activity [14]. Zhou et al. describe various methods for cytotoxicity analysis of medical devices [120].

3.1.8. Blood Compatibility

For any biomedical device or formulation, especially those intended to be introduced in-vivo through intravenous route and blood vessels, embolizing agents must have blood compatibility and should not cause hemolysis and blood coagulation. For blood compatibility analysis, the capsule formulation must undergo five stages of screening tests, which include thrombosis (blood clotting index, coagulation analysis, and platelets), hemolysis rate (nonhemolytic (0–2%), slightly hemolytic (2–5%), or hemolytic (>5%)) [121], and immunology testing [122].

3.1.9. Flowability

Flow properties of the dried micro- or nano-capsule powder is an important parameter that establishes the powder quality. Usually, flow properties are analyzed by calculating the bulk and the tapped densities of a powdered sample. The procedure involves transferring a measured amount (m) of the powdered sample into a calibrated measuring cylinder and noting the bulk volume (V_L) occupied by the powder to calculate the bulk density, ρ_b by $\frac{m}{V_L}$. After this, the cylinder with the m amount of powdered sample is manually tapped for a certain amount of time to reach the tapped volume V_T for calculating the tapped density, ρ_T by $\frac{m}{V_T}$. The flowability of the power is then indirectly predicted using

$$\text{Carr's index } (\%) = \frac{\rho_T - \rho_b}{\rho_T} \times 100$$

$$\text{Hausner Ratio } = \frac{\rho_T}{\rho_b}$$

Carr's index ratings up to 10% are deemed excellent, between 10–15% are good, 16–20% are poor, 32–37% are very poor, and greater than 38% are abysmal. A Hausner ratio $\leq$1.25 indicates that the powdered sample is free-flowing, while a ratio $\geq$1.25 indicates poor flowability [12].

3.1.10. Pore Size and Porosity

Depending upon the pore diameter size, micro- and nano-spheres can be microporous (<2 nm), mesoporous (2–50 nm), or macroporous (>200 nm). The pore size can be measured during morphological analysis using SEM, TEM, or confocal laser scanning microscopy. Porosity is the ratio between the pore volume and total volume of the microsphere. It can be calculated using a variety of methods [64].

3.2. Function-Specific Characteristics

3.2.1. Drug Release and Kinetics

To understand the release behavior of the drug from a sustained-release capsule formulation, it is essential to study its release kinetics in-vitro. This is usually done by dispersing the drug-loaded capsule formulations in a release media under constant stirring and by measuring the drug concentrations in the release media at set time intervals. Conditions, such as the selection of proper release media, pH, temperature, and stirring speed, must be maintained and monitored throughout the in-vitro release experiments. The in-vitro release media is generally composed of the route- and target-specific biomimicking fluids at various pH values and bodily temperature (~37 °C). For example, orally administered capsule formulations are tested in-vitro in the gastrointestinal-mimicking release media. However, simulating exact in-vivo conditions is difficult.

D'Souza reviewed various in-vitro drug release study methods [123], including 'sample and separate', 'continuous flow', and 'dialysis method'. The 'sample and separate' method involves retrieving a certain amount of sample from the release media at certain time-intervals, separating the retrieved sample from capsules (via filtration, ultrafiltration, centrifugation, ultra-centrifugation, or their combination), and, finally, measuring the drug concentration in the filtrate or/and the evaluating the filtered capsules. This method, although straightforward, poses many challenges, including the clogging of filters during filtration and absorption of the drug molecules into the filters. We also faced similar challenges during drug release studies from organic-core BSA microcapsules [85]. In addition, we observed that BSA microcapsules ruptured several times during sample ultrafiltration, which resulted in the premature drug release in the filtrate leading. In a continuous flow method, the release media flows through a column containing immobilized drug-loaded capsule formulation, and the effluent is collected and monitored by detectors. Several types of apparatus are available for the continuous flow method. However, it is a costly method and requires complicated set-up assembly. The dialysis method is straightforward. Generally, the sample is placed in a dialyzing membrane and suspended into the release media. Samples are retrieved from the release media and analyzed. The method is simple and advantageous over the 'sample and separate' and 'continuous flow' methods, with the exception that a few drugs can bind to the dialysis membrane, affecting their concentration in the release media. In addition, the behavior of the dialysis membrane in the release media must be monitored prior to their employment for drug release studies. Finally, the drug release concentration in the release media vs. time profile is generated and compared to theoretical and computational models to predict the drug release behavior from the capsules and ascertain the underlaying release mechanisms.

The drug release process typically involves the migration of drug molecules from their initial location in the capsule to the external surface of the capsule and then, eventually, into an in-vitro release media or at the in-vivo target site. The movement and release of the drug via this route are facilitated by various mechanisms, which are briefly discussed below [30,124]. In-vivo drug release is usually governed by a combination of two or more of these mechanisms, depending upon the type and design of the capsules or the spheres.

Diffusion. This process involves the mass transfer of the molecules of a substance (solute) from one part of a system or solution to another, driven by the solute concentration gradient [125]. In other words, it is the movement of solute molecules from their higher concentration to their lower concentration in a solution, as long as this concentration gradient is maintained. After the concentration difference is equalized, the system reaches a state of equilibrium where no more solute diffusion from one part of the system to another takes place. This mass transfer of molecules is facilitated by thermal and Brownian motion, which results in random and repeated collisions between molecules. Usually, in a gradient of solute concentration, not all the solute molecules have a preference to move in one direction. Hence, while studying mass transfer by diffusion, a solution is divided into volume groups of solute molecules [30]. One group of molecules may move in one direction, while another group in the reverse direction. If the concentration of the first volume group is more than the second one, overall, more particles will move from the first group to the second, leading to a net flow of molecules from their higher concentration in group one to their lower concentration in group two. For releasing from a polymeric capsule, the drug molecules must diffuse from their initial position (inside the core drug reservoir or matrix, or the polymer matrix) to the outer surface of the polymer matrix and, eventually, into the release media.

Erosion. Drug release, by polymeric capsules, sometimes involves the erosion or disintegration of the polymer matrix by the kinetic degradation of the appropriate links between polymer–polymer molecules or polymer–APC molecules, due to the hydrolysis of bonds [126]. The hydrolysis of a bond depends upon the local environment (acidic or basic). In a drug reservoir system, erosion-controlled drug release occurs when the polymer matrix degrades, releasing the APC that it physically encapsulates. In the case of a matrix system, the APC is usually chemically linked to either the polymeric shell or the core and is released after the breakage of those chemical links, accompanying the degradation of the matrix. Erosion can occur at the surface [127], or in bulk [128], of the capsules. When water invasion is slow and the hydrolysis of polymeric bonds is rapid, surface erosion occurs, which reduces system dimensions [30]. In a matrix-type system, the surface erosion of the polymeric SDDS is accompanied by the release of the APC molecules. When water invades the SDDS more rapidly throughout the system than the hydrolysis of the surface bonds, several polymeric chains are broken, leading to the bulk erosion of the system. During the bulk erosion, the drug is initially released from the system through the surface and pores. This initial release is followed by a dormant stage (almost no drug release), where broken polymer chains, triggered by water invasion, form crystallites that are resilient against hydrolysis. Finally, the drug is released rapidly, due to the accelerated degradation of the polymer and polymeric crystallites, due to autocatalysis.

Osmosis. Osmosis involves the movement of solvent (biological fluid) from its higher concentration (i.e., lower concentration of the solute) to its lower concentration (i.e., the higher concentration of the solute) through a semi-permeable membrane, which allows the transport of smaller solvent molecules into the system but prevents bigger solute molecules from leaving the system. The process is controlled by osmotic pressure, which develops when two solutions of different solute concentrations are separated by a semi-permeable membrane [30]. The higher the osmotic pressure, the higher the chemical potential, which leads to an increased rate of transport of the solvent molecules through the semi-permeable membrane. In osmotically-driven drug release, the polymer matrix of the capsules or spheres acts as a semi-permeable membrane. Due to the built-up osmotic pressure, the water outside the DDC/S starts to permeate the capsule polymer matrix, resulting in its hydration and swelling. Eventually, due to the permeation of water molecules into the matrix, the drug (solute) concentration inside the SDDC starts lowering, which results in a decrease in the osmotic pressure. The hydration and swelling of the polymer matrix result in the matrix becoming partially permeable to the drug molecules, which decreases the osmotic pressure and drives the drug molecules to slowly escape the system through the now partially permeable polymer matrix of the SDDC [129]. The process is repeated alternatively on both sides of the polymer matrix on account of osmotic pressure and chemical potential, leading to a slow and controlled release of the drug. The rate of osmotic flow depends upon the concentration and nature of the drug, temperature, and hydraulic permeability of the polymer matrix.

Swelling. Swelling of the polymeric membrane of an SDDC usually depends upon the hydrophilic behavior of the polymer or water–molecule interaction [130]. When the polymer is surrounded by water, the polymeric network expands because water enters the DDC rapidly, as opposed to polymer dissolution, which is slow. This leads to the swelling of the polymeric shell. The mechanism is similar to swelling, in the case of osmotically driven drug release from an SDDC. The primary parameters that control swelling are ionic content, cross-link content, and hydrophilic content of the polymeric shell.

Partitioning. DDCs are usually made of one or more polymers of different affinities and polarities from the APC they contain [30]. Hydrophilic drugs partition themselves in the aqueous phase hydrophilic moieties of the DDC, whereas the hydrophobic drugs tend to reside in the organic phase or hydrophobic moieties of the DDC. In order to be released from the DDC, the drug molecules travel through mediums of different affinities (hydrophilic or hydrophobic polymers) at different rates, depending upon their relative concentration and affinity to each phase. This affinity is defined as a partition coefficient, which is the ratio of drug solubilities in the two phases and describes the relative frequency with which the drug moves from one medium to another.

3.2.2. Swelling Ratio

The diameter of the micro- and nano-capsules is measured before and after the swelling of the capsules. During swelling experiments, the capsules are dispersed in an aqueous media under stirring at varying pH and temperatures conditions [14]. Their diameters are measured at each interval of time, and the swelling ratio (%) is calculated using the equation:

$$\frac{D_t - D_0}{D_0} \times 100 \qquad (1)$$

where, D_0 is the initial diameter and D_t is the diameter of the capsules after swelling at time (t). It is vital to build a swelling ratio profile prior to in-vivo testing, in order to understand the swelling behavior of micro- and nano-capsules, especially for their utility as embolizing agents operating at different diameters of blood vessels, as well as osmosis-controlled drug release systems.

3.2.3. Cell Survival Number

For determining the efficacy of the capsule as a protective enclosure to probiotic bacterial cells against the harsh gastric environment, in-vitro incubation of cell-encapsulating capsules and free cells in a simulated gastric fluid (SGF) is carried out for a set period to evaluate the cell survival number [11].

3.2.4. In-Vivo Bioavailability

Capsules prepared for aiding the solubility characteristics of the encapsulated drug are tested, in comparison to the unencapsulated free drug, for its in-vivo oral bioavailability. The procedure involves live subjects (such as male or female rats in a similar weight range), divided into test and control groups. A certain amount (by weight of the live subjects) of drug-encapsulating capsules and the free drug are administered orally in the test and the control groups, respectively. Fixed volumes of blood samples are then drawn from the test and control groups at fixed time intervals ($t_0, t_1, \ldots, t_n$), through the experimentally preferred vein type (for example, the retro-orbital, the saphenous vein, or the tail vein in rats) [131]. Blood samples from a second control group of live subjects, to which no drug is administered, can also be studied for conducting an accurate evaluation. The collected blood samples from each group are analyzed for the blood plasma drug concentrations. Pharmacological analyses are carried out by generating the mean plasma concentrations of drug vs. time profile and analyzing the maximum plasma concentration (C_{max}) at the time (t_{max}) and area under the curve (AUC), to evaluate drug bioavailability from free drug and capsule-encapsulated drug [12].

3.2.5. Dissolution Profile of the Capsules

The dissolution profile of a capsule formulation is built based on in-vitro experiments, which usually involve incubating the capsules in water/simulated gastric fluids over a definite period [12]. In such as case, the dissolution behavior is evaluated by observing the change in the absorbance intensity and optical density with time at the absorbance frequency of the capsule-forming polymer. The dissolution profile of capsule formulations reflects the capsule erosion over time in the release media and, as a result, indicates its biodegradation and elimination from the body, and affects the release behavior of the encapsulated APC.

4. Recent Advances in Protein-Based Spherical Capsules towards Biomedical Applications

In the past few decades, various types of animal- and plant-based proteins and peptides have been studied for their use as drug and growth factor carriers, embolizing agents, and cell culture platforms, in order to enable sustained drug release, protection from the biological environment, enhanced bioavailability, targeted delivery, pH- and temperature-responsive release, embolization of blood vessels, better cell integration into the body and toxicity moderation. Table 1 lists various natural polymers (biopolymers), including proteins, that have been utilized to develop spherical capsules in biomedical applications. The interest in protein-based drug carriers stems from several of their advantages, namely higher biocompatibility and lower toxicity, biodegradability, high drug-binding capacity leading to a good drug-loading efficiency, possibility of straightforward and cheaper production due to their abundance in nature, the feasibility of structural modifications, due to the presence of several functional groups, non-immunogenicity, etc. Protein-based systems in the form of hydrogels, micro and nanoparticles, micro- and nanocapsules, implants, microneedles, bio-adhesives, fibers, rods, and films have been developed and tested for various applications in cancer therapy, nutritional therapy, diabetes, bone diseases, neurological conditions, and stem cell therapy. Spheres and capsules made of animal- and plant-proteins having liquid (organic or aqueous), hollow and solid cores have been developed for the above applications (see Tables 5 and 6).

In addition, the past decade has seen a considerable evolution of composite capsules made of two or more proteins, protein–polymer composite capsules, and composite capsules of protein conjugated with other materials, such as ceramics and metallic nanoparticles. Protein capsules have also been functionalized using other polymers to develop drug delivery formulations for targeted delivery. Herein, we focus on the advances made in the past decade towards developing micrometric and nanometric capsules with liquid, solid, and hollow core encapsulated by shells made of functionalized proteins and protein–protein composites, protein–polymer composites, protein composites with other materials, and multiwalled capsules. Our discussion revolves around protein capsules and spheres developed for biomedical applications, under the designing aspects discussed hitherto.

4.1. Liquid-Core Protein-Shell Capsules

With the purpose of drug protection and its sustained-release, microcapsules made of the proteins, bovine- and human serum albumin (BSA and has), encapsulating various hydrophilic and hydrophobic drugs have been developed using various synthesis techniques, including sonochemical synthesis. One such work, by Shimanovich et al., involved the sonochemical encapsulation of ribonucleic acid (RNA) molecules in the BSA microspheres, having an organic core, to study the possibility of using protein microspheres for delivering RNA to Trypanosoma brucei parasites (causes sleeping sickness) and mammalian cells (human U2OS cancer cells). The aim of encapsulation in the microspheres was to protect the RNA molecules from the outer cellular environment and enable their controlled release from the microspheres [16]. Various organic solvents, such as dodecane, soya bean oil, canola oil, and olive oil, were tested to form the organic core of the microspheres, among which the soya bean oil was found to be the most biocompatible to the Trypanosoma brucei parasites and U2OS cancer cells. The RNA molecules were successfully encapsulated inside the BSA microspheres, with 60% encapsulation efficiency, causing no damage to the RNA molecules during encapsulation. The RNA molecules were initially found to be localized in the hydrophobic organic core of the microspheres but delocalized themselves into the hydrophilic BSA crust within ca. 24 h after the formation of the spheres. The average size of the RNA-loaded BSA microspheres (RNA@BSAMS) ranged from 0.5 μm to 2.5 μm, which depended on the size of the encapsulated RNA molecules, whereas the surface charge on the RNA-BSAMS was around −40 meV. RNA@BSAMS were observed to degrade slowly over five months (at 4 °C), while gradually releasing the RNA molecules, thus ensuring the slow and controlled release of RNA by the BSA microspheres. A successful in vitro uptake of RNA@BSAMS by Trypanosoma brucei parasites and U2OS cancer cells was observed spontaneously without the help of additional mediators. The RNA@BSAMS were stable in the cellular environment of the two types of cells, thus proving that RNA remained protected inside the BSA microspheres.

In the follow-up work, Shimanovich et al. functionalized RNA-loaded BSA microspheres by coating their surface with polymers, either polyvinyl alcohol (PVA) or polyethyleneimine (PEI), to enable targeted delivery of the RNA to the cell nucleus of Trypanosoma brucei parasites and U2OS cancer cells [35]. They observed changes in the surface charge from −40 meV to 0 meV upon coating with PVA and to +20 meV upon coating with PEI. The enhanced cell uptake of the coated microsphere, which was four times larger than the uncoated microspheres, was attributed to the changes in the surface charge. Moreover, unlike the uncoated microspheres, which were localized near the cell membrane, the microspheres coated with PVA localized themselves near the cell nucleus, and the ones coated with PEI were able to penetrate the cell nucleus, thus enabling targeted delivery into it. In another study, Grinberg et al. demonstrated significant inhibition of pancreatic cancer cells (AsPC1) proliferation using antibody-modified BSA microcapsules, loaded with the FDA-approved anti-cancer drug gemcitabine [20]. It is known that the pancreatic cells have an overexpression of EGFR (epidermal growth factor receptor), and one of the strategies of inhibiting their growth is to inhibit the EGFR signaling pathway. For implementing this strategy, anti-EGFR-modified BSA microcapsules, loaded with gemcitabine (BSA-Gem-EGFR), were synthesized sonochemically. The core of the microcapsules was made of dodecane, and gemcitabine was found embedded in the BSA shell matrix. The average size of the obtained antibody-modified microcapsules was ~1.1 μm with the maximum 30% loading capacity for gemcitabine. They demonstrated that BSA microcapsules alone, when incubated with AsPC1 cells, do not show any inhibition in cell proliferation. However, BSA-Gem-EGFR displayed significant inhibition of proliferation of AsPC-1 cells (up to 31%), as compared to controlled gemcitabine-free (cell inhibition up to 15%) and unmodified gemcitabine-loaded BSA microspheres (cell inhibition up to 25%). The strategy used in that work may be effective for treating cancer cells exhibiting an overexpression of EGFR.

Qian et al. prepared protein-lipid nanocapsules of fluorescently tagged-BSA (FITC-BSA) shells with double emulsion features, wherein oily shell containing PLGA-linolic acid encapsulated a protein-containing aqueous core for the co-delivery of lipophilic paclitaxel and hydrophilic transcription factor p53 for cancer theragnostic [25]. Prepared by the double emulsion technique, the obtained nanocapsules were ~180 nm in diameter, with a zeta potential of −36.4 mV. Paclitaxel was loaded within the oily shell, containing PLGA and linolic acid, whereas p53 resided in the aqueous core of the nanocapsules. Paclitaxel and p53 synergistically induced ca. 100% apoptosis in the HeLa cells, significantly higher than either paclitaxel or p53 alone. The BSA-FITC shell of the nanocapsules could enable the observation of apoptotic cells under a fluorescence microscope. Such a formulation with therapeutic and diagnostic ability has excellent potential in biomedical applications.

Organic core HSA-shell capsules have also been prepared. Gaber et al. developed HSA-based nanocapsules with an oily core, containing a combination of hydrophobic drugs, exemestane, and hesperetin for the targeted breast cancer therapy [10]. A two-stage polymer coating method was applied to make the HSA nanocapsules, wherein an oil-in-water emulsion, containing exemestane in the oil phase and hesperetin added later to the oil phase, was prepared to form a cationic nanoemulsion. The negatively charged HSA shell was deposited on the oily core by adding the aqueous solution of HSA dropwise to the cationic nanoemulsion under stirring. 3-Aminophenylboronic acid (APBA) conjugated HSA was used by this method to prepare functionalized nanocapsules containing exemestane and hesperetin. The HSA nanocapsules and APBA–HSA nanocapsules were both obtained, in the size of ca. 172, which is suitable for delivery into cancer cells. The average zeta potentials were 20.7 ± 1.3 mV and 16.5 ± 2.8 mV, respectively. In-vitro drug release studies showed a biphasic release profile, indicating a diffusion-controlled system. Increased cell internalization of drug-loaded HSA and APBA–HSA capsules was observed in MCF-7 cell lines, compared to free drugs. The ABPA–HSA nanocapsules were successfully able to passively target the hypervascular breast tumor and actively target the overexpressed receptors in this tissue. In vivo studies showed a significant reduction in tumor volume, decreased cell proliferation, and accelerated necrosis when drug-loaded APBA–HSA nanocapsules were introduced. The study successfully utilized the superior synergistic effect of the two hydrophobic drugs, exemestane, and hesperetin, by their targeted delivery into the tumor cells using APBA functionalized HSA nanocapsules. Various other studies have also demonstrated the successful targeted delivery of drug-loaded HSA capsules by functionalizing HSA crust using various biomolecules. Rollet et al. prepared folic acid (FA) functionalized HSA nanocapsules to demonstrate cell-specific internalization by folate receptor (FRβ) macrophages, which are known to be expressed by chronically activated macrophages responsible for inflammation and tissue degradation in Rheumatoid Arthritis patients [22]. The HSA capsules (having an organic core of dodecane) were prepared sonochemically and functionalized with folic acid post-synthesis. The obtained FA-HSA nanocapsules of size ~440 nm and surface charge around −20 mV were able to successfully internalize in positive FRβ macrophages. It was observed that the FA modified HSA nanocapsules were taken up three-fold higher in concentration by FRβ-positive macrophages than in macrophages not expressing FRβ, thus paving the way for the targeted delivery into inflammation-causing macrophages during Rheumatoid Arthritis. In a recent study by Skoll et al., wheat germ agglutinin-functionalized HSA nanocapsules, with a core composed of biocompatible plant oils, were sonochemically prepared for the targeted delivery into urothelial cancer cells [105]. Various oils such as almond oil, rapeseed oil, olive oil, and linseed oil were incorporated in the HSA nanocapsules to form the organic core and analyzed for their effect on the size and stability of the nanocapsules. HSA nanocapsules with olive oil core, obtained in size range of 830–900 nm, proved to possess long-term stability (in the suspension and after freezing). Studies on Human urothelial-5637 cell lines indicated a significantly higher uptake of wheat germ agglutinin functionalized HSA nanocapsules as compared to the unfunctionalized HSA capsules.

4.2. Spherical Protein Capsules with a Solid Core

Protein capsules of hydrophobic and hydrophilic solid cores have been developed to encapsulate hydrophobic and hydrophilic payload. As mentioned in Section 2, hydrophobic payloads have usually been encapsulated to enhance their bioavailability by altering their water solubility. Parthasarathi et al. prepared vitamin E ((+)-α-tocopherol)-loaded whey protein isolate (WPI) microcapsules, altering the solubility of vitamin E, using a combination of spray drying and freeze-drying techniques, which they designated as spray freeze-drying method [12]. The microcapsules consisted of an organic core of solidified sunflower oil containing vitamin E. The synthesis methodology involved firstly the formation of a nanoemulsion of vitamin E and sunflower oil (in an aqueous solution of a surfactant) using a microfluidizer, which was then homogenized with the shell material, WPI, where the WPI molecules were rapidly absorbed on the emulsion interface to form a continuous shell layer around the oil droplets. The homogenized mixture of the organic nanoemulsion and the shell material was then spray-dried to form solid microcapsules with an average size of 145.3 μm, which were later freeze-dried. Vitamin E was encapsulated inside the microcapsules with 89.3% efficiency and seemed to be localized in the core matrix formed of solidified sunflower oil. In-vivo testing on male Wister rats, facilitated by orally administering vitamin E-loaded WPI microcapsules, revealed better pharmacokinetics, with an almost two-fold increase in the oral bioavailability of vitamin E. Increase in the bioavailability of water-insoluble anti-fungal agents has also been observed, due to their encapsulation by microcapsules. For this purpose, Li et al. developed itraconazole-loaded gelatin microcapsules using the spray-drying technique, during which itraconazole was dissolved in a mixture of dichloromethane and ethanol, which was added to an aqueous solution of gelatin and citric acid. This solution was then spray dried to give gelatin microcapsules, containing a solid core of citric acid-containing itraconazole [21]. During the preparation, itraconazole changed from the insoluble crystalline form to the soluble amorphous form, which was an important factor, apart from its acidic microenvironment, due to the presence of citric acid, contributing to a 10-fold enhancement in the solubility of the drug. In-vivo studies on 6–9 week old male Sprague–Dawley rats revealed higher concentrations of the drug in the blood plasma after the oral administration of itraconazole-loaded gelatin microcapsules, compared to the commercial product.

Microcapsules of protein shells containing hydrophilic polymeric solid core have been developed for the protection of various hydrophilic drugs, as well as probiotic cells. In a study by Laelorspoen et al., alginate microspheric core containing Lactobacillus acidophilus, a probiotic bacteria, were synthesized using the electro-spraying technique and then coated with citric acid-modified zein protein shell layer [11]. The rationale behind zein coating was to protect probiotic cells contained within the matrix of the solid alginate protein core from the harsh gastric environment, given that zein, being highly hydrophobic, possesses good resistivity against degradation due to gastric acids. The sizes of the obtained microcapsules depended upon the concentration of citric acid and ranged from 543 to

650 μm. The effects of electro-spraying voltage and citric acid concentration on the viability of encapsulated probiotic cells were evaluated. It was observed that an increase in the electro-spraying voltage reduced the size of the obtained microcapsules, which, in turn, adversely influenced the survival number of the encapsulated probiotic cells. Moreover, an increase in the citric acid concentration, during the zein protein coating, resulted in a pH decrease in the microenvironment of the cells within the capsules, reducing the number of viable probiotic cells. This is because probiotic cells cannot survive at a pH lower than 2. In-vitro studies in simulated gastric conditions (pH 1.2) revealed a five-fold increase in the cell survival number of the probiotic cells encapsulated within alginate-zein core–shell microcapsules, compared to free probiotic cells. Moreover, zein coating over alginate core proved to be highly effective in protecting the probiotic cells, establishing its supremacy over previously reported [132], uncoated probiotic cell-alginate microspheres.

Along with their application as drug delivery carriers, protein-based microcapsules with a solid core of other natural polymers have been developed for their use as embolizing agents in cancer therapy, due to their biodegradable and biocompatible nature. In a recent work by Chen et al., adriamycin hydrochloride-loaded sodium alginate-core encapsulated within silk fibroin protein-shell microcapsules were prepared as transcatheter arterial chemoembolizing agents for the treatment of hepatocellular carcinoma, using the emulsified cross-linking method [14]. The method involved emulsifying the aqueous sodium alginate and silk fibroin solution, followed by gelling and cross-linking, facilitated by lowering the pH to 3.5 and adding the cross-linker glutaraldehyde. A stable sodium alginate sphere was formed during the process, due to the acetal reaction, initiated by the chemical interaction of glutaraldehyde with the carboxyl and hydroxyl groups of sodium alginate. Silk fibroin molecules were then deposited on the outer surface of the alginate spheres via iconic and hydrogen bonding. The average size of 142 μm silk fibroin microcapsules were obtained, suitable to act as embolizing agents. The microcapsules showed good blood compatibility and almost no cytotoxicity towards vascular endothelial cells, thus proving safe for intravenous administration of microcapsules. The degradability rate of the microcapsules was found to reach 20.8% in three weeks. It indicated a good degradability trend for the slow release of a chemotherapy drug during the embolization and recanalization of blood vessels. The release mechanism of adriamycin hydrochloride was swelling-controlled, characterized by rapid initial release-kinetics, due to the initial uptake of water by the microcapsules leading to a rapid dissolution of the drug. However, the release kinetics eventually slowed down, due to microcapsule swelling after the water intake. In-vivo embolization studies on the rat ear model revealed that the microcapsules could embolize the arteries in 3 weeks, leading to ischemic necrosis in rats' ears. Their study, thus, showed that sodium alginate–silk fibroin core–shell microcapsules could effectively serve two purposes: vascular embolization and sustained-drug release, proving their potential as effective embolizing agents.

Silk fibroin microcapsules have been utilized to encapsulate solid drug-loaded synthetic polymeric-spheres, to enable sustained-release of various drugs. Qiao et al. prepared silk fibroin shell-PLGA-core microcapsules to study the sustained-release of simvastatin (SIM), a cholesterol-reducing drug known for its osteoinductive properties to enable alveolar ridge preservation after tooth extraction [24]. The hydrophobic PLGA core was made porous with a dimpled surface to improve its deposition onto the affected area. Silk-fibroin coating was performed to reduce the initial burst-release of SIM from the microcapsules, as observed in conventional uncoated PLGA microspheres. The preparation process involved synthesizing porous PLGA microspheres, using the emulsion/solvent evaporation technique, wherein an organic solution of SIM-PLGA was emulsified in an aqueous solution, followed by solvent evaporation. The maximum encapsulation efficiency of SIM in the PLGA core was found to be 85%, directly related to the concentration of PLGA. Silk fibroin coating on the PLGA microspheres involved the incubation of PLGA microspheres in an aqueous solution of silk fibroin, followed by glutaraldehyde cross-linking of silk fibroin.

In-vitro SIM release studies revealed that initial burst release was reduced to 13.2% in silk fibroin shell-coated PLGA microspheres, compared to the uncoated PLGA microspheres.

4.3. Porous/Hollow Core Protein Capsules

The past decade has seen a dramatic rise in systematic studies concerning the utilization of collagen-based microspheres in stem cell therapy, proving their great potential in tissue regeneration applications. Collagen-PLGA hollow core microcapsules functionalized by MnO_2 nanoparticles (PLGA-Col-MnO_2) were prepared by Tapeinos et al. to act as scavengers of overexpressed reactive oxygen species (ROS) such as hydrogen peroxide (responsible for oxidative stress in cells, which leads to damaged cellular protein) lipids, membranes and DNA [18]. PLGA hollow core microspheres were prepared by emulsification, followed by collagen-shell coating and incorporation of MnO_2 nanoparticles (~15 nm). It was shown that MnO_2 is capable of completely decomposing H_2O_2 into water and oxygen without forming an intermediate hydroxyl radical while turning itself into the easily excretable Mn^{+2} ions. In this study, the embedding of MnO_2 nanoparticles in the collagen-coating of the PLGA microspheres ensured their stability, facilitated better circulation in the bloodstream, prevented their easy removal by macrophages, and preserved their ability to scavenge H_2O_2 to release oxygen. In-vitro studies on the two oxidative stress-induced immortalized cell lines, 3T3 and MCF7, revealed that the microspheres could prevent H_2O_2-induced cell apoptosis by scavenging on H_2O_2 and releasing oxygen, which was cell-specific and was directly affected by PLGA-Col-MnO_2 microsphere concentration.

Collagen has been transformed into capsules and spheres to deliver substances, such as drugs, growth factors, progenitor cells, etc., for bone cancer therapy, tissue engineering, and bone regeneration applications. Nagai et al. synthesized injectable collagen microspheres loaded with recombinant human vascular endothelial growth factor (rhVEGF) aiming to protect rhVEGF from early degradation in the body and demonstrate its sustained-release to promote angiogenesis [28]. The collagen microspheres were synthesized using the emulsification technique and impregnated with rhVEGF post-synthesis. Collagen microspheres of sizes 1–30 μm possessing a positive surface charge of 8.86 mV (in phosphate-buffer saline) and 3.15 mV (in the culture medium) were obtained. Sustained-release of rhVEGF from the collagen microspheres was observed over four weeks due to the slow degradation of the microspheres. The released rhVEGF maintained its bioactivity and was able to induce capillary formation in human umbilical vein endothelial cells. In another study by Yang et al., collagen microspheres loaded (during synthesis) with steroidal saponins, a glycoside with osteoinductive properties, were synthesized using the emulsion/solvent evaporation technique [133]. They aimed to demonstrate the sustained-release of steroidal saponins from the collagen microspheres and evaluate the osteogenic properties of the composed formulation. The release of steroidal saponins from collagen microspheres was erosion-controlled, facilitated by the degradation of the microspheres in the PBS buffer. In-vitro release studies on pre-osteoblastic MC3T3-E1 cells revealed an increased and sustained expression of alkaline phosphatase (ALP), an enzyme that induces the formation of osteoblasts for bone regeneration. Liu et al. used collagen microspheres as carriers for the sustained-release of basic Fibroblast Growth Factor (bFGF) [31]. The collagen microspheres were cross-linked using different concentrations of carbodiimide to avoid their fast biodegradation and initial burst release of bFGF. The average diameter of the resultant microspheres ranged from 600–3000 nm. The authors observed a significant increase in the stability of bFGF-loaded collagen microspheres, which reduced initial burst release. Sustained-release of bFGF has also been demonstrated by porous silk fibroin (SF) microspheres prepared by Qu et al., in the size range 95–260 μm and the pore size of 1.5–7.0 μm using high voltage electrostatic differentiation, followed by lyophilization [39]. They observed a sustained biphasic release of bFGF from the porous SF microspheres when bFGF was loaded into the microspheres during synthesis, compared to its absorption into the microspheres post-synthesis. Moreover, the culture of mouse embryonic lung fibroblast cells L929 on the bFGF loaded SF microspheres exhibited significant cell proliferation in

5–9 days with very high cell viability and number compared to their culture on bFGF unloaded SF microspheres.

Hollow/porous microcapsules/spheres have also been studied as scaffolds for growing stem cells and facilitating progenitor cell delivery to a variety of damaged tissues for their regeneration. Such a need arises because stem cells introduced directly to the lesion-affected tissue do not survive long enough to undergo cell differentiation and promote tissue regeneration. For this purpose, Yao et al. utilized collagen microspheres for culturing oligodendrocyte progenitor cells (OPC) to study if the microspheres support cell progenitor growth and differentiation [13]. The collagen microspheres, synthesized using the water-in-oil emulsion technique, were obtained in sizes ranging from 73–192 μm and could support the growth and differentiation of OPC (derived from 2 rats) into oligodendrocytes. When co-cultured with dorsal root ganglion (taken from a 15-day old rat embryo), the oligodendrocytes grown from OPC-collagen microspheres could form neurite myelin sheath and initiate other processes in the dorsal root ganglion. In a series of studies by Chan et al., various types of cells, such as mesenchymal stem cell (MSC), mesenchymal stromal cells, osteoarthritis chondrocytes, and neuroblastoma cells, were microencapsulated in collagen microspheres to study their survival, growth, and differentiation, along with the potential usage of collagen microspheres as in-vitro 3D culture platforms [27,134,135]. In one of those studies, two different sets of MSC-loaded collagen microspheres (MSC@CM), one with undifferentiated MSC@CM and the other with differentiated MSC@CM, to check the effect of cell density and differentiation on cartilage repair [27]. It was observed that undifferentiated MSC@CM implanted at the affected area led to the formation of thicker but softer cartilage, whereas differentiated MSC@CM promoted the growth of stiffer but thinner cartilage. Additionally, the introduction of higher cell density into the affected area favors cartilage regeneration. In another series of studies, Cardier et al. introduced bone marrow MSC@CM into the platelet-rich blood (PRB) clots (MSC@CM-PRB), to induce bone regeneration in non-union lesions and fractures. New bone formation could be observed at the nonunion fracture areas, after three to five months of implanting MSC@CM-PRB into three patients (aged 27, 43, and 81). Moreover, no signs of in-situ abnormalities were observed. The patients' non-union fractures were healed entirely after 14 months to three years, restoring full functionality and the ability to walk [136,137].

Similar to collagen, gelatin microcapsules and silk fibroin microspheres have also been transformed to serve as cell and tissue delivery scaffolds. In a recent work by Hou et al., hyaluronic acid-graft-amphiphilic gelatin hollow microcapsules (HA-AGMC), with shell-embedded superparamagnetic iron oxide nanoparticles (SPIO nps), were prepared to serve as chondrocytes 3D-culture platforms, to form cartilage tissue-mimicking pellets for the correction of articular cartilage damage [23]. The synthesis first involved the grafting of hyaluronic acid (HA) onto the amino groups of amphiphilic gelatin (AG) and the formation of SPIO nanoparticles. The double emulsion technique was employed to prepare the microcapsules, wherein the aqueous solution of HA-g-AG was added to SPIO chloroform solution to form (w/o) single emulsion, followed by the addition of HA-g-AG aqueous solution to form w/o/w double emulsion, then solvent evaporation, dialysis, and freeze-drying. The microcapsules produced had a hydrophilic hollow core, encapsulated by a hydrophobic bilayer, composed of the amphiphilic gelatin. The hydrophobic SPIO nps resided in the shell matrix, with a loading efficiency of 92.2%, and HA covered the inner and the outer surface of the microcapsules. The microcapsules formed were highly biocompatible to the chondrocytes. HA served to connect the microcapsules to the chondrocytes because of the presence of its receptors (CD44) on the chondrocyte membrane. It was revealed that the HA-AGMC concentration of 170 μg/mL, incubated with chondrocytes for 14 days, resulted in the formation of the largest tissue pellet size of 200 μm, with the attachment efficiency of 90%, higher cell density, and viability in the cartilage tissue-mimetic, as well as a stronger connection of the microcapsules to the chondrocyte extracellular-membrane, as compared to the control groups. Due to the presence of SPIO in the HA-AGMC, the pallets could be subjected to biophysical stimulation (via static magnetic field (MF) and magnet derived sheer stress (S)) for the gene expression of Aggrecan, type I and II collagen, and SOX9, which are essential regulators of chondrogenesis and chondrocyte promotion. It was observed that their expression was dramatically up-regulated when the MF and S were applied to HA-AGMC, containing cartilage tissue pallets. Four weeks of implantation of HA-AGMC cartilage tissue-mimicking pellets in the osteoarthritic male New Zealand rabbits revealed improved retention, biofunctionality, better growth, and ordering of chondrocytes. The presence of SPIO-loaded HA-AGMC in the cartilage tissue-mimicking pallet, as well as the application of the magnetic field, exhibits better growth and ordering of chondrocytes in the pallet, and similar strategies can be applied for tissue repair in a variety of conditions/disease states. In another study, Fang et al. prepared strontium-loaded silk fibroin porous (pore diameter ~ 25 μm) microcarriers as the potential osteoinductive platforms to enable the sustained-release of osteogenesis-promoting strontium ions and the attachment, proliferation, and differentiation of mesenchymal stem cells (BMSCs) [26]. These porous microcarriers were prepared using the w/o emulsion-phase separation technique, followed by freeze-drying and strontium mineralization. They allowed the controlled release of strontium ions and attachment, proliferation, and differentiation of seeded BMCs, which was in contrast with unloaded SF microspheres, due to the limited osteoinductive property of SF. Similar to collagen and gelatin, SF-based microspheres can, thus, be used as potential osteoinductive scaffolds for stem cell growth and differentiation, to prepare injectable tissue engineering vehicles.

4.4. Multiwalled Core–Shell Protein Capsules

Zheng et al. modified the conventional LBL technique to successfully fabricate hollow SF nanocapsules with efficient encapsulation of doxorubicin (a cationic antitumor drug), as well as chlorin e6 (an anionic photosensitizer) drugs, with efficiencies of 80% and 90%, respectively, to study their sustained-release [8]. A sacrificial core of polystyrene (~250 nm) was used as a stencil to create the hollow SF nanocapsules. Alternating layers of positively charged aminopropyl triethoxysilane (APTES) were introduced in between the SF layers to promote the growth of negatively charged SF layers. SF nanocapsules with positive or negative surface charge were obtained depending upon whether the last SF layer of the capsules was subjected to APTES treatment and if the loading of chlorin e6 (Ce6) or doxorubicin (DOX), respectively, was required. The encapsulation efficiency of DOX and Ce6 reached 80% and 90%, respectively, as the layers of SF were increased, thus proving that their synthesis strategy was efficient in loading cationic and anion drugs. Burst release of DOX and Ce6 was observed from the SF nanocapsules at pH 6.5, but the slow release was evident at pH 7.4. In-vitro cytotoxicity analysis of unloaded SF nanocapsules on L929 cells and MCF-7 cells revealed a 90% higher viability than the control, proving that the nanocapsules were biocompatible. When treated with DOX- or Ce6-loaded SF nanocapsules, MCF-7 breast cancer cell lines experienced higher apoptosis, compared to the free DOX or Ce6.

Mashoofnia et al. reported multilayered nanocapsules fabrication by the LbL self-assembly of alginate polyanion and BSA polycation for the pH-responsive release of betamethasone disodium phosphate (BSP), a synthetic glucocorticoid with metabolic, immunosuppressive, and anti-inflammatory activity [17]. SiO_2 was used as a sacrificial core for the deposition of polyelectrolyte layers. They found that efficient complexation between the two polyelectrolytes for their LbL self-assembly was obtained at an alginate/BSA mixing ratio of 1:4 at pH 4. The 7- and 9-times assembly of alginate and BSA layers resulted in nanocapsules of 170 and 188 nm, respectively. The thickness of each layer was found to be ~5–6 nm. The loaded drug was sustainably released at pH 7.4, due to decreased electrostatic interactions between the alginate and the BSA layers. MTT assay analysis of MCF-7 cell lines indicated that the nanocapsules were biocompatible and suitable for drug delivery applications.

Protein–polyphenol multiwalled microcapsules have been prepared for the encapsulation and protection of hydrophilic and hydrophobic drugs. Polyphenols possess antioxidant properties, which are crucial to protect the encapsulated drugs and prolong their lifetime. Shutava et al. prepared protein –polyphenol microcapsules of alternative gelatin–epigallocatechin gallate (EGCG) layers, using the LbL technique [36], wherein EGCG polyphenol was used for its anti-cancer and antioxidant activity. It was found that the interaction between the gelatin and EGCG layers was predominantly hydrophobic, and the total EGCG content was up to 30% w/w. In another study, Lomova et al. used the LbL technique to prepare multilayered capsules of BSA protein and polyphenol Tannic acid (TA) to load hydrophilic model drug, tetramethylrhodamine-isothiocyanate labeled BSA (TRITC-BSA) and hydrophobic model drug, 3,4,9,10-tetra-(hectoxy-carbonyl)-perylene (THCP) [33]. For encapsulating the hydrophilic TRITC-BSA, a sacrificial TRITC-BSA loaded-$CaCO_3$ microparticle core was first coated with poly-L-arginine hydrochloride (PARG), followed by six bilayers of TA/BSA, to give a core/shell particle. PARG was used to provide a stronger interaction between the $CaCO_3$ core and shell TA/BSA layers. TRITC-BSA resided as a solid core in the microcapsules. To load the hydrophobic THCP into the microcapsules, THCP was dissolved in sunflower oil and emulsified with the aqueous solution of TRITC-BSA, wherein the latter stabilized on the THCP containing oil droplets, followed by the absorption of TA and BSA layers. THCP was encapsulated as an oily core in the microcapsules. Enzyme-catalyzed degradation of the capsules was employed using α-chymotrypsin, which enabled the sustained-release of the encapsulated model drugs (Tables 5 and 6).

Table 5. Recent advances in the biomedical applications of protein-based solid/liquid/hollow capsules.

Protein	Shell Composition	Core Type	Shell-Bound API	Core API	Biomedical Function	Type of Therapy	Ref.
Albumins	BSA	Liquid, organic (soybean oil)	RNA	-	Controlled release of RNA and its protection from the outer cellular environment	Gene expression and function	[16]
	PVA and PEI functionalized-BSA protein	Liquid, organic (soybean oil)	RNA	-	Targeted delivery of RNA to the cell nucleus, controlled release, and protection from the outer cellular environment	Gene expression and function	[35]
	Anti-EGFR-modified BSA	Liquid, organic (dodecane)	Gemcitabine	-	Sustained-release of Gemcitabine and EGFR blocking	Pancreatic-cancer therapy	[20]
	FITC-BSA bound liquid organic shell filled with PLGA-linolic acid	Liquid Aqueous	Paclitaxel	Transcription factor p53	Sustained-release synergistic apoptotic effect of hydrophilic and hydrophobic drugs on HeLa cells	Cancer theragnostic	[25]
	Multiwalled, BSA polycation–alginate polyanion layered alternatively	Hollow	Betamethasone disodium phosphate (BSP)	-	Sustained-release of BSP having metabolic, immunosuppressive, and anti-inflammatory activity	Rheumatoid arthritis, Crohn's disease, etc.	[17]
	Multiwalled, BSA-Tannic acid layered alternatively	Solid, hydrophilic tetramethylrho damine-isothiocyanate labeled BSA (TRITC-BSA)	-	TRITC-BSA	-	-	[33]
	Multiwalled, BSA-Tannic acid layered alternatively	Liquid, organic (sunflower oil)	TRITC-BSA	3,4,9,10-tetra-(hectoxy-carbonyl)-perylene (THCP)	Co-encapsulation of hydrophobic and hydrophilic drugs for sustained-release and their protection by polyphenol Tannic Acid	All types of therapies	[33]

Table 5. *Cont.*

Protein	Shell Composition	Core Type	Shell-Bound API	Core API	Biomedical Function	Type of Therapy	Ref.
Albumins	3-aminophenylboronic acid functionalized-HSA	Liquid, organic	-	Exemestane and Hesperetin	Cell-specific internalization and Targeted delivery into MCF-7 cell lines and sustained-release	Breast-cancer therapy	[10]
	Folic acid-functionalized HSA	Liquid, organic (dodecane)	Folic acid	-	Cell-specific internalization and Targeted delivery into folic-receptor macrophages	Rheumatoid arthritis	[22]
Whey Protein Isolate (WPI)	WPI	Solid Hydrophobic (sunflower oil)	-	vitamin E ((+)-α-tocopherol)	Enhanced bioavailability of water-insoluble vitamin E	Nutritional therapy	[12]
Collagen	MnO_2 functionalized-collagen-PLGA	Hollow	-	-	Prevention of oxidative stress-induced protein-, lipid- or DNA damage and cell apoptosis	Cancer therapy, cardiovascular and neurological disorders treatment	[18]
Silk Fibroin	Silk fibroin protein	Solid Hydrophilic (alginate)	Adriamycin hydrochloride	-	Transcatheter arterial chemoembolizing by the microcapsules and controlled release of adriamycin hydrochloride	Liver cancer therapy	[14]
	Silk fibroin protein	Solid Hydrophobic (PLGA)	-	Simvastatin	sustained-release of cholesterol-reducing and osteoinductive simvastatin	Bone regeneration	[24]
	Multiwalled, silk fibroin-APTES layered alternatively	Hollow	chlorin e6 (Ce6) and doxorubicin (DOX)	-	Sustained-release of anti-tumor drug DOX and photosensitizer Ce6	Chemophoto therapy	[8]
Zein	Citric acid-modified zein	Solid Hydrophilic (alginate)	-	Lactobacillus acidophilus	Protection of probiotic L. acidophilus from the gastric environment	Nutritional therapy	[11]
Gelatin	Gelatin	Solid hydrophilic (citric acid)	-	Itraconazole	Enhanced bioavailability of water-insoluble itraconazole	Treatment of mycotic infections	[21]
	Hyaluronic acid-graft gelatin hydrophobic shell embedding SPIO	Hollow (hydrophilic)	-	-	Chondrocyte cells 3D-culture platforms to form cartilage tissue-mimicking pellets, magnetic field, and magnetic stress-induced gene expression	Tissue repair (correction of articular cartilage damage)	[23]
	Multiwalled gelatin–epigallocatechin gallate (EGCG) LbL	Hollow	-	-	EGCG layers introduce antioxidant properties to the microcapsules to prolong the lifetime and enhance the effectiveness of encapsulated APIs	Cancer therapy and more	[36]

Table 6. Recent advances in the biomedical applications of porous protein microspheres.

Protein	Composition	Biomedical Cargo	Biomedical Function	Type of Therapy	Ref.
Collagen	Collagen microspheres	Recombinant human vascular endothelial growth factor (rhVEGF)	Sustained-release of signal protein rhVEGF	Cardiac muscle repair	[28]
		Steroidal saponins	Sustained-release of Steroidal saponins	Osteogenesis and bone regeneration	[133]
		Oligodendrocyte progenitor cells (OPC)	Culturing OPC and their delivery to lesion-affected tissue for the repair of the neurite myelin sheath	Tissue regeneration	[13]
		Mesenchymal stem cells, mesenchymal stromal cells, osteoarthritis chondrocytes, and neuroblastoma cells	3D cell culture platform for stem cell culture, differentiation, and delivery	Stem cell therapy	[134,135]
		Bone marrow mesenchymal stromal cells	Integration into platelet-rich blood clots and implantation at the nonunion lesion site	Bone regeneration for nonunion fractures	[136,137]
Silk Fibroin	Porous silk fibroin (SF) microspheres	Basic fibroblast growth factor (bFGF)	Sustained-release of bFGF and lowering of biodegradability	Tissue repair	[39]
	Strontium loaded porous SF microspheres	Strontium and mesenchymal stem cell (MSC)	Sustained-release of osteogenic strontium and the culture of MSC	Bone regeneration	[26]

5. Concluding Remarks and Future Perspectives

Natural, polymer-based APC carriers, especially proteins and polysaccharides, have been utilized widely in biomedical applications, mainly due to features such as biodegradability, biocompatibility, functionalization capability, low-immunogenicity, and blood compatibility. Amongst polysaccharides, chitosan (and derivatives), cellulose (and derivatives), and alginate have been the most commonly utilized shell candidates in core–shell capsules, whereas BSA, HSA, collagen, gelatin, silk fibroin, and zein proteins rule the polypeptide family. Keratin, resilin, and gliadin are some of the less explored polypeptides as shell-forming polymers for core–shell capsules. Recent research trends indicate an accelerated rate in the development of APC carriers with various core–shell structural configurations. Compared to non-porous and porous sphere structures, core–shell capsule configurations have been proven superior, as they enable the introduction of multifunctionalities, higher cargo loading capacity, and encapsulation efficiency. A variety of hydrophobic and hydrophilic cargo can be simultaneously encapsulated and entrapped in the single- and multiwalled core–shell capsules. Porous spheres, on the other hand, were proven to be advantageous as platforms for cell culture. The majority of research and development in cell carrier platforms seems to utilize porous microspheres, possibly due to the availability of a larger surface area for culture within the pores, as well as the polymer matrix surface.

Several approaches and techniques have been utilized to synthesize porous spheres and core–shell capsules. The majority of these techniques have been utilized for many decades and have undergone slight modifications over the years to meet the system-specific needs. It is also evident that these techniques largely utilize templating approach, wherein either solid or emulsion templates are prepared and used to guide the formation of core–shell-type, as well as sphere-type structural configurations. Solid templating is the easiest and most direct approach for synthesizing solid- and hollow-core, as well as single- and multiwalled capsules. There are many reports on utilizing the solid templating approach for solid- and hollow-core–shell capsule synthesis. Liquid-core capsules have also been indirectly prepared using solid templating. However, the number of such studies is relatively low. Emulsion templating is another commonly used method of synthesis, especially for oily core capsules and porous spheres. An important difference between sphere and capsule synthesis using emulsion templating is that the latter generally requires the formation of o/w emulsion, as opposed to w/o emulsion, especially when the polymer is hydrophilic and soluble in water. Ultrasonication-assisted emulsification has also been established as one of the leading capsule synthesis approaches, due to its facile and time-efficient methodology. Several studies have utilized the ultrasonication approach for the synthesis of liquid-core capsules. However, the proportion of studies involving the ultrasonic synthesis of oily-core capsules is higher than that of the aqueous-core capsules. Thus, ultrasonication can be further explored towards the synthesis of aqueous-core capsules.

The stability of polymeric carriers, especially core–shell single- and multiwalled capsules, has been a concern. For their long-term stability, covalent crosslinking has been exploited. However, as mentioned earlier, a balance between covalent and non-covalent interactions must be achieved to ensure capsule stability but must enable stimuli-responsive cargo release. Storage is another concern when it comes to core–shell capsules. Liquid-core and hollow core capsules are highly prone to structural deformations and bursting when dried for storage and characterization as dry powders. Liquid-core capsules can be stored as colloidal solutions, instead, to avoid these issues. However, it must be noted that if the colloidal conditions are not appropriately maintained, the capsules suffer aggregation, which results in the loss of shell functionalities. More efforts are needed in the direction of purification and proper storage. In addition, alternative, less invasive approaches for preparing characterization samples of liquid-core capsules are needed.

Author Contributions: Conceptualization, K.S., Z.P. and A.G.; methodology, investigation, resources, writing—original draft preparation, K.S.; writing—review and editing, Z.P. and A.G.; supervision, Z.P. and A.G. All authors have read and agreed to the published version of the manuscript.

Funding: This research received no external funding.

Institutional Review Board Statement: Not applicable.

Informed Consent Statement: Not applicable.

Data Availability Statement: Not applicable.

Conflicts of Interest: The authors declare no conflict of interest.

References

1. Ekladious, I.; Colson, Y.L.; Grinstaff, M.W. Polymer–drug conjugate therapeutics: Advances, insights and prospects. *Nat. Rev. Drug Discov.* **2019**, *18*, 273–294. [CrossRef] [PubMed]
2. Park, K. Controlled drug delivery systems: Past forward and future back. *J. Control. Release* **2014**, *190*, 3–8. [CrossRef] [PubMed]
3. Folkman, J.; Long, D.M. The use of silicone rubber as a carrier for prolonged drug therapy. *J. Surg. Res.* **1964**, *4*, 139–142. [CrossRef]
4. Ringsdorf, H. Structure and properties of pharmacologically active polymers. *J. Polym. Sci. Polym. Symp.* **1975**, *51*, 135–153. [CrossRef]
5. Gregoriadis, G. Drug entrapment in liposomes. *FEBS Lett.* **1973**, *36*, 292–296. [CrossRef]
6. Kramer, P.A. Albumin Microspheres as Vehicles for Achieving Specificity in Drug Delivery. *J. Pharm. Sci.* **1974**, *63*, 1646–1647. [CrossRef]
7. Gurny, R.; Peppas, N.A.; Harrington, D.D.; Banker, G.S. Development of Biodegradable and Injectable Latices for Controlled Release of Potent Drugs. *Drug Dev. Ind. Pharm.* **1981**, *7*, 1–25. [CrossRef]
8. Zheng, H.; Duan, B.; Xie, Z.; Wang, J.; Yang, M. Inventing a facile method to construct Bombyx mori (*B. mori*) silk fibroin nanocapsules for drug delivery. *RSC Adv.* **2020**, *10*, 28408–28414. [CrossRef]
9. Cheng, C.; Teasdale, I.; Brüggemann, O. Stimuli-Responsive Capsules Prepared from Regenerated Silk Fibroin Microspheres. *Macromol. Biosci.* **2014**, *14*, 807–816. [CrossRef] [PubMed]
10. Gaber, M.; Hany, M.; Mokhtar, S.; Helmy, M.W.; Elkodairy, K.A.; Elzoghby, A.O. Boronic-targeted albumin-shell oily-core nanocapsules for synergistic aromatase inhibitor/herbal breast cancer therapy. *Mater. Sci. Eng. C* **2019**, *105*, 110099. [CrossRef]

11. Laelorspoen, N.; Wongsasulak, S.; Yoovidhya, T.; Devahastin, S. Microencapsulation of Lactobacillus acidophilus in zein–alginate core–shell microcapsules via electrospraying. *J. Funct. Foods* **2014**, *7*, 342–349. [CrossRef]
12. Parthasarathi, S.; Anandharamakrishnan, C. Enhancement of oral bioavailability of vitamin E by spray-freeze drying of whey protein microcapsules. *Food Bioprod. Process.* **2016**, *100*, 469–476. [CrossRef]
13. Yao, L.; Phan, F.; Li, Y. Collagen microsphere serving as a cell carrier supports oligodendrocyte progenitor cell growth and differentiation for neurite myelination in vitro. *Stem Cell Res. Ther.* **2013**, *4*, 109. [CrossRef]
14. Chen, G.; Wei, R.; Huang, X.; Wang, F.; Chen, Z. Synthesis and assessment of sodium alginate-modified silk fibroin microspheres as potential hepatic arterial embolization agent. *Int. J. Biol. Macromol.* **2020**, *155*, 1450–1459. [CrossRef]
15. Ortiz, M.; Jornada, D.S.; Pohlmann, A.R.; Guterres, S.S. Development of Novel Chitosan Microcapsules for Pulmonary Delivery of Dapsone: Characterization, Aerosol Performance, and In Vivo Toxicity Evaluation. *AAPS PharmSciTech* **2015**, *16*, 1033–1040. [CrossRef]
16. Shimanovich, U.; Tkacz, I.D.; Eliaz, D.; Cavaco-Paulo, A.; Michaeli, S.; Gedanken, A. Encapsulation of RNA Molecules in BSA Microspheres and Internalization into Trypanosoma Brucei Parasites and Human U2OS Cancer Cells. *Adv. Funct. Mater.* **2011**, *21*, 3659–3666. [CrossRef]
17. Mashoofnia, A.; Mohamadnia, Z.; Kompany-Zareh, M. Application of Multivariate and Spectroscopic Techniques for Investigation of the Interactions between Polyelectrolyte Layers in Layer-by-Layer Assembled pH-Sensitive Nanocapsules. *Macromol. Chem. Phys.* **2021**, *222*, 2100107. [CrossRef]
18. Tapeinos, C.; Larrañaga, A.; Sarasua, J.-R.; Pandit, A. Functionalised collagen spheres reduce H_2O_2 mediated apoptosis by scavenging overexpressed ROS. *Nanomed. Nanotechnol. Biol. Med.* **2018**, *14*, 2397–2405. [CrossRef] [PubMed]
19. Zhang, S.; Zhou, Y.; Nie, W.; Song, L.; Li, J.; Yang, B. Fabrication of uniform "smart" magnetic microcapsules and their controlled release of sodium salicylate. *J. Mater. Chem. B* **2013**, *1*, 4331. [CrossRef]
20. Grinberg, O.; Gedanken, A.; Mukhopadhyay, D.; Patra, C.R. Antibody modified Bovine Serum Albumin microspheres for targeted delivery of anticancer agent Gemcitabine. *Polym. Adv. Technol.* **2013**, *24*, 294–299. [CrossRef]
21. Li, D.X.; Park, Y.-J.; Oh, D.H.; Joe, K.H.; Lee, J.H.; Yeo, W.H.; Yong, C.S.; Choi, H.-G. Development of an itraconazole-loaded gelatin microcapsule with enhanced oral bioavailability: Physicochemical characterization and in-vivo evaluation. *J. Pharm. Pharmacol.* **2010**, *62*, 448–455. [CrossRef]
22. Rollett, A.; Reiter, T.; Nogueira, P.; Cardinale, M.; Loureiro, A.; Gomes, A.; Cavaco-Paulo, A.; Moreira, A.; Carmo, A.M.; Guebitz, G.M. Folic acid-functionalized human serum albumin nanocapsules for targeted drug delivery to chronically activated macrophages. *Int. J. Pharm.* **2012**, *427*, 460–466. [CrossRef] [PubMed]
23. Hou, K.-T.; Liu, T.-Y.; Chiang, M.-Y.; Chen, C.-Y.; Chang, S.-J.; Chen, S.-Y. Cartilage Tissue-Mimetic Pellets with Multifunctional Magnetic Hyaluronic Acid-Graft-Amphiphilic Gelatin Microcapsules for Chondrogenic Stimulation. *Polymers* **2020**, *12*, 785. [CrossRef] [PubMed]
24. Qiao, F.; Zhang, J.; Wang, J.; Du, B.; Huang, X.; Pang, L.; Zhou, Z. Silk fibroin-coated PLGA dimpled microspheres for retarded release of simvastatin. *Colloids Surf. B Biointerfaces* **2017**, *158*, 112–118. [CrossRef] [PubMed]
25. Fu, A.; Wu, J.; Zhang, E.; Zhang, Y.; Qian, K. Biodegradable double nanocapsule as a novel multifunctional carrier for drug delivery and cell imaging. *Int. J. Nanomed.* **2015**, *10*, 4149–4157. [CrossRef]
26. Fang, J.; Wang, D.; Hu, F.; Li, X.; Zou, X.; Xie, J.; Zhou, Z. Strontium mineralized silk fibroin porous microcarriers with enhanced osteogenesis as injectable bone tissue engineering vehicles. *Mater. Sci. Eng. C* **2021**, *128*, 112354. [CrossRef] [PubMed]
27. Li, Y.Y.; Cheng, H.W.; Cheung, K.M.C.; Chan, D.; Chan, B.P. Mesenchymal stem cell-collagen microspheres for articular cartilage repair: Cell density and differentiation status. *Acta Biomater.* **2014**, *10*, 1919–1929. [CrossRef]
28. Nagai, N.; Kumasaka, N.; Kawashima, T.; Kaji, H.; Nishizawa, M.; Abe, T. Preparation and characterization of collagen microspheres for sustained release of VEGF. *J. Mater. Sci. Mater. Med.* **2010**, *21*, 1891–1898. [CrossRef]
29. Zhi, Z.L.; Song, L.; Pickup, J.J. Nanolayer encapsulation of insulin-chitosan complexes improves efficiency of oral insulin delivery. *Int. J. Nanomed.* **2014**, *9*, 2127–2136. [CrossRef]
30. Bruschi, M.L. (Ed.) Main mechanisms to control the drug release. In *Strategies to Modify the Drug Release from Pharmaceutical Systems*; Elsevier: San Diego, CA, USA, 2015; pp. 37–62.
31. Liu, T.; Dan, N.; Dan, W. The effect of crosslinking agent on sustained release of bFGF–collagen microspheres. *RSC Adv.* **2015**, *5*, 34511–34516. [CrossRef]
32. Alvarez-Lorenzo, C.; Blanco-Fernandez, B.; Puga, A.M.; Concheiro, A. Crosslinked ionic polysaccharides for stimuli-sensitive drug delivery. *Adv. Drug Deliv. Rev.* **2013**, *65*, 1148–1171. [CrossRef] [PubMed]
33. Lomova, M.V.; Brichkina, A.I.; Kiryukhin, M.V.; Vasina, E.N.; Pavlov, A.M.; Gorin, D.A.; Sukhorukov, G.B.; Antipina, M.N. Multilayer Capsules of Bovine Serum Albumin and Tannic Acid for Controlled Release by Enzymatic Degradation. *ACS Appl. Mater. Interfaces* **2015**, *7*, 11732–11740. [CrossRef]
34. Toublan, F.J.-J.; Boppart, S.; Suslick, K.S. Tumor Targeting by Surface-Modified Protein Microspheres. *J. Am. Chem. Soc.* **2006**, *128*, 3472–3473. [CrossRef]
35. Shimanovich, U.; Eliaz, D.; Zigdon, S.; Volkov, V.; Aizer, A.; Cavaco-Paulo, A.; Michaeli, S.; Shav-Tal, Y.; Gedanken, A. Proteinaceous microspheres for targeted RNA delivery prepared by an ultrasonic emulsification method. *J. Mater. Chem. B* **2013**, *1*, 82–90. [CrossRef] [PubMed]

36. Shutava, T.G.; Balkundi, S.S.; Lvov, Y.M. (−)-Epigallocatechin gallate/gelatin layer-by-layer assembled films and microcapsules. *J. Colloid Interface Sci.* **2009**, *330*, 276–283. [CrossRef]
37. Van den Mooter, G. The use of amorphous solid dispersions: A formulation strategy to overcome poor solubility and dissolution rate. *Drug Discov. Today Technol.* **2012**, *9*, e79–e85. [CrossRef] [PubMed]
38. Sapkal, S.B.; Adhao, V.S.; Thenge, R.R.; Darakhe, R.A.; Shinde, S.A.; Shrikhande, V.N. Formulation and Characterization of Solid Dispersions of Etoricoxib Using Natural Polymers. *Turk. J. Pharm. Sci.* **2020**, *17*, 7–19. [CrossRef] [PubMed]
39. Qu, J.; Wang, L.; Niu, L.; Lin, J.; Huang, Q.; Jiang, X.; Li, M. Porous Silk Fibroin Microspheres Sustainably Releasing Bioactive Basic Fibroblast Growth Factor. *Materials* **2018**, *11*, 1280. [CrossRef]
40. Gentilini, R.; Munarin, F.; Bloise, N.; Secchi, E.; Visai, L.; Tanzi, M.C.; Petrini, P. Polysaccharide-based hydrogels with tunable composition as 3D cell culture systems. *Int. J. Artif. Organs* **2018**, *41*, 213–222. [CrossRef]
41. De Oliveira, A.C.; Sabino, R.M.; Souza, P.R.; Muniz, E.C.; Popat, K.C.; Kipper, M.J.; Zola, R.S.; Martins, A.F. Chitosan/gellan gum ratio content into blends modulates the scaffolding capacity of hydrogels on bone mesenchymal stem cells. *Mater. Sci. Eng. C* **2020**, *106*, 110258. [CrossRef]
42. Liang, Y.-J.; Yu, H.; Feng, G.; Zhuang, L.; Xi, W.; Ma, M.; Chen, J.; Gu, N.; Zhang, Y. High-Performance Poly(lactic-co-glycolic acid)-Magnetic Microspheres Prepared by Rotating Membrane Emulsification for Transcatheter Arterial Embolization and Magnetic Ablation in VX 2 Liver Tumors. *ACS Appl. Mater. Interfaces* **2017**, *9*, 43478–43489. [CrossRef] [PubMed]
43. Kim, H.; Lee, G.-H.; Ro, J.; Kuh, H.-J.; Kwak, B.-K.; Lee, J. Recoverability of freeze-dried doxorubicin-releasing chitosan embolic microspheres. *J. Biomater. Sci. Polym. Ed.* **2013**, *24*, 2081–2095. [CrossRef] [PubMed]
44. Katsumori, T.; Miura, H.; Arima, H.; Hino, A.; Tsuji, Y.; Masuda, Y.; Nishimura, T. Tris-acryl gelatin microspheres versus gelatin sponge particles in uterine artery embolization for leiomyoma. *Acta Radiol.* **2017**, *58*, 834–841. [CrossRef]
45. Sommer, C.M.; Do, T.D.; Schlett, C.L.; Flechsig, P.; Gockner, T.L.; Kuthning, A.; Vollherbst, D.F.; Pereira, P.L.; Kauczor, H.U.; Macher-Göppinger, S. In vivo characterization of a new type of biodegradable starch microsphere for transarterial embolization. *J. Biomater. Appl.* **2018**, *32*, 932–944. [CrossRef]
46. Choi, H.; Choi, B.; Yu, B.; Li, W.; Matsumoto, M.M.; Harris, K.R.; Lewandowski, R.J.; Larson, A.C.; Mouli, S.K.; Kim, D.-H. On-demand degradable embolic microspheres for immediate restoration of blood flow during image-guided embolization procedures. *Biomaterials* **2021**, *265*, 120408. [CrossRef]
47. Guo, L.; Qin, S. Studies on preparations and properties of drug-eluting embolization microspheres made from oxidated alginate and carboxymethyl chitosan. *Int. J. Polym. Mater. Polym. Biomater.* **2019**, *68*, 844–849. [CrossRef]
48. Nicolas, J.; Mura, S.; Brambilla, D.; Mackiewicz, N.; Couvreur, P. Design, functionalization strategies and biomedical applications of targeted biodegradable/biocompatible polymer-based nanocarriers for drug delivery. *Chem. Soc. Rev.* **2013**, *42*, 1147–1235. [CrossRef]
49. Piñón-Segundo, E.; Llera-Rojas, V.G.; Leyva-Gómez, G.; Urbán-Morlán, Z.; Mendoza-Muñoz, N.; Quintanar-Guerrero, D. The emulsification-diffusion method to obtain polymeric nanoparticles. In *Nanoscale Fabrication, Optimization, Scale-Up and Biological Aspects of Pharmaceutical Nanotechnology*; Elsevier: Kidlington, UK, 2018; pp. 51–83.
50. Lengyel, M.; Kállai-Szabó, N.; Antal, V.; Laki, A.J.; Antal, I. Microparticles, Microspheres, and Microcapsules for Advanced Drug Delivery. *Sci. Pharm.* **2019**, *87*, 20. [CrossRef]
51. Lensen, D.; Vriezema, D.M.; van Hest, J.C.M. Polymeric Microcapsules for Synthetic Applications. *Macromol. Biosci.* **2008**, *8*, 991–1005. [CrossRef] [PubMed]
52. Deng, S.; Gigliobianco, M.R.; Censi, R.; Di Martino, P. Polymeric Nanocapsules as Nanotechnological Alternative for Drug Delivery System: Current Status, Challenges and Opportunities. *Nanomaterials* **2020**, *10*, 847. [CrossRef]
53. Choudhury, N.; Meghwal, M.; Das, K. Microencapsulation: An overview on concepts, methods, properties and applications in foods. *Food Front.* **2021**, *2*, 1–17. [CrossRef]
54. Kozlovskaya, V.; Baggett, J.; Godin, B.; Liu, X.; Kharlampieva, E. Hydrogen-Bonded Multilayers of Silk Fibroin: From Coatings to Cell-Mimicking Shaped Microcontainers. *ACS Macro Lett.* **2012**, *1*, 384–387. [CrossRef]
55. Yang, Y.; Zhu, H.; Wang, J.; Fang, Q.; Peng, Z. Enzymatically Disulfide-Crosslinked Chitosan/Hyaluronic Acid Layer-by-Layer Self-Assembled Microcapsules for Redox-Responsive Controlled Release of Protein. *ACS Appl. Mater. Interfaces* **2018**, *10*, 33493–33506. [CrossRef] [PubMed]
56. Yitayew, M.Y.; Tabrizian, M. Hollow Microcapsules Through Layer-by-Layer Self-Assembly of Chitosan/Alginate on *E. coli*. *MRS Adv.* **2020**, *5*, 2401–2407. [CrossRef]
57. Szafraniec-Szczęsny, J.; Janik-Hazuka, M.; Odrobińska, J.; Zapotoczny, S. Polymer Capsules with Hydrophobic Liquid Cores as Functional Nanocarriers. *Polymers* **2020**, *12*, 1999. [CrossRef] [PubMed]
58. Singh, A.; Bajpai, J.; Tiwari, A.; Bajpai, A.K. Designing casein-coated iron oxide nanostructures (CCIONPs) as superparamagnetic core–shell carriers for magnetic drug targeting. *Prog. Biomater.* **2015**, *4*, 39–53. [CrossRef] [PubMed]
59. Chua, P.-H.; Neoh, K.-G.; Kang, E.-T.; Wang, W. Surface functionalization of titanium with hyaluronic acid/chitosan polyelectrolyte multilayers and RGD for promoting osteoblast functions and inhibiting bacterial adhesion. *Biomaterials* **2008**, *29*, 1412–1421. [CrossRef]
60. Deng, C.; Dong, W.-F.; Adalsteinsson, T.; Ferri, J.K.; Sukhorukov, G.B.; Möhwald, H. Solvent-filled matrix polyelectrolyte capsules: Preparation, structure and dynamics. *Soft Matter* **2007**, *3*, 1293. [CrossRef] [PubMed]

61. Dong, Y.; Lan, T.; Wang, X.; Zhang, Y.; Jiang, L.; Sui, X. Preparation and characterization of soy protein microspheres using amorphous calcium carbonate cores. *Food Hydrocoll.* **2020**, *107*, 105953. [CrossRef]
62. Zhang, W.; Wang, X.; Wang, J.; Zhang, L. Drugs adsorption and release behavior of collagen/bacterial cellulose porous microspheres. *Int. J. Biol. Macromol.* **2019**, *140*, 196–205. [CrossRef]
63. Fan, J.-B.; Huang, C.; Jiang, L.; Wang, S. Nanoporous microspheres: From controllable synthesis to healthcare applications. *J. Mater. Chem. B* **2013**, *1*, 2222. [CrossRef]
64. Yuan, W.; Cai, Y.; Chen, Y.; Hong, X.; Liu, Z. Porous microsphere and its applications. *Int. J. Nanomed.* **2013**, *8*, 1111–1120. [CrossRef] [PubMed]
65. Zhao, Q.; Han, B.; Wang, Z.; Gao, C.; Peng, C.; Shen, J. Hollow chitosan-alginate multilayer microcapsules as drug delivery vehicle: Doxorubicin loading and in vitro and in vivo studies. *Nanomed. Nanotechnol. Biol. Med.* **2007**, *3*, 63–74. [CrossRef] [PubMed]
66. Yilmaz, M.D. Layer-by-layer hyaluronic acid/chitosan polyelectrolyte coated mesoporous silica nanoparticles as pH-responsive nanocontainers for optical bleaching of cellulose fabrics. *Carbohydr. Polym.* **2016**, *146*, 174–180. [CrossRef] [PubMed]
67. Sukhishvili, S.A.; Granick, S. Layered, Erasable Polymer Multilayers Formed by Hydrogen-Bonded Sequential Self-Assembly. *Macromolecules* **2002**, *35*, 301–310. [CrossRef]
68. Manna, U.; Bharani, S.; Patil, S. Layer-by-Layer Self-Assembly of Modified Hyaluronic Acid/Chitosan Based on Hydrogen Bonding. *Biomacromolecules* **2009**, *10*, 2632–2639. [CrossRef] [PubMed]
69. Liu, P. Stabilization of layer-by-layer engineered multilayered hollow microspheres. *Adv. Colloid Interface Sci.* **2014**, *207*, 178–188. [CrossRef] [PubMed]
70. Tiwari, S.; Mishra, B. Multilayered membrane-controlled microcapsules for controlled delivery of isoniazid. *Daru* **2011**, *19*, 41–46.
71. Pal, K.; Paulson, A.T.; Rousseau, D. Biopolymers in Controlled-Release Delivery Systems. In *Handbook of Biopolymers and Biodegradable Plastics*; Elsevier: London, UK, 2013; pp. 329–363.
72. Mu, B.; Lu, C.; Liu, P. Disintegration-controllable stimuli-responsive polyelectrolyte multilayer microcapsules via covalent layer-by-layer assembly. *Colloids Surf. B Biointerfaces* **2011**, *82*, 385–390. [CrossRef]
73. Wang, C.; Luo, W.; Li, P.; Li, S.; Yang, Z.; Hu, Z.; Liu, Y.; Ao, N. Preparation and evaluation of chitosan/alginate porous microspheres/Bletilla striata polysaccharide composite hemostatic sponges. *Carbohydr. Polym.* **2017**, *174*, 432–442. [CrossRef]
74. Huang, L.; Xiao, L.; Jung Poudel, A.; Li, J.; Zhou, P.; Gauthier, M.; Liu, H.; Wu, Z.; Yang, G. Porous chitosan microspheres as microcarriers for 3D cell culture. *Carbohydr. Polym.* **2018**, *202*, 611–620. [CrossRef]
75. Fang, J.; Zhang, Y.; Yan, S.; Liu, Z.; He, S.; Cui, L.; Yin, J. Poly(l-glutamic acid)/chitosan polyelectrolyte complex porous microspheres as cell microcarriers for cartilage regeneration. *Acta Biomater.* **2014**, *10*, 276–288. [CrossRef] [PubMed]
76. Moinard-Chécot, D.; Chevalier, Y.; Briançon, S.; Beney, L.; Fessi, H. Mechanism of nanocapsules formation by the emulsion–diffusion process. *J. Colloid Interface Sci.* **2008**, *317*, 458–468. [CrossRef]
77. Trojanowska, A.; Nogalska, A.; Valls, R.G.; Giamberini, M.; Tylkowski, B. Technological solutions for encapsulation. *Phys. Sci. Rev.* **2017**, *2*, 171–202. [CrossRef]
78. Grinstaff, M.W.; Suslick, K.S. Air-filled proteinaceous microbubbles: Synthesis of an echo-contrast agent. *Proc. Natl. Acad. Sci. USA* **1991**, *88*, 7708–7710. [CrossRef] [PubMed]
79. Suslick, K.S.; Grinstaff, M.W. Protein microencapsulation of nonaqueous liquids. *J. Am. Chem. Soc.* **1990**, *112*, 7807–7809. [CrossRef]
80. Panigrahi, R.; Srivastava, S.K. Ultrasound assisted synthesis of a polyaniline hollow microsphere/Ag core/shell structure for sensing and catalytic applications. *RSC Adv.* **2013**, *3*, 7808. [CrossRef]
81. Rajabinejad, H.; Patrucco, A.; Caringella, R.; Montarsolo, A.; Zoccola, M.; Pozzo, P.D. Preparation of keratin-based microcapsules for encapsulation of hydrophilic molecules. *Ultrason. Sonochem.* **2018**, *40*, 527–532. [CrossRef]
82. Wong, M.; Suslick, K.S. Sonochemically Produced Hemoglobin Microbubbles. *MRS Proc.* **1994**, *372*, 89. [CrossRef]
83. Sharma, K.; Sadhanala, H.K.; Mastai, Y.; Porat, Z.; Gedanken, A. Sonochemically Prepared BSA Microspheres as Adsorbents for the Removal of Organic Pollutants from Water. *Langmuir* **2021**, *37*, 9927–9938. [CrossRef]
84. Mutalikdesai, A.; Nassir, M.; Saady, A.; Hassner, A.; Gedanken, A. Sonochemically modified ovalbumin enhances enantioenrichment of some amino acids. *Ultrason. Sonochem.* **2019**, *58*, 104603. [CrossRef]
85. Sharma, K.; Saady, A.; Jacob, A.; Porat, Z.; Gedanken, A. Entrapment and release kinetics study of dyes from BSA microspheres forming a matrix and a reservoir system. *J. Mater. Chem. B* **2020**, *8*, 10154–10161. [CrossRef]
86. Saady, A.; Varon, E.; Jacob, A.; Shav-Tal, Y.; Fischer, B. Applying styryl quinolinium fluorescent probes for imaging of ribosomal RNA in living cells. *Dyes Pigment.* **2020**, *174*. [CrossRef]
87. Gedanken, A. Preparation and Properties of Proteinaceous Microspheres Made Sonochemically. *Chem. Eur. J.* **2008**, *14*, 3840–3853. [CrossRef] [PubMed]
88. Silva, R.; Ferreira, H.; Azoia, N.G.; Shimanovich, U.; Freddi, G.; Gedanken, A.; Cavaco-Paulo, A. Insights on the mechanism of formation of protein microspheres in a biphasic system. *Mol. Pharm.* **2012**, *9*, 3079–3088. [CrossRef]
89. Xu, H.; Zeiger, B.W.; Suslick, K.S. Sonochemical synthesis of nanomaterials. *Chem. Soc. Rev.* **2013**, *42*, 2555–2567. [CrossRef] [PubMed]
90. Silva, R.; Ferreira, H.; Cavaco-Paulo, A. Sonoproduction of Liposomes and Protein Particles as Templates for Delivery Purposes. *Biomacromolecules* **2011**, *12*, 3353–3368. [CrossRef] [PubMed]

91. Suslick, K.S.; Grinstaff, M.W.; Kolbeck, K.J.; Wong, M. Characterization of sonochemically prepared proteinaceous microspheres. *Ultrason. Sonochem.* **1994**, *1*, S65–S68. [CrossRef]
92. Abismaïl, B.; Canselier, J.; Wilhelm, A.; Delmas, H.; Gourdon, C. Emulsification by ultrasound: Drop size distribution and stability. *Ultrason. Sonochem.* **1999**, *6*, 75–83. [CrossRef]
93. Shimanovich, U.; Bernardes, G.J.L.; Knowles, T.P.J.; Cavaco-Paulo, A. Protein micro- and nano-capsules for biomedical applications. *Chem. Soc. Rev.* **2014**, *43*, 1361–1371. [CrossRef]
94. Atkin, R.; Davies, P.; Hardy, J.; Vincent, B. Preparation of Aqueous Core/Polymer Shell Microcapsules by Internal Phase Separation. *Macromolecules* **2004**, *37*, 7979–7985. [CrossRef]
95. Dolçà, C.; Ferrándiz, M.; Capablanca, L.; Franco, E.; Mira, E.; López, F.; García, D. Microencapsulation of Rosemary Essential Oil by Co-Extrusion/Gelling Using Alginate as a Wall Material. *J. Encapsulation Adsorpt. Sci.* **2015**, *5*, 121–130. [CrossRef]
96. Wu, X.-B.; Peng, C.-H.; Huang, F.; Kuang, J.; Yu, S.-L.; Dong, Y.-D.; Han, B.-S. Preparation and characterization of chitosan porous microcarriers for hepatocyte culture. *Hepatobiliary Pancreat. Dis. Int.* **2011**, *10*, 509–515. [CrossRef]
97. Shepherd, S.J.; Issadore, D.; Mitchell, M.J. Microfluidic formulation of nanoparticles for biomedical applications. *Biomaterials* **2021**, *274*, 120826. [CrossRef]
98. Mendes, A.C.; Baran, E.T.; Lisboa, P.; Reis, R.L.; Azevedo, H.S. Microfluidic Fabrication of Self-Assembled Peptide-Polysaccharide Microcapsules as 3D Environments for Cell Culture. *Biomacromolecules* **2012**, *13*, 4039–4048. [CrossRef] [PubMed]
99. Duncanson, W.J.; Zieringer, M.; Wagner, O.; Wilking, J.N.; Abbaspourrad, A.; Haag, R.; Weitz, D.A. Microfluidic synthesis of monodisperse porous microspheres with size-tunable pores. *Soft Matter* **2012**, *8*, 10636. [CrossRef]
100. Paşcalu, V.; Soritau, O.; Popa, F.; Pavel, C.; Coman, V.; Perhaita, I.; Borodi, G.; Dirzu, N.; Tabaran, F.; Popa, C. Curcumin delivered through bovine serum albumin/polysaccharides multilayered microcapsules. *J. Biomater. Appl.* **2016**, *30*, 857–872. [CrossRef]
101. Zaeim, D.; Sarabi-Jamab, M.; Ghorani, B.; Kadkhodaee, R.; Liu, W.; Tromp, R.H. Microencapsulation of probiotics in multi-polysaccharide microcapsules by electro-hydrodynamic atomization and incorporation into ice-cream formulation. *Food Struct.* **2020**, *25*, 100147. [CrossRef]
102. Qu, J.; Wang, L.; Hu, Y.; Wang, L.; You, R.; Li, M. Preparation of Silk Fibroin Microspheres and Its Cytocompatibility. *J. Biomater. Nanobiotechnol.* **2013**, *4*, 84–90. [CrossRef]
103. Shuai, Y.; Yang, S.; Li, C.; Zhu, L.; Mao, C.; Yang, M. In situ protein-templated porous protein–hydroxylapatite nanocomposite microspheres for pH-dependent sustained anticancer drug release. *J. Mater. Chem. B* **2017**, *5*, 3945–3954. [CrossRef] [PubMed]
104. Skoll, K.; Ritschka, M.; Fuchs, S.; Wirth, M.; Gabor, F. Characterization of sonochemically prepared human serum albumin nanocapsules using different plant oils as core component for targeted drug delivery. *Ultrason. Sonochem.* **2021**, *76*, 105617. [CrossRef] [PubMed]
105. Mu, X.-T.; Ju, X.-J.; Zhang, L.; Huang, X.-B.; Faraj, Y.; Liu, Z.; Wang, W.; Xie, R.; Deng, Y.; Chu, L.-Y. Chitosan microcapsule membranes with nanoscale thickness for controlled release of drugs. *J. Membr. Sci.* **2019**, *590*, 117275. [CrossRef]
106. Rocha-Selmi, G.A.; Theodoro, A.C.; Thomazini, M.; Bolini, H.M.A.; Favaro-Trindade, C.S. Double emulsion stage prior to complex coacervation process for microencapsulation of sweetener sucralose. *J. Food Eng.* **2013**, *119*, 28–32. [CrossRef]
107. Meng, Q.; Zhong, S.; He, S.; Gao, Y.; Cui, X. Constructing of pH and reduction dual-responsive folic acid-modified hyaluronic acid-based microcapsules for dual-targeted drug delivery via sonochemical method. *Colloid Interface Sci. Commun.* **2021**, *44*, 100503. [CrossRef]
108. Li, X.; van der Gucht, J.; Erni, P.; de Vries, R. Core–Shell Microcapsules from Unpurified Legume Flours. *ACS Appl. Mater. Interfaces* **2021**, *13*, 37598–37608. [CrossRef] [PubMed]
109. Lan, Y.; Ohm, J.-B.; Chen, B.; Rao, J. Microencapsulation of hemp seed oil by pea protein isolate−sugar beet pectin complex coacervation: Influence of coacervation pH and wall/core ratio. *Food Hydrocoll.* **2021**, *113*, 106423. [CrossRef]
110. Hoshyar, N.; Gray, S.; Han, H.; Bao, G. The effect of nanoparticle size on in vivo pharmacokinetics and cellular interaction. *Nanomedicine* **2016**, *11*, 673–692. [CrossRef] [PubMed]
111. Kothamasu, P.; Kanumur, H.; Ravur, N.; Maddu, C.; Parasuramrajam, R.; Thangavel, S. Nanocapsules: The weapons for novel drug delivery systems. *BioImpacts* **2012**, *2*, 71–81. [CrossRef]
112. Palomo, M.E.; Ballesteros, M.P.; Frutos, P. Solvent and plasticizer influences on ethylcellulose-microcapsules. *J. Microencapsul.* **1996**, *13*, 307–318. [CrossRef]
113. Dunne, M.; Corrigan, O.I.; Ramtoola, Z. Influence of particle size and dissolution conditions on the degradation properties of polylactide-co-glycolide particles. *Biomaterials* **2000**, *21*, 1659–1668. [CrossRef]
114. Visscher, G.E.; Pearson, J.E.; Fong, J.W.; Argentieri, G.J.; Robison, R.L.; Maulding, H.V. Effect of particle size on thein vitro andin vivo degradation rates of poly(DL-lactide-co-glycolide) microcapsules. *J. Biomed. Mater. Res.* **1988**, *22*, 733–746. [CrossRef]
115. Saravanan, M.; Rao, K.P. Pectin–gelatin and alginate–gelatin complex coacervation for controlled drug delivery: Influence of anionic polysaccharides and drugs being encapsulated on physicochemical properties of microcapsules. *Carbohydr. Polym.* **2010**, *80*, 808–816. [CrossRef]
116. Jégat, C.; Taverdet, J.L. Stirring speed influence study on the microencapsulation process and on the drug release from microcapsules. *Polym. Bull.* **2000**, *44*, 345–351. [CrossRef]
117. Kristmundsdóttir, T.; Ingvarsdóttir, K. Influence of emulsifying agents on the properties of cellulose acetate butyrate and ethylcellulose microcapsules. *J. Microencapsul.* **1994**, *11*, 633–639. [CrossRef]

118. Valot, P.; Baba, M.; Nedelec, J.-M.; Sintes-Zydowicz, N. Effects of process parameters on the properties of biocompatible Ibuprofen-loaded microcapsules. *Int. J. Pharm.* **2009**, *369*, 53–63. [CrossRef] [PubMed]
119. Pomeranz, Y.; Meloan, C.E. Determination of Moisture. In *Food Analysis*; Springer: Boston, MA, USA, 1994; pp. 575–601.
120. Li, W.; Zhou, J.; Xu, Y. Study of the in vitro cytotoxicity testing of medical devices. *Biomed. Rep.* **2015**, *3*, 617–620. [CrossRef] [PubMed]
121. Urbán, P.; Liptrott, N.J.; Bremer, S. Overview of the blood compatibility of nanomedicines: A trend analysis of in vitro and in vivo studies. *WIREs Nanomed. Nanobiotechol.* **2019**, *11*, e1546. [CrossRef] [PubMed]
122. Raghavendra, G.M.; Varaprasad, K.; Jayaramudu, T. Biomaterials. In *Nanotechnology Applications for Tissue Engineering*; William Andrew: Oxford, UK, 2015; pp. 21–44.
123. D'Souza, S. A Review of In Vitro Drug Release Test Methods for Nano-Sized Dosage Forms. *Adv. Pharm.* **2014**, *2014*, 304757. [CrossRef] [PubMed]
124. Siepmann, J.; Siepmann, F. Mathematical modeling of drug delivery. *Int. J. Pharm.* **2008**, *364*, 328–343. [CrossRef]
125. Siepmann, J.; Siepmann, F. Modeling of diffusion controlled drug delivery. *J. Control. Release* **2012**, *161*, 351–362. [CrossRef]
126. Faisant, N.; Siepmann, J.; Richard, J.; Benoit, J.P. Mathematical modeling of drug release from bioerodible microparticles: Effect of gamma-irradiation. *Eur. J. Pharm. Biopharm.* **2003**, *56*, 271–279. [CrossRef]
127. Lee, P.I. Modeling of drug release from matrix systems involving moving boundaries: Approximate analytical solutions. *Int. J. Pharm.* **2011**, *418*, 18–27. [CrossRef] [PubMed]
128. Lao, L.L.; Peppas, N.A.; Boey, F.Y.C.; Venkatraman, S.S. Modeling of drug release from bulk-degrading polymers. *Int. J. Pharm.* **2011**, *418*, 28–41. [CrossRef] [PubMed]
129. Keraliya, R.A.; Patel, C.; Patel, P.; Keraliya, V.; Soni, T.G.; Patel, R.C.; Patel, M.M. Osmotic Drug Delivery System as a Part of Modified Release Dosage Form. *ISRN Pharm.* **2012**, *2012*, 528079. [CrossRef] [PubMed]
130. Arifin, D.Y.; Lee, L.Y.; Wang, C.H. Mathematical modeling and simulation of drug release from microspheres: Implications to drug delivery systems. *Adv. Drug Deliv. Rev.* **2006**, *58*, 1274–1325. [CrossRef] [PubMed]
131. Van Herck, H.; Baumans, V.; Brandt, C.J.W.M.; Boere, H.A.G.; Hesp, A.P.M.; van Lith, H.A.; Schurink, M.; Beynen, A.C. Blood sampling from the retro-orbital plexus, the saphenous vein and the tail vein in rats: Comparative effects on selected behavioural and blood variables. *Lab. Anim.* **2001**, *35*, 131–139. [CrossRef]
132. Iyer, C.; Kailasapathy, K.; Peiris, P. Evaluation of survival and release of encapsulated bacteria in ex vivo porcine gastrointestinal contents using a green fluorescent protein gene-labelled *E. coli*. *LWT Food Sci. Technol.* **2004**, *37*, 639–642. [CrossRef]
133. Yang, C.; Wang, J. Preparation and characterization of collagen microspheres for sustained release of steroidal saponins. *Mater. Res.* **2014**, *17*, 1644–1650. [CrossRef]
134. Yeung, P.; Sin, H.S.; Chan, S.; Chan, G.C.F.; Chan, B.P. Microencapsulation of Neuroblastoma Cells and Mesenchymal Stromal Cells in Collagen Microspheres: A 3D Model for Cancer Cell Niche Study. *PLoS ONE* **2015**, *10*, e0144139. [CrossRef]
135. Yeung, P.; Cheng, K.H.; Yan, C.H.; Chan, B.P. Collagen microsphere based 3D culture system for human osteoarthritis chondrocytes (hOACs). *Sci. Rep.* **2019**, *9*, 12453. [CrossRef]
136. Wittig, O.; Romano, E.; González, C.; Diaz-Solano, D.; Marquez, M.E.; Tovar, P.; Aoun, R.; Cardier, J.E. A method of treatment for nonunion after fractures using mesenchymal stromal cells loaded on collagen microspheres and incorporated into platelet-rich plasma clots. *Int. Orthop.* **2016**, *40*, 1033–1038. [CrossRef] [PubMed]
137. Wittig, O.; Diaz-Solano, D.; Cardier, J. Viability and functionality of mesenchymal stromal cells loaded on collagen microspheres and incorporated into plasma clots for orthopaedic application: Effect of storage conditions. *Injury* **2018**, *49*, 1052–1057. [CrossRef] [PubMed]

polymers MDPI

Article

Application of Redox-Responsive Hydrogels Based on 2,2,6,6-Tetramethyl-1-Piperidinyloxy Methacrylate and Oligo(Ethyleneglycol) Methacrylate in Controlled Release and Catalysis

Miriam Khodeir, He Jia, Alexandru Vlad and Jean-François Gohy *

Institute of Condensed Matter and Nanosciences (IMCN), Université catholique de Louvain, Place L. Pasteur 1, B-1348 Louvain-la-Neuve, Belgium; miriam.khodeir92@gmail.com (M.K.); he.jia@uclouvain.be (H.J.); alexandru.vlad@uclouvain.be (A.V.)
* Correspondence: jean-francois.gohy@uclouvain.be

Abstract: Hydrogels have reached momentum due to their potential application in a variety of fields including their ability to deliver active molecules upon application of a specific chemical or physical stimulus and to act as easily recyclable catalysts in a green chemistry approach. In this paper, we demonstrate that the same redox-responsive hydrogels based on polymer networks containing 2,2,6,6-tetramethyl-1-piperidinyloxy (TEMPO) stable nitroxide radicals and oligoethylene glycol methyl ether methacrylate (OEGMA) can be successfully used either for the electrochemically triggered release of aspirin or as catalysts for the oxidation of primary alcohols into aldehydes. For the first application, we take the opportunity of the positive charges present on the oxoammonium groups of oxidized TEMPO to encapsulate negatively charged aspirin molecules. The further electrochemical reduction of oxoammonium groups into nitroxide radicals triggers the release of aspirin molecules. For the second application, our hydrogels are swelled with benzylic alcohol and tert-butyl nitrite as co-catalyst and the temperature is raised to 50 °C to start the oxidation reaction. Interestingly enough, benzaldehyde is not miscible with our hydrogels and phase-separate on top of them allowing the easy recovery of the reaction product and the recyclability of the hydrogel catalyst.

Keywords: hydrogels; redox-responsive polymers; TEMPO; encapsulation-release; catalysis

Citation: Khodeir, M.; Jia, H.; Vlad, A.; Gohy, J.-F. Application of Redox-Responsive Hydrogels Based on 2,2,6,6-Tetramethyl-1-Piperidinyloxy Methacrylate and Oligo(Ethyleneglycol) Methacrylate in Controlled Release and Catalysis. *Polymers* **2021**, *13*, 1307. https://doi.org/10.3390/polym13081307

Academic Editor: M. Ali Aboudzadeh and Shaghayegh Hamzehlou

Received: 18 March 2021
Accepted: 14 April 2021
Published: 16 April 2021

1. Introduction

Hydrogels are three-dimensional, hydrophilic, polymeric networks capable of swelling in aqueous medium and resembling to some extent living tissues [1]. They may be chemically stable or they may degrade or dissolve under specific conditions [2]. When the polymer networks are held together via polymer chain entanglements, crystallites or non-covalent interactions including hydrophobic interactions, Van der Waals interactions, hydrogen bonds and ionic forces, they are referred to as physical gels [3]. The so-called permanent or chemical hydrogels are obtained from covalently-crosslinked networks and generally present a better homogeneity than physical gels [2]. Since the seminal publication of Wichterle and Lim [4], chemical hydrogels have gained attention owing to their unique characteristics, especially when they are originating from stimuli-responsive networks that can dynamically and reversibly alter their structure and properties in response to changes in the environment [5]. Hydrogels are utilized in many areas and more specifically in the biomedical field where they can act as drug protectors, targetable carriers for bioactive drugs etc. [2,3,6,7]. As far as biomedical applications are concerned, hydrogels derived from bio-based polymers are particularly attracting since those polymers are generally biocompatible and biodegradable and often show a high level of biomimicry, a highly desired characteristic for in vivo applications [8]. Moreover, in order to reach the desired properties with ever increasing complexity, polymer-based hydrogels may be mixed with other polymeric or non-polymeric components to form composite hydrogel systems, including e.g., polymer nanoparticles, electrospun fibers, nanocarbons, etc. [8–10].

In the present contribution, we focus on stimuli-responsive chemical hydrogels. Typical stimuli include variations in temperature [11], pH [11], applied stress [12], magnetic and electromagnetic field [13], ionic strength [14], light [15] and the presence of bioactive species [16]. Redox-responsive gels have been scarcely reported. Typically, species which may undergo reversible oxidation-reduction reactions are good candidates to achieve redox responsiveness in hydrogels [17]. The design of redox active networks usually involves the incorporation of redox responsive groups either in the polymer main chain, in the side groups or as cross-linking moieties [17,18]. From a practical point of view, redox stimuli may be applied chemically or electrochemically. This last possibility is particularly promising since the addition of reactants to the hydrogel is not required in order to observe the redox-responsive behavior. However, the hydrogel should display a sufficient electric conductivity in order to be electrochemically addressed.

As far as redox groups are concerned, 2,2,6,6-tetramethyl-1-piperidinyloxy (TEMPO) is a very interesting candidate. TEMPO is a stable nitroxide radical that can be easily oxidized into an oxoammonium cation or reduced into an aminoxyl group [19]. Moreover, TEMPO derivatives are largely used in chemistry as catalysts [20], and in the biomedical field as imaging enhancers in electron spin resonance techniques or as a radical scavenger of reactive oxygen species and are, therefore, considered valuable candidates for anti-oxidant therapies [21]. Moreover, the redox equilibrium associated with nitroxide radicals, and especially TEMPO, has been recently used for the production of energy storage devices [22,23]. For the energy storage application, scientists developed a poly(methacrylate) bearing TEMPO moieties as side groups. In our recent previous works [24,25], we have described and characterized new redox-responsive hydrogels based on polymer networks containing the TEMPO-containing methacrylate and oligoethylene glycol methyl ether methacrylate (OEGMA). TEMPO groups can undergo reversible oxidation-reduction reactions that lead to redox activity while OEGMA groups are hydrophilic and allow water swelling in order to obtain hydrogels. In the present contribution, the use of those hydrogels for two different applications is presented.

The first application deals with the release of a guest molecule, namely aspirin, in response to an electrochemical redox trigger. The advantage of the electrochemical stimulus is that it can be localized in time, it does not require the addition of reagents and the trigger parameters, such as current intensity and reaction time, and it can easily by modulated to adequately comply with the system [26,27]. Therefore, the amount of released guest molecules can be controlled and realized on demand.

The second application is based on the well-known use of TEMPO groups as catalysts for the oxidation of alcohols to obtain ketones, aldehydes or carboxylic acids [28,29]. The oxidation of alcohols using TEMPO-based catalysts is often efficient, fast, selective, realized in mild conditions and can tolerate sensitive functional groups [28,29]. However, the difficulty of catalyst recycling as well as the need for organic solvents and transition metal co-oxidants are limiting the application of TEMPO-based catalysts [30]. Here, we report a methodology using our hydrogels containing TEMPO groups as catalysts for the oxidation of benzyl alcohol into benzaldehyde in aqueous medium. The applied methodology is inspired by the previous work of Karimi et al. [31]. Such an application follows some of the principles of green chemistry since water is used as solvent and no metallic co-catalysts are needed. Moreover, it allows a very easy purification of the product of the reaction since the latter phase separates from the hydrogel. Therefore, the TEMPO-containing hydrogel can be easily recycled after the reaction to be used again.

2. Materials and Methods

2.1. Materials

All chemicals were purchased from Sigma-Aldrich (Overijse, Belgium), Acros (Geel, Belgium) and TCI (Zwijndrecht, Belgium). 2,2,6,6-Tetramethyl-1-piperidyl methacrylate

(TMPM, 98%, TCI), oligo(ethylene glycol) methyl ether methacrylate (OEGMA, average molar mass of 300 g/mol, Sigma-Aldrich) and di(ethylene glycol) dimethacrylate ($OEGMA_2$, 98%, Sigma-Aldrich) were purified on a AlO_x-filtration column prior use in order to remove the inhibitor. The 2,2′-azobisisobutyronitrile (AIBN, 98% purity, Sigma Aldrich) initiator was recrystallized twice from methanol (Acros, 99%) prior use. Acetylsalicylic acid (ASA, 99.9%, Sigma-Aldrich), isopropanol (IPA, Acros, 99.5%), methanol (Acros, 99.8%), diethyl ether (Acros, 99%), $NaHCO_3$ (Sigma Aldrich, 99%), $Na_2WO_4.2H_2O$ (Sigma-Aldrich, 99%), H_2O_2 (Sigma-Aldrich, 30 wt% solution in water), ethylene diamine tetraacetic acid disodium salt dihydrate (EDTA, Sigma Aldrich, 99%), NaClO (Sigma Aldrich, 99%), HBF_4 (Sigma Aldrich, 48 wt% solution in water), $NaClO_4$ (Sigma Aldrich, 98%), acetonitrile (ACN, 99.9%, Sigma Aldrich), formic acid (98%, Sigma Aldrich), benzylic alcohol (Sigma Aldrich, 99%) and tert-butyl nitrite (Sigma Aldrich, 90%) were used as received.

2.2. Synthesis of Hydrogels

The investigated poly(2,2,6,6-tetramethyl-1-piperidinyloxo ammonium methacrylate-random-oligo(ethylene glycol) methyl ether methacrylate), further abbreviated as P($TEMPO^+$-*r*-OEGMA), hydrogels have been synthesized via a methodology described in our previous work [24]. Briefly, TMPM was dissolved in isopropanol into a round-bottom flask. The required amount of OEGMA, $OEGMA_2$ and the initiator AIBN (0.5 eq.) were then added and stirred (Figure 1). The molar ratio of TMPM/(TMPM+OEGMA) (abbreviated as X_{TEMPO}) was set to 0.2 and the cross-linker molar ratio $OEGMA_2$/(TMPM+OEGMA) (abbreviated as X_{CL}) was set to 0.03. The solution was then degassed by three freeze pump–thaw cycles and filled with argon before stirring in an oil bath at 70 °C overnight to lead to a transparent material. The TMPM units were then oxidized into TEMPO nitroxide radicals using Na_2WO_4 (0.25 eq.), EDTA (0.15 eq.) and H_2O_2 (5 eq.) in methanol (see Figure 1). The mixture was then stirred at 60 °C overnight. An orange colored gel was obtained, washed four times with distilled water and methanol (1:1, *v:v*) and dried in vacuum at 40 °C overnight. An orange sticky material corresponding to dried P(TEMPO-*r*-OEGMA) was finally obtained. For the oxidation of TEMPO into $TEMPO^+$ (Figure 1), the dried P(TEMPO-*r*-OEGMA) material was swollen in distilled H_2O (31.25 eq.). HBF_4 (1 eq.) was then slowly added at room temperature, followed by the addition of NaClO (0.5 eq.) at 0 °C and additional stirring for 1 h at 0 °C. The oxidized P($TEMPO^+$-*r*-OEGMA) obtained was washed with ice-cold 5 wt% $NaHCO_3$ aqueous solution and ice-cold diethyl ether. The yellow material obtained was finally dried overnight at 40 °C in vacuum. The final hydrogels were obtained by swelling the P($TEMPO^+$-*r*-OEGMA) hydrogels in distilled water for 48 h to be sure to reach the swelling equilibrium. The excess of water was removed and the remaining hydrogel was used for further experiments.

2.3. Electrochemical Measurements

All electrochemical experiments were performed at room temperature using 0.1 M $NaClO_4$ as supporting electrolyte on a Biologic VMP300. 5 μL of sample were dropped on the working electrode (carbon screen printed electrode DRP-110-U75, Metrohm, Antwerpen, Belgium) and, after 60 s, cyclic voltammetry was performed in the potential range that comprised between 0.2 and 0.6 V at a scan rate of 10 mV/s. Measurements were repeated 5 times to ensure reproducibility.

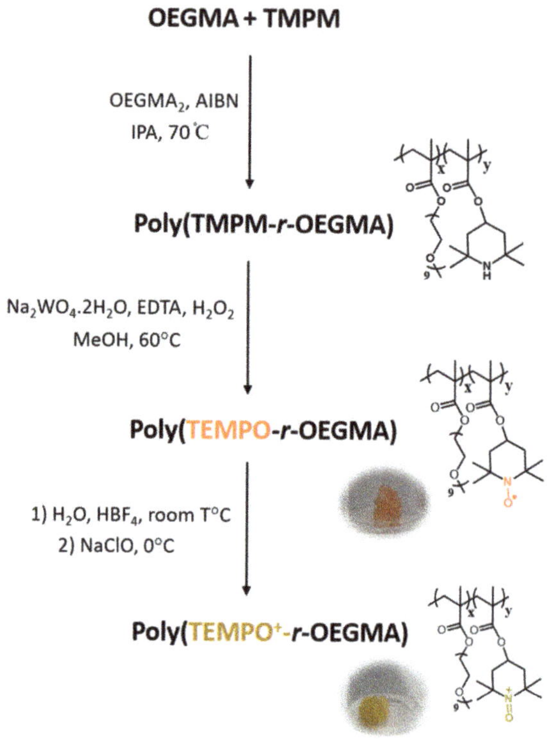

Figure 1. Synthesis of the poly(2,2,6,6-tetramethyl-1-piperidinyloxo ammonium methacrylate-*random*-oligo(ethylene glycol) methyl ether methacrylate) (P(TEMPO-*r*-OEGMA)) and oxidized P(TEMPO-*r*-OEGMA) (P(TEMPO$^+$-*r*-OEGMA)) hydrogels.

2.4. *Ultra-High-Performance Liquid Chromatography (UHPLC)–Electrospray Ionization (ESI) Mass Spectrometry Analysis*

An ultra-high-performance liquid chromatography (UHPLC) system (ThermoFisher Scientific, Merelbeke, Belgium) consisting of a binary pump, an automatic injector, a column oven and an Agilent (Machelen, Belgium) 1290 series ultraviolet (UV) detector was used. The separation was carried out on an Eclipse plus C18 rapid resolution high definition (RRHD) column (100 × 2.1 mm, 1.8 µm) at a flow rate of 0.2 mL/min and using an aqueous solution containing 1.0% of formic acid and 5% ACN (solvent A) and ACN containing 0.1% formic acid (solvent B). The elution program was started at 90% solvent A for 1 min, then 90% solvent B for 6 min at a flow rate of 0.2 mL/min with UV detection at 280 nm. The injection volume was kept at 5 µL for the standard and all the other samples. The detection in mass spectrometry was carried out by an electrospray ionization (ESI) jet stream Agilent (Machelen, Belgium) 6150B mass spectrometer. The analysis was carried out at 200 °C with a capillary voltage of 1500 eV and a voltage nozzle at 2000 eV. The parameters were optimized for the detection of acetylsalicylic acid as follows: nebulization pressure at 50 psi, drying gas at 4 L/min and power of fragmentary set at 75 eV. The retention time of aspirin was observed at 5.52 ± 0.1 min for MS and 5.32 ± 0.1 min for UV. An aspirin standard (Sigma Aldrich, Overijse, Belgium) was dissolved in ACN containing 0.1% formic acid. A calibration curve ranging from 15 µM (2.7 ppm) to 301 µM (54.2 ppm) was injected. The aspirin was followed in LC-MS in SIM (-) mode for m/z of 179 and 225 and in UV at 280 nm. 30 µL of samples were diluted with 100 µL of water (dilution factor 4.33). Measurements were repeated 3 times to ensure reproducibility.

2.5. Gas Chromatography (GC)–Mass Spectrometry (MS) Analysis

The chromatographic separation of benzaldehyde and benzylic alcohol was performed using a Thermo Scientific (Merelbeke, Belgium) TRACE 1310 GC (gas chromatograph) coupled with a Thermo Scientific (Merelbeke, Belgium) single quadrupole ISQ™ QD™ mass spectrometer (MS). The GS was equipped with a RESTEK™ Rxi®-5Sil MS column (L = 30 m, d_c = 0.25 mm, and d_f = 0.25 μm). The GC temperature program started at a temperature of 60 °C, ramped to 300 °C (20 °C/min, held 5 min) with a constant flow of 2 mL/min, resulting in an overall analysis time of 17 min. Injection mode with split ratio of 1/10 was used. The MS was operated with source of mass at 305 °C and full scan mode from 40 to 400 m/z. Measurements were repeated 3 times to ensure reproducibility.

2.6. Fourier-Transform Infrared (FTIR)

Infrared spectra were collected on a Thermo Scientific (Merelbeke, Belgium) Nicolet 6700 Fourier transform infrared (FTIR) spectrometer. For the kinetic study, a probe 6.3 mm AgX DiComp (Au, Diamond, C22) was connected to a ReactIR 15 from Mettler Toledo (Zaventem, Belgium). The probe was introduced in an Easymax reactor from Mettler Toledo (Zaventem, Belgium). Measurements were repeated 3 times to ensure reproducibility.

3. Results

3.1. Synthesis of Hydrogels

P(TEMPO$^+$-*r*-OEGMA) hydrogels with a molar fraction of TEMPO$^+$ monomer equal to 0.2 and a molar fraction of crosslinker equal to 0.03 were synthesized (Figure 1). Briefly, a P(TMPM-*r*-OEGMA) precursor hydrogel was first prepared by conventional radical copolymerization of TMPM, OEGMA and OEGMA$_2$ (OEGMA$_2$ playing the role of chemical crosslinking agent since it contains two polymerizable double bonds). This was followed by the oxidation of the secondary amine of TMPM units with H_2O_2 and Na_2WO_4 in methanol to obtain TEMPO units. Finally, the nitroxide radical units of TEMPO were oxidized into oxoammonium groups (TEMPO$^+$) with NaClO in the presence of HBF_4. This reaction can be macroscopically followed by the change of color of the hydrogel from orange to yellow (Figure 1) and is confirmed by FTIR spectroscopy with a characteristic N–O vibration at 1540 cm^{-1} for P(TEMPO-*r*-OEGMA) and a characteristic N=O vibration at 1570 cm^{-1} for P(TEMPO$^+$-*r*-OEGMA).

3.2. Encapsulation-Release of Aspirin from P(TEMPO$^+$-r-OEGMA) Hydrogels

In order to demonstrate the encapsulation abilities of P(TEMPO$^+$-*r*-OEGMA) hydrogels, we have designed a proof-of-concept experiment by taking opportunity of the presence of positively charged units in the oxidized P(TEMPO$^+$-*r*-OEGMA) hydrogels to encapsulate a negatively charged drug, namely ASA, commonly known as aspirin. In a first step, the equilibrium swelling of the P(TEMPO$^+$-*r*-OEGMA) hydrogel was realized with aspirin (0.1 g/L) dissolved in 0.1 M aqueous $NaClO_4$ solution (at this pH close to 7 the aspirin is mainly negatively charged since its carboxylic group is in the carboxylate anion form). After the swelling step, the hydrogel was rinsed with water and a small amount of the 1 M $NaClO_4$ solution was deposited on top of the aspirin loaded P(TEMPO$^+$-*r*-OEGMA) hydrogel that was further equilibrated for 12 h (Figure 2a). Afterwards, the supernatant of the hydrogel has been analyzed by mass spectrometry and no aspirin molecules were detected confirming their strong electrostatic encapsulation in the hydrogel and the fact that no aspirin molecules are diffusing out of the hydrogel with time (Figure 2a).

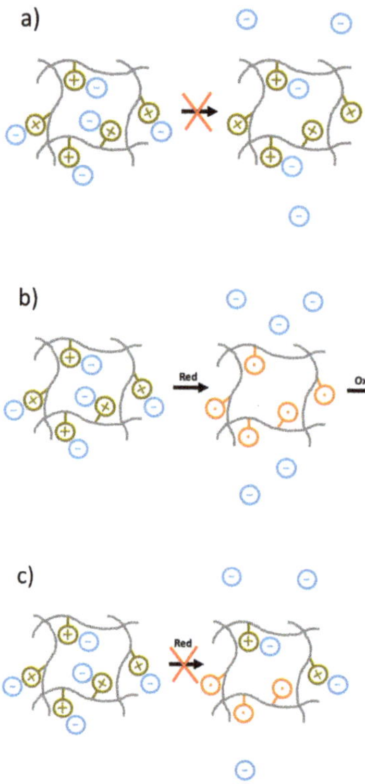

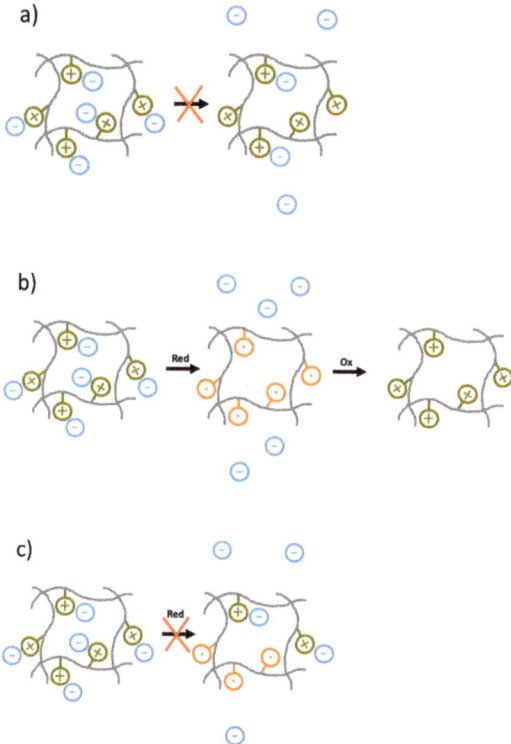

Figure 2. Sketches for encapsulation-release of aspirin from P(TEMPO$^+$-*r*-OEGMA) hydrogels (grey lines stand for polymer chains, negatively charged aspirin molecules are represented in blue, positively charged oxoammonium are represented in yellow-greenish and nitroxide radicals are in orange). (**a**) Firmly attached aspirin do not diffuse out of the positively charged oxidized hydrogel. (**b**) The whole aspirin molecules are released from the hydrogel upon electrochemical reduction and the re-oxidized hydrogel contains no longer aspirin molecules. (**c**) Incomplete release (e.g., due to incomplete reduction of oxoammonium cations) is not observed.

In order to electrochemically address the aspirin-loaded P(TEMPO$^+$-*r*-OEGMA) hydrogel, a small piece (2 mm^2) of the hydrogel was cut and brought in direct contact with a printed carbon electrode to perform the electrochemical reduction of TEMPO$^+$ into TEMPO (Figure S1). Cyclic voltammetry measurements (CV) for the P(TEMPO$^+$-*r*-OEGMA) hydrogel loaded with aspirin show clearly the reversible reduction of the oxoammonium cations into nitroxide radicals (Figure 3).

In order to demonstrate the release of aspirin molecules during the reduction step, the supernatant aqueous solution above the hydrogel was analyzed by mass spectrometry and the presence of aspirin molecules released from the hydrogel was detected (Figure 2b, and green mass spectrum in Figure S2). In a second step, the electrochemical oxidation of the P(TEMPO-*r*-OEGMA) hydrogel into P(TEMPO$^+$-*r*-OEGMA) was performed and followed by CV (Figure 3). Once again, the supernatant aqueous solution above the hydrogel was analyzed by mass spectrometry but this time no aspirin molecules were detected (Figure 2b and blue mass spectrum in Figure S2). In order to demonstrate that the scenario depicted in Figure 2b is operating and that no aspirin molecules remain trapped into the hydrogel because of e.g., incomplete reduction of oxoammonium cations (see sketch depicted in Figure 2c), UHPLC has been used to determine the amount of aspirin released in the supernatant of the P(TEMPO$^+$-*r*-OEGMA) hydrogels after the electrochemical reduction.

The determined concentration of aspirin molecules was 1372 µM (247 ppm), for a total surface of 6.9 mm^2 (surface of the electrode) and a cut gel surface of 2 mm^2. The further calculation indicated that 100% of aspirin molecules were released in the whole supernatant, proving the efficiency of the electrochemically triggered release process in agreement with the sketch depicted in Figure 2b.

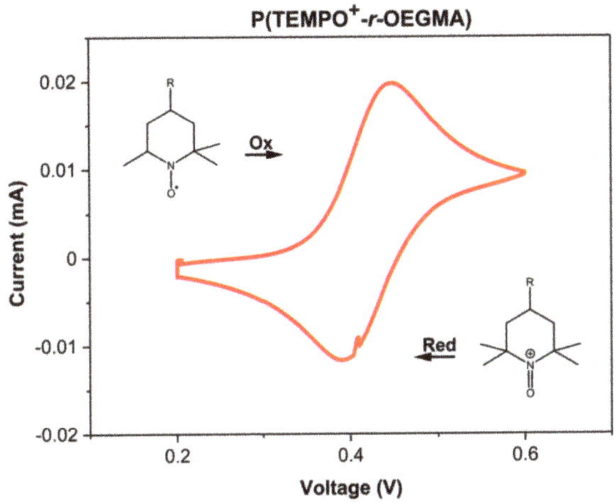

Figure 3. Cyclic voltammetry of P(TEMPO+-*r*-OEGMA) hydrogels. Analyses conducted in 0.1 M $NaClO_4$ in distilled water at a carbon electrode. Scan rate = 5 mV s^{-1}.

3.3. *P(TEMPO-r-OEGMA) Hydrogels as Catalytic Scaffolds for the Oxidation of Alcohols*

The catalytic activity of P(TEMPO-*r*-OEGMA) hydrogels has been determined by monitoring the oxidation of benzylic alcohol under aerobic condition (Figure 4a). Practically, the reaction was started by mixing in a 10 mL round-bottom flask the dry gel (153 mg) and all the reactants (benzylic alcohol, 0.52 mL; tert-butyl nitrite (TBN) as a metal-free co-catalyst, 5 mol% and water, 1.5 mL) together and stirring for 2 h at room temperature. In a second step, the temperature was raised to 50 °C to start the oxidation reaction for 4 h. Then, the reaction was stopped by let it cool down to room temperature for 30 min and the supernatant of the gel was analyzed.

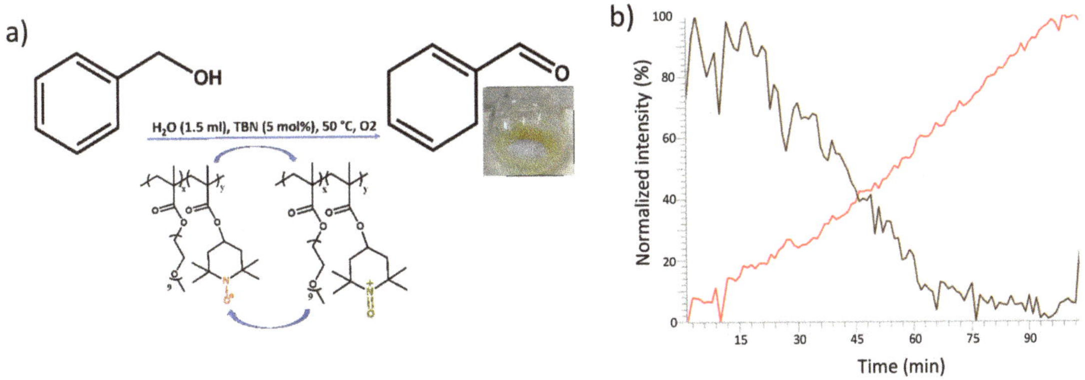

Figure 4. (**a**) Reaction conditions for benzyl alcohol oxidation. (**b**) Evolution of the normalized integration of the infrared spectra of benzylic alcohol and benzaldehyde during the oxidation reaction of benzyl alcohol (black curve) into benzaldehyde (red curve) over time.

GC-MS analysis of the products of the reaction confirmed quantitative benzylic alcohol conversion into benzaldehyde. It was found that 100% of benzyl alcohol was converted into benzaldehyde (Figure S3). The reaction conditions and corresponding conversion rates of different experiments for the oxidation of benzyl alcohol to benzaldehyde in water based on TEMPO catalysts have been summarized in Table 1.

Table 1. Different reaction conditions for the oxidation of benzyl alcohol into benzaldehyde in water with 2,2,6,6-tetramethyl-1-piperidinyloxy (TEMPO)-based catalysts.

Entry	TEMPO [mol%]	Co-Catalyst	Condition	T (°C)	t (hour)	Conversion (%)	Reference
1	0.2	TBN *	O_2	50	1.5	100	This work
2	0.2	TBN	O_2	50	4	100	[31]
3	0.3	$NaNO_2$, DBDMH *	Air (0.9 MPa)	80	1.5	99.8	[28]
4	5	Cu, K_2CO_3	Air	40	24	100	[32]
5	0.68	Co	Air	RT	0.5	98	[33]

* DBDMH: 1,3-dibromo-5,5-dimethylhydantoin, TBN: tert-butyl nitrite

Compared with other representative experimental conditions, without the addition of transition metal co-catalysts, our P(TEMPO-*r*-OEGMA) hydrogels still maintain competitive catalytic efficiency and conversion rate. Such a high catalytic efficiency of the P(TEMPO-*r*-OEGMA) hydrogel could be mainly due to its composition allowing an important and uniform uptake of benzylic alcohol inside the hydrogel. In this respect, a homogenous system is observed when the P(TEMPO-*r*-OEGMA) hydrogel is added with the reactants and equilibrated. Benzylic alcohol was absorbed inside the matrix and as the reaction starts to proceed, a supernatant appears on top of the hydrogel consisting of benzaldehyde that phase separates from the hydrogel. This last characteristic feature is very interesting since it allows the easy recovery of the product of the reaction and the easy recycling of the TEMPO catalyst.

In order to obtain more information about the kinetics of the oxidation reaction of benzylic alcohol into benzaldehyde, an in situ infrared monitoring of the reaction was realized. Practically, the reaction was performed using the same experimental conditions as described above in the presence of an infrared probe immersed in the reaction medium. The recording of the infrared probe was started when temperature was raised to 50 °C to start the oxidation reaction. The formation of benzaldehyde is followed in the infrared spectra by the peaks at 1700, 1167 and 1204 cm^{-1} corresponding to vibrations associated with the aldehyde function. On the other hand, the peaks corresponding to the benzylic alcohol molecule in the area between 1605 and 1660 cm^{-1} are linearly vanishing with time. The recorded data allow the deduction of two main trends corresponding to a linear disappearance of benzylic alcohol and a concomitant linear formation of benzaldehyde over time (Figure 4). Moreover, they also indicate the complete conversion of benzylic alcohol into benzaldehyde after 90 min (Figure 4). Those results indicate a zero order for the kinetics of the reaction as is often the case for diffusion limited reactions due to the presence of a gel [34,35].

4. Conclusions

Redox-responsive hydrogels represent a highly interesting and versatile class of materials for loading and release applications in the biomedical field. In this paper, a new class of redox-responsive hydrogel is developed by combining TEMPO stable nitroxide radicals and OEGMA. TEMPO groups can undergo reversible oxidation-reduction reactions that lead to redox activity while OEGMA groups are hydrophilic and allow water swelling to obtain hydrogels. In this contribution, we have taken advantage of the positive charge present on the oxidized form of TEMPO (oxoammonium cations in $TEMPO^+$) to electrostatically complex a negatively charged aspirin molecule. We have demonstrated that aspirin is tightly bound to $TEMPO^+$ and cannot diffuse out the hydrogel. However, when $TEMPO^+$

groups are electrochemically reduced into TEMPO radicals, the electrostatic interaction between aspirin and $TEMPO^+$ disappears which allows the release of the aspirin molecule from the hydrogel. Furthermore, by using UHPLC, we have demonstrated that aspirin can be quantitatively released from the hydrogel. Finally, it should be pointed out that a very simple carbon printed electrode has been used to electrochemically trigger our hydrogels, proving that this set-up could be easily utilized for real-life applications. Finally, we have demonstrated the efficiency of our hydrogels as TEMPO-based catalysts for the oxidation of benzylic alcohols to the corresponding aldehydes under aerobic conditions. Moreover, the final products are easily separated from the hydrogel starting materials without using toxic organic solvents or complicated and costly processes.

Supplementary Materials: The following are available online at https://www.mdpi.com/article/10.3390/polym13081307/s1, Figure S1: Experimental set-up for the electrochemical reduction of P($TEMPO^+$-*r*-OEGMA) hydrogels, swollen in aqueous $NaClO_4$ and loaded with aspirin using the printed carbon electrode method (PCE); Figure S2: Mass spectroscopy analysis of the supernatant for the oxidation step (blue curve), reduction step (green curve) and the aspirin reference molecule (orange curve); Figure S3: GC-MS spectrum of the solution obtained after oxidation of benzylic alcohol into benzaldehyde.

Author Contributions: Conceptualization, J.-F.G.; methodology, M.K., A.V., H.J. and J.-F.G.; investigation, M.K. and H.J.; resources, J.-F.G.; data curation, M.K. and H.J.; writing—original draft preparation, M.K.; writing—review and editing, H.J. and J.-F.G.; supervision, J.-F.G.; project administration, J.-F.G.; funding acquisition, J.-F.G. All authors have read and agreed to the published version of the manuscript.

Funding: This research was funded by the UCLouvain International Relation Offices (ADRI) and by the UCLouvain Concerted Research Action BATTAB grant number 14/19-057.

Institutional Review Board Statement: Not applicable.

Informed Consent Statement: Not applicable.

Data Availability Statement: The data presented in this study are available on request from the corresponding author.

Conflicts of Interest: The authors declare no conflict of interest.

References

1. Das, N. Preparation methods and properties of hydrogel: A review. *Int. J. Pharm. Pharm. Sci.* **2013**, *5*, 112–117.
2. Hoffman, A.S. Hydrogels for biomedical applications. *Adv. Drug Deliv. Rev.* **2012**, *64*, 18–23. [CrossRef]
3. Chai, G.; Jiao, Y.; Yu, X. Hydrogels for biomedical applications: Their characteristics and the mechanisms behind them. *Gels* **2017**, *3*, 6. [CrossRef] [PubMed]
4. Wichterle, O.; Lim, D. Hydrophilic gels in biologic use. *Nature* **1960**, *185*, 117–118. [CrossRef]
5. Echeverria, C.; Fernandes, S.; Godinho, M.; Borges, J.; Soares, P. Functional stimuli-responsive gels: Hydrogels and microgels. *Gels* **2018**, *4*, 54. [CrossRef]
6. Liu, F.; Urban, M.W. Recent advances and challenges in designing stimuli-responsive polymers. *Prog. Polym. Sci.* **2010**, *35*, 3–23. [CrossRef]
7. Stuart, M.A.C.; Huck, W.T.S.; Genzer, J.; Müller, M.; Ober, C.; Stamm, M.; Sukhorukov, G.B.; Szleifer, I.; Tsukruk, V.V.; Urban, M.; et al. Emerging applications of stimuli-responsive polymer materials. *Nat. Mater.* **2010**, *9*, 101–113. [CrossRef]
8. Buwalda, S.J. Bio-based composite hydrogels for biomedical applications. *Multifunct. Mater.* **2020**, *3*, 022001. [CrossRef]
9. Rohani Rad, E.; Vahabi, H.; Formela, K.; Saeb, M.R.; Thomas, S. Injectable poloxamer/graphene oxide hydrogels with well-controlled mechanical and rheological properties. *Polym. Adv. Technol.* **2019**, *30*, 2250–2260. [CrossRef]
10. Rastin, H.; Zhang, B.; Mazinani, A.; Hassan, K.; Bi, J.; Tung, T.; Losic, D. 3D bioprinting of cell-laden electroconductive MXene nanocomposite bioinks. *Nanoscale* **2020**, *12*, 16069–16080. [CrossRef]
11. Rezanejade, G.; Ghavami, S.; Sadat, S. A review on pH and temperature responsive gels in drug delivery. *J. Chem Rev.* **2020**, *2*, 80–89. [CrossRef]
12. Gupta, T.; Pradhan, A.; Bandyopadhyay-Ghosh, S.; Ghosh, S.B. Thermally exfoliated graphene oxide reinforced stress responsive conductive nanocomposite hydrogel. *Polym. Adv. Technol.* **2019**, *30*, 2392–2401. [CrossRef]
13. Reddy, N.N.; Mohan, Y.M.; Varaprasad, K.; Ravindra, S.; Joy, P.A.; Raju, K.M. Magnetic and electric responsive hydrogel–magnetic nanocomposites for drug-delivery application. *J. Appl. Polym. Sci.* **2011**, *122*, 1364–1375. [CrossRef]

14. Zhou, S.; Wu, B.; Zhou, Q.; Jian, Y.; Le, X.; Lu, H. Ionic strength and thermal dual-responsive bilayer hollow spherical hydrogel actuator. *Macromol. Rapid Commun.* **2020**, *41*, 1900543. [CrossRef]
15. Tao, N.; Zhang, D.; Li, X.; Lou, D.; Sun, X.; Wei, C.; Li, J.; Yang, J.; Liu, Y.N. Near-infrared light-responsive hydrogels via peroxide-decorated MXene-initiated polymerization. *Chem. Sci.* **2019**, *10*, 10765–10771. [CrossRef]
16. Ahn, S.K.; Kasi, R.M.; Kim, S.C.; Sharma, N.; Zhou, Y. Stimuli-responsive polymer gels. *Soft Matter* **2008**, *4*, 1151–1157. [CrossRef]
17. Sui, X.; Feng, X.; Hempenius, M.A.; Vancso, G.J. Redox active gels: Synthesis, structures and applications. *J. Mater. Chem. B* **2013**, *1*, 1658–1672. [CrossRef]
18. Zoetebier, B.; Hempenius, M.A.; Vancso, G.J. Redox-responsive organometallic hydrogels for in situ metal nanoparticle synthesis. *Chem. Commun.* **2015**, *51*, 636–639. [CrossRef]
19. Blinco, J.P.; Hodgson, J.L.; Morrow, B.J.; Walker, J.R.; Will, G.D.; Coote, M.L.; Bottle, S.E. Experimental and theoretical studies of the redox potentials of cyclic nitroxides. *J. Org. Chem.* **2008**, *73*, 6763–6771. [CrossRef]
20. Hansen, K.A.; Blinco, J.P. Nitroxide radical polymers—A versatile material class for high-tech applications. *Polym. Chem.* **2018**, *9*, 1479–1516. [CrossRef]
21. Yoshitomi, T.; Kuramochi, K.; Binh Vong, L.; Nagasaki, Y. Development of nitroxide radicals–containing polymer for scavenging reactive oxygen species from cigarette smoke. *Sci. Technol. Adv. Mater.* **2014**, *15*, 035002. [CrossRef] [PubMed]
22. Janoschka, T.; Morgenstern, S.; Hiller, H.; Friebe, C.; Wolkersdörfer, K.; Häupler, B.; Hager, M.D.; Schubert, U.S. Synthesis and characterization of TEMPO- and viologen-polymers for water-based redox-flow batteries. *Polym. Chem.* **2015**, *6*, 7801–7811. [CrossRef]
23. Vlad, A.; Singh, N.; Rolland, J.; Melinte, S.; Ajayan, P.M.; Gohy, J.F. Hybrid supercapacitor-battery materials for fast electrochemical charge storage. *Sci. Rep.* **2014**, *4*, 4315. [CrossRef]
24. Khodeir, M.; Ernould, B.; Brassinne, J.; Ghiassinejad, S.; Jia, H.; Antoun, S.; Friebe, C.; Schubert, U.S.; Kochovski, Z.; Lu, Y.; et al. Synthesis and characterisation of redox hydrogels based on stable nitroxide radicals. *Soft Matter* **2019**, *15*, 6418–6426. [CrossRef] [PubMed]
25. Khodeir, M.; Antoun, S.; Van Ruymbeke, E.; Gohy, J.F. Temperature and redox-responsive hydrogels based on nitroxide radicals and oligoethyleneglycol methacrylate. *Macromol. Chem. Phys.* **2020**, *221*, 1900550. [CrossRef]
26. Cross, E.R. The electrochemical fabrication of hydrogels: A short review. *SN Appl. Sci.* **2020**, *2*, 397. [CrossRef]
27. Kleber, C.; Lienkamp, K.; Rühe, J.; Asplund, M. Electrochemically controlled drug release from a conducting polymer hydrogel (PDMAAp/PEDOT) for local therapy and bioelectronics. *Adv. Healthc. Mater.* **2019**, *8*, 1801488. [CrossRef]
28. Liu, R.; Dong, C.; Liang, X.; Wang, X.; Hu, X. Highly efficient catalytic aerobic oxidations of benzylic alcohols in water. *J. Org. Chem.* **2005**, *70*, 729–731. [CrossRef]
29. Bobbitt, J.M. Oxoammonium salts. 6. 4-Acetylamino-2,2,6,6-tetramethylpiperidine-1-oxoammonium perchlorate: A stable and convenient reagent for the oxidation of alcohols. Silica gel catalysis. *J. Org. Chem.* **1998**, *63*, 9367–9374. [CrossRef]
30. Gharehkhani, S.; Zhang, Y.; Fatehi, P. Lignin-derived platform molecules through TEMPO catalytic oxidation strategies. *Prog. Energy Combust. Sci.* **2019**, *72*, 59–89. [CrossRef]
31. Karimi, B.; Farhangi, E. A Highly recyclable magnetic core-shell nanoparticle-supported TEMPO catalyst for efficient metal- and halogen-free aerobic oxidation of alcohols in water. *Chem. Eur. J.* **2011**, *17*, 6056–6060. [CrossRef]
32. Bocian, A.; Gorczyński, A.; Marcinkowski, D.; Witomska, S.; Kubicki, M.; Mech, P.; Bogunia, M.; Brzeski, J.; Makowski, M.; Pawluć, P.; et al. New benzothiazole based copper(II) hydrazone Schiff base complexes for selective and environmentally friendly oxidation of benzylic alcohols: The importance of the bimetallic species tuned by the choice of the counterion. *J. Mol. Liq.* **2020**, *302*, 112590. [CrossRef]
33. Mahmoudi, B.; Rostami, A.; Kazemnejadi, M.; Hamah-Ameen, B.A. Catalytic oxidation of alcohols and alkyl benzenes to carbonyls using Fe_3O_4@SiO_2@(TEMPO)-co-(Chlorophyll-Co^{III}) as a bi-functional, self-co-oxidant nanocatalyst. *Green Chem.* **2020**, *22*, 6600–6613. [CrossRef]
34. Dash, S.; Murthy, P.N.; Nath, L.; Chowdhury, P. Kinetic modeling on drug release from controlled drug delivery systems. *Acta Pol. Pharm.* **2010**, *67*, 217–223.
35. Vyavahare, N.R.; Kulkarni, M.G.; Mashelkar, R.A. Zero order release from glassy hydrogels. *J. Memb. Sci.* **1990**, *54*, 205–220. [CrossRef]

polymers

Review

Recent Developments in Ion-Sensitive Systems for Pharmaceutical Applications

Michał Rudko, Tomasz Urbaniak and Witold Musiał *

Department of Physical Chemistry and Biophysics, Pharmaceutical Faculty, Wroclaw Medical University, Borowska 211, 50-556 Wroclaw, Poland; michalzdzislawrudko@gmail.com (M.R.); tomasz.urbaniak@umed.wroc.pl (T.U.)
* Correspondence: witold.musial@umed.wroc.pl

Abstract: Stimuli-responsive carriers of pharmaceutical agents have been extensively researched in recent decades due to the possibility of distinctively precise targeted drug delivery. One of the potentially beneficial strategies is based on the response of the medical device to changes in the ionic environment. Fluctuations in ionic strength and ionic composition associated with pathological processes may provide triggers sufficient to induce an advantageous carrier response. This review is focused on recent developments and novel strategies in the design of ion-responsive drug delivery systems. A variety of structures i.e., polymeric matrices, lipid carriers, nucleoside constructs, and metal-organic frameworks, were included in the scope of the summary. Recently proposed strategies aim to induce different pharmaceutically beneficial effects: localized drug release in the desired manner, mucoadhesive properties, increased residence time, or diagnostic signal emission. The current state of development of ion-sensitive drug delivery systems enabled the marketing of some responsive topical formulations. Concurrently, ongoing research is focused on more selective and complex systems for different administration routes. The potential benefits in therapeutic efficacy and safety associated with the employment of multi-responsive systems will prospectively result in further research and applicable solutions.

Keywords: ion-sensitive systems; drug delivery; smart pharmaceutical systems; biocompatible medical devices

Citation: Rudko, M.; Urbaniak, T.; Musiał, W. Recent Developments in Ion-Sensitive Systems for Pharmaceutical Applications. *Polymers* **2021**, *13*, 1641. https://doi.org/10.3390/polym13101641

Academic Editor: M. Ali Aboudzadeh

Received: 26 April 2021
Accepted: 16 May 2021
Published: 18 May 2021

1. Introduction

The precise delivery of an active pharmaceutical ingredient to the site of action in a controlled manner is one of the primary aims of research conducted in the field of pharmaceutical technology. With an increasing number of pharmacologically active macromolecules, i.a., proteins, the design of an accurate drug delivery system (DDS) capable of effective drug protection gains great attention. Numerous methods were evaluated to provide these beneficial features [1].

One of the extensively researched concepts in the field of drug delivery is focused on stimuli-responsive systems, also termed "smart" carriers. Various chemical and physical stimulants may serve as triggers for drug release. Most of the proposed carriers respond to the changes in temperature and pH value; nonetheless, systems sensitive to the magnetic field, electromagnetic radiation, redox potential, and enzyme presence were reported [2]. In this review, we aim to summarize recent advancements in the development of ion-responsive formulations. Changes in homeostatic ionic strength and composition may be exploited as the trigger, for example, for drug release, phase transition, or diagnostic signal emission. The ions present in physiological fluids and on the surface of mucous membranes are potential stimuli for mucoadhesive and topical formulations. Additionally, a number of medical conditions are accompanied by a change in ionic concentration or may be caused by the presence of exogenic toxic ions present in the environment. Increased Ca^{2+} serum concentrations were linked to various vascular and bone diseases, Zn^{2+} concentrations are considerably higher in nervous tissue, particularly in synaptic vesicles, and changes in Fe^{3+} concentration may indicate anemia and several different diseases [3–5]. Furthermore, due to ongoing technological progress, the rise in potentially harmful heavy metal pollution is observed [6]. The development of safe carriers for chelating agents, which are released in response to particular toxic ions, may find an application in industry. Due to the critical role of ion concentrations in different body compartments, the homeostatic mechanisms do not allow significant fluctuations. Therefore, the proposed DDS should be capable of selective response to minor variations in physiological concentration.

Various vehicles exhibiting ion sensitivity were proposed, including prodrugs, synthetic polymeric architectures, metal-organic frameworks, polynucleotide structures, or liquid crystals. In this review, we summarize the most recent and distinguished ion-sensitive DDSs divided into two groups, according to their selectivity response toward ions. Within these two main categories, proposed DDSs were classified according to their structure, which is frequently related to the sensitivity mechanism and potential applications (Figure 1).

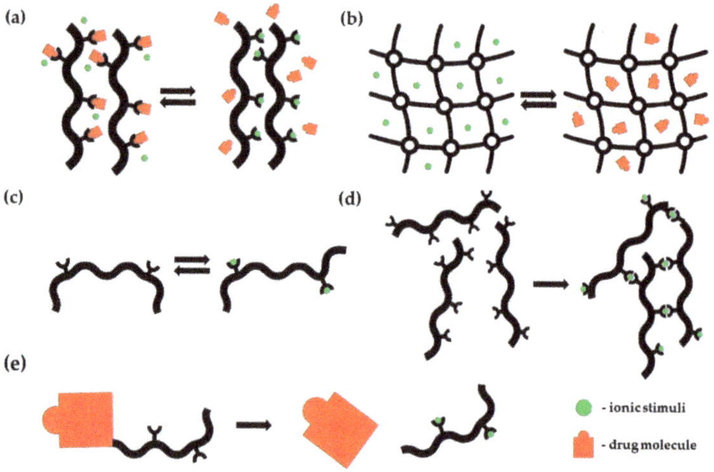

Figure 1. Possible mechanisms of ion-dependent response of pharmaceutical systems: (**a**) ion exchange in polymer resins; (**b**) ion exchange in porous materials; (**c**) conformational change; (**d**) ionic crosslinking; (**e**) prodrug activation.

Among currently available approaches, non-selective topically administered in situ gelling DDSs based on biopolymers are researched most accurately, while more elaborated and selective structures are most often in the initial stage of development. Additionally, multi stimuli-responsive systems aiming at particular microenvironments in the human body gained considerable interest. The possibility of higher therapy efficacy and improved safety offered by an ion-responsive DDS will likely result in further progress and the development of clinically applicable strategies.

2. Non-Selective Ion-Sensitive Formulations

There is a considerably large group of formulations that respond to ions generally, without any specific sensitivity towards particular ions or valence. Polymer-based formulations included in this category react to ions in a defined concentration range. Ion-dependent responses include most often gelling and swelling associated with the supramolecular rearrangement of polymeric architectures. The process of ion-dependent gelling is mainly utilized in ocular and nasal formulations. In this part, non-selective ion responsive systems were described and categorized according to the ion-sensitive agent (Table 1).

2.1. Gellan Gum

Gellan Gum (GG) is a linear polysaccharide composed of two D-glucose molecules, one L-rhamnose, and one D-glucuronic acid, in a structural unit. In the native form of GG, one acetyl moiety per two repeating units is bound to a glucose molecule. Additionally, there are commercially available deacylated variants described as low acyl GG. The deacylation degree significantly influences the gelation process and mechanical properties of GG gel [7]. A high acetylation degree results in soft gel formation, whereas a low acetylation degree leads to the formation of the stiff gel structure. Due to D-glucuronic acid presence in the structural unit, GG is polyanionic in its deprotonated form. The mechanism of ion-assisted gelling is linked to the formation of cation-induced crosslinking of a double-helical polysaccharide structure, a process sensitive to temperature changes. GG has the ability to form gels after exposition to different metal ions and, in a less pronounced manner, to hydrogen ions. Gels obtained in the presence of divalent ions such as Ca^{2+}, Mg^{2+} are less viscous and form at significantly lower concentrations in comparison to monovalent cations, e.g., Na^{+} or K^{+}. The minimum gelling concentration of monovalent ions is 100 mM, while for divalent ions, it is equal to 5 mM [7]. The concentrations required for GG gelation in contact with physiological fluids can be found in the mucosal nasal fluid [8], blood [9], and tear film [10]. The experimental nasal formulations are frequently modified by the supplementation with the ion-sensitive GG. The presence of GG prolongs drug residence time in the nasal mucosa, and thus it may enhance the beneficial effects of the intranasal administration route, e.g., avoidance of first-pass effect and decrease in systemic side effects. A combination of modern drug carriers such as nanoparticles with GG may increase the concentration of drug in targeted organs e.g., in the brain. In many cases, formulations utilizing GG as a gelling agent in the nasal cavity proved to be safe and effective. The most often applied concentrations of GG were in the range of 0.3% to 0.5% (*w/v*), whereas the 0.5% (*w/v*) concentration was preferred due to the plausible balance between viscosity, mucoadhesive properties, and impact on drug release profile.

2.1.1. Nasal Formulation

The rapid elimination of poorly soluble drug suspension from the nasal cavity due to mucociliary beating is a significant hindrance for traditional nasal formulations. Application of in situ gelling systems may result in prolonged contact between drug molecules and the vascularized tissues capable of drug absorption. Moreover, low drug diffusion rates, accompanied by a high gel viscosity may result in prolonged drug release. Mometasone, a gel-based carrier designed by Xin-guo Jiang et al., took advantage of these characteristics A nasal DDS based on xanthan gum and 0.5% (*w/v*) GG resulted in a favorable response in animal models compared to the traditional suspension-based system [11]. The nasal drug administration route proved to be beneficial in the alleviation of symptoms linked to motion sickness. Mucoadhesivity is one of the features enabling a bioavailability increase and an overall better therapeutic response in the nasal application. In situ gelling formulation based on 0.3% deacetylated GG and 0.15 carbopol 934 P resulted in a significant mucoadhesive force allowing a prolonged exposition of the dimenhydrinate loaded system to the absorptive tissues. The evaluated formulation showed a slightly negative effect on the nasal epithelium in the animal model [12]. Incorporation of an ion-sensitive gelling agent to the colloidal DDS may prolong residence time. This approach was employed in the design of the curcumin-loaded microemulsion system for nasal drug delivery. Deacetylated GG in the concentration of 0.3% was found to be a suitable ion-sensitive gelling agent for described self-emulsifying system [13]. Nasal drug application is considered a possibly favorable brain tissue delivery path [14]. Therefore, there were attempts to design intranasal systems aiming to increase drug concentration in the brain structures via the prolonged exposure of nasal mucosa to the drug carrier. The formulation including solid lipid nanoparticles with paeonol suspended in 0.4% deacetylated GG as an in situ gelling agent, was employed to alleviate the first-pass effect and to enable brain delivery via the olfactory nerve pathway [15]. Resveratrol loaded nanosuspension, based on 0.6% (*w/v*) deacetylated

GG, was employed to exploit an analogous delivery route. The drug concentration in brain tissue increased in the animal model compared to the intravascular administration, and the effect was prolonged. Moreover, the amount of resveratrol distributed in other organs was lower than in brain tissue and was eliminated significantly faster [16]. Employment of 0.5% (*w/v*) GG and 0.15% (*w/v*) xanthan gum enabled brain favoring donepezil delivery. Some advantages were revealed, including high donepezil concentrations in the brain tissue and a significant decrease in drug concentration in the liver and other vital organs compared to a marketed oral formulation [17]. In conclusion, intranasal systems combining in situ gelling properties with contemporary drug carriers may be promising systems for brain drug delivery.

2.1.2. Ocular Formulations

Topical drug formulations are frequently used in the treatment of ocular diseases. The most notable topical ocular drug form are eye drops, with specific local activity and uncomplicated administration. Several hindrances limit the extensive application of eye drops, including the short residence time, resulting in limited absorption and penetration rates of the active compound. Additionally, the eye drops that flow into the lacrimal canaliculus may result in an escalation of side effects. The introduction of in situ gelling agents, i.e., GG, may help to overcome the issues. Some eye-related disorders require prolonged or nearly zero-order drug release. Glaucoma, the incurable disease leading to vision impairment and blindness in the final stage, may be treated with several drugs that can mitigate and even stop disease progression. Most often, they are administered in eye drop form, but due to the short residence time, they require a frequent application to provide sufficient therapeutic concentration. In order to improve eye drop performance, GG was employed in a few strategies. The most straightforward approach to obtain GG-based in situ gelling eye drops was a combination of the drug brinzolamide and GG with a 95% degree of deacetylation. GG in the concentration of 0.5% (*w/v*) was determined as the most beneficial due to both desirable stiffness and optimal residence time of the in situ formed gel. In vitro trials showed that, compared to market eye drops, formulations containing GG release brinzolamide slower and without a burst of release during the first two hours of exposition to artificial tear fluid. Measurements of intraocular pressure in the animal model showed that in situ gelling formulation provided prolonged drug activity without ocular irritation [18]. Gayatri et al. described the complex formulations including GG: carbopol 934 P as a pH-sensitive mucoadhesive agent and benzododecinium bromide as a preservative and corneal penetration enhancer. The experimental formulation was developed via Box-Behnken design and compared to the commercially available TIMOPTIC-XE®, which also contained GG as an in situ gelling agent. Compared to the marketed product, the optimized formulation had a comparable release profile in vitro with less pronounced concentration fluctuations. The same observations were made for the measurements of intraocular pressure in animal models; the formulation was stable and well-tolerated [19]. Another approach is based on liposomes as drug carriers combined with deacetylated GG, in order to address low active pharmaceutical agent bioavailability. The deacetylated GG concentration of 0.4% (*w/v*) was found to be optimal due to the most beneficial viscosity and release rate. Supplementation with deacetylated GG resulted in a less pronounced burst release and an overall decrease in drug release rate in comparison to an aqueous suspension of liposomal timolol maleate. Measurement of residence time via fluorescence imagining proved that in situ gelling liposomal timolol eye drops exhibited the longest contact time with eye surface. A comparison of the intraocular pressure after exposition to the evaluated system and reference timolol eye drops showed a more pronounced response in the case of the in situ gelling formulation [20]. In ocular infections, topical DDSs can simultaneously decrease systemic impact and locally increase therapy efficiency. The potency of some antibiotics depends on the time of their presence in a site of action at the proper concentration. Therefore, beneficial in situ gelling systems are employed in experimental antimicrobial formulations for ocular administration. Asgar Ali et al. utilized pefloxacin as an antibiotic agent combined with a GG solution. An increased concentration in GG resulted in a slower release rate and a more linear drug liberation. As a result, GG also influenced the antimicrobial properties of the formulation. In vitro tests conducted via the cup plate method showed that optimized formulation had a bigger area of inhibition than marketed pefloxacin eye drops [21]. An analogous approach was employed in the GG-based ocular system loaded with moxifloxacin and ketorolac as an anti-inflammatory agent. Dissolution tests of formulations employing GG in concentration range 0.1–0.25% (*w/v*) confirmed the influence of ion-sensitive polysaccharides on the release of both drugs [22]. A combination of liposomal drug carriers with an in situ gelling ion-sensitive matrix was proposed as an efficient delivery approach for the lipophilic antifungal drug—natamycin. A comparison with the marketed drug suspension confirmed the advantages of the evaluated system, a superior corneal permeability, as well as prolonged residence time [23]. Due to the distinctive anatomical structure of the human eye, systemic delivery of lipophilic pharmaceutically active agents of high molecular weight is extremely difficult. Therefore, local administration is often employed as a strategy to achieve high intraocular concentration. An example of such an approach was described by Chetoni et al. Application of cyclosporine-A loaded micelles entrapped in ion-sensitive in situ gelling matrix resulted in improved drug solubility and enhanced residence time in tear fluid [24]. A distinct example of a dry ocular DDS composed of GG-pullulan electrospun nanofibers was proposed. Immediate gel formation after lens application in the animal model was achieved. Model drug residence time on the ocular surface was significantly longer compared to a drug applied in the form of eye drops [25].

2.1.3. Other Applications

Prior discussed approaches demonstrated the two most commonly described examples of GG's use in pharmaceutical applications—ocular and intranasal. However, there are reports of this in situ gelling agent employment in other ion-rich environments. Wound dressings are one of the medical supplies which may potentially benefit from GG employment. Carboxymethyl chitosan-based wound dressing enriched with GG-derived microparticles exhibited prolonged release of the antibacterial drugs—tetracycline and silver sulfadiazine [26]. Another study focused on burn wounds treated with a collagen-GG crosslinked network, employed as a carrier for cells promoting regeneration. The obtained interpenetrating network improved early wound closure, reduced inflammation, and promoted regeneration for third-degree burn wounds [27]. GG is also utilized in formulations for hard tissue regeneration enhancement. P. Matricardi et al. developed an ion-responsive formulation for bone and cartilage defects treatment. The described system consisted of two solutions, which, after contact, formed a bioadhesive hydrogel via ion-activation. The primary solution consisted of hyaluronic acid and CaCl2, whereas the secondary solution was an aqueous GG solution. Hyaluronic acid with Ca^{2+} was applied to the defect cavity, followed by dropwise application of the GG solution. Initially, a two-layered environment inside the cavity was formed; subsequently, diffusion of the formulation components promoted further hydrogel formation. The hydrogel provided a good environment for osteoblasts proliferation; the formulation had plausible adhesiveness and durability [28]. A clotrimazole-loaded formulation for anti-fungal dental application was obtained via a combination of GG as a gelling agent and Ca^{2+} ions in the form of a citrate complex. In the oral cavity, Ca^{2+} was released from the citrate complex due to a slightly acidic environment. Thus, the Ca^{2+} level sufficient for gelation was obtained. The employment of GG enhanced the residence time of clotrimazole on the surface of the oral cavity improving the anti-fungal activity of the formulation [29]. Medical procedures requiring intrauterine device insertion are often avoided due to associated significant pain. Employment of local anesthetics can mitigate pain during intrauterine procedures. An intravaginal formulation composed of GG as an ion-responsive agent was proposed as a carrier of lidocaine for local analgesia. The investigated system exhibited good biocompatibility in the animal model. The formulation was employed during different intravaginal medical procedures performed on women volunteers and provided significant pain alleviation [30].

2.2. Alginates

Alginates are polysaccharides occurring naturally in brown algae and soil bacteria. The backbone of the alginate chain consists of α-L-guluronic acid and β-D-mannuronic acid residues linked with 1-4 bonds. The arrangement of moieties present in the alginate molecule is variable and depends on the source organism or even tissue of origin. The molecule of alginate consists mainly of randomly distributed guluronic and mannuronic monomers, with locally ordered fragments. Both arrangement and monomer ratio strongly influence polymer chain stiffness and results in varied physical and mechanical properties. Guluronic acid units, especially if present in homopolymeric sequences, are capable of interacting with multivalent cations, notably Ca^{2+}. As a result, alginates are capable of ion-dependent gel formation via electrostatic crosslinking. Therefore, there is an applicative potential for the alginate-based in situ gelling system in divalent cation-rich environments, e.g., in the ocular surface, nasal cavity, or in the vascular bed. Moreover, alginates exhibit temperature-independent gelation, which allows thermal processing of obtained ion-sensitive formulations [31].

Table 1. Summary of non-selective ion-responsive systems for pharmaceutical applications.

Ion-Sensitive Component	Incorporated Substance	Application	Ion-Induced Response	Ref.
Gellan gum	Momentasone	Allergic rhinitis	Prolonged residence in nasal cavity	[11]
	Dimenhydrate	Motion sickness	Alternative administration route	[12]
	Curcumin	n/a	Nose-to-brain delivery	[13]
	Paeonol	Neuroprotection	Nose-to-brain delivery	[15]
	Resveratrol	Neurodegenerative diseases	Enhanced pharmacokinetic profile	[16]
	Donepezil	Alzheimer's disease	Alternative route of administration	[17]
	Brinzolamide	Glaucoma	Enhanced pharmacokinetic profile	[18]
Gellan gum	Momentasone	Glaucoma	Drug release control	[19]
	Dimenhydrate	Glaucoma	Prolonged residence time	[20]
	Curcumin	Bacterial infection	Enhanced antibacterial activity	[21]
	Paeonol	Bacterial infection	Enhanced pharmacokinetic profile	[22]
	Resveratrol	Fungal infection	Permeability and residence time enhance	[23]
	Cyclosporine-A	Dry eye disease, choroid inflammation	Enhanced solubility and residence time	[24]
	Tetracycline, Silver sulfadiazine	Wound dressing	Sustained drug release	[25]
	Collagen	Wound dressing	Wound regeneration improvement	[26]
	Hyaluronic acid	Bone and cartilage regeneration	Increase in proliferation of osteoblasts	[28]
	Clotrimazole	Dental fungal infection	Prolonged residence on mucous membrane	[29]
	Lidocaine	Local analgesia	Pain alleviation during medical intervention	[30]
Alginates	Gatifloxacin	Bacterial infection	Sustained drug release	[32]
	Ofloxacin	Bacterial infection	Sustained drug release	[33]
	Nepafenac	Anti-inflammatory	Enhanced permeability	[34]
	Paracetamol	Pain and fever therapy	Prolonged release	[35]
	Rifampicin	Tuberculosis infection	Delayed release	[36]
	IL-2	Immunomodulation	Formation of matrix for cell colonization	[37]
Poly (styrene-divinyl benzene) sulfonic acid	Betaxolol	Glaucoma or ocular hypertension treatment	Ion-dependent release	[38]

Table 1. *Cont.*

Ion-Sensitive Component	Incorporated Substance	Application	Ion-Induced Response	Ref.
Carboxymethyl chitosan	Interferon α-2b	Antitumor	Sustained release and lung accumulation	[39]
Carboxymethyl cellulose	Lysozyme	n/a	Gel swelling and protein uptake	[40]
Dextran-poly (acrylic acid) copolymer	Ibuprofen	n/a	Controlled drug release and gel swelling	[41]
Eudragit RS/LS	Diltiazem	n/a	Controlled drug release	[42]
Methacrylate	n/a	n/a	Gel swelling, water uptake	[43]
Acrylic acid grafted polyvinylidene fluoride	Propranolol, caffeine, sodium salicylate	n/a	Controlled drug release	[44]
MOF	Procainamide	n/a	Controlled drug release	[45]

2.2.1. Ocular Formulations

Alginates are employed in the preparation of in situ gelling ocular formulations based on ion-sensitive behavior, similar to GG. Ion-induced alginate gelation is one of the features that can address the issue of low drug bioavailability and short drug residence time. Significant attention in the field of ocular infections was paid to antimicrobial drugs from the group of quinolones. The use of quinolone-loaded sodium alginate as an ion-sensitive in situ gelling agent combined with cellulose derivatives was reported. Sodium alginate containing 40% of guluronic and 60% mannuronic acid, combined with HPMC as a viscosity-enhancing agent, formed an ion-responsive formulation. A higher alginate concentration resulted in the increase in formulation gelling capacity; the same effect was observed in systems containing a constant alginate concentration and increasing content of 90 kDa HPMC. A profile of gatifloxacin release from the optimized formulation was characterized by a less pronounced burst release effect and provided sustained drug release for 8 h. The formulation was well tolerated in the animal model; no ocular damage and iatrogenic abnormalities were observed [32]. An analogous ion-sensitive system based on sodium alginate and HPC was developed as a potential ofloxacin carrier. An optimized formulation containing 1.5% (*w/v*) alginate and 0.5% (*w/v*) HPC exhibited, similarly to the abovementioned, an 8 h sustained release of the drug [33]. Alginate-based ocular formulation designed to carry the lipophilic anti-inflammatory drug nepafenac was investigated [34]. The active pharmaceutical ingredient was complexed with hydroxypropyl-β-cyclodextrin to improve water solubility and incorporated into alginate and an HPMC solution. An optimized formulation containing 0.3% (*w/v*) alginate was compared to the marketed Nevanac®, an ophthalmic suspension of nepafenac. The concentration of nepafenac in the cornea was significantly higher in the case of the investigated formulation compared to reference suspension.

2.2.2. Other Applications

A sodium alginate-based oral in situ gelling DDS loaded with paracetamol capable of forming a gel in the stomach was reported [35]. The administered solution contained Ca^{2+} ions complexed in citrate form to ensure gelation in the acidic stomach fluid, as a result of Ca^{2+} release via H^+ substitution. The formed gel provided paracetamol release for 6 h and a release profile close to the reference suspension form. An alginate-based system for prolonged rifampicin release was proposed for administration in the form of poly (lactic-co-glycolic acid) microspheres suspended in sodium alginate solution. The described formulation was administered by endotracheal intubation in the animal model. The employment of alginate resulted in a significant delay in drug delivery and 24 h adhesion. Presented results indicate that the described formulation could be useful for the interventional treatment of tuberculosis [36]. A two-component alginate injectable formulation capable of in vivo matrix formation was reported. Alginate was employed as a gelling agent alongside a Ca^{2+} reservoir in the form of microspheres for inducing in vivo gelation. Microspheres in an alginate solution may be applied as carriers for pharmacologically active substances e. g., interleukins. After formation, gels were infiltrated by host cells. The authors speculate that such a formulation could be employed for cell incorporation [37].

2.3. Other Carriers

DDSs of another origin were also utilized in order to obtain non-selective ion-sensitive products potentially applicable in various drug delivery strategies. A complex ion-sensitive DDS for ocular application in the form of a contact lens was evaluated. The main barrier in drug delivery via contact lenses is drug leak during storage. This disadvantage can be addressed by the application of ion-sensitive macromolecular architectures. The described lenses were made out of a silicone outer layer and a poly(styrene-divinylbenzene) sulfonic acid resin. The resin was dispersed in a copolymeric matrix and served as a betaxolol reservoir. A prepared DDS, stored in distilled water, due to a lack of ions capable of drug substitution, did not release the drug during a prolonged period. However, after exposition to the artificial tear fluid, the drug was liberated with an observable burst release in the first 4 h and a subsequent slower discharge in the following 6-days [38]. Previously discussed ion-sensitive systems were designed mainly for topical applications. Nevertheless, some ion-dependent formulations may be beneficial in parenteral administration. Y. Xu et al. described ion-sensitive microparticles targeting lung cancer tissue [39]. The particles were electrostatically loaded with interferon α-2b. Nanoporous microspheres were obtained using carboxymethyl chitosan and subsequently loaded via electrostatic interaction with the chemotherapeutic agent. Due to the ionic nature of the interaction between the drug and the carrier, the protein release occurred via an ion-exchange mechanism. Physiologically occurring cations substituted positively charged interferon and resulted protein release from the carrier in the target tissue. The obtained formulation provided a sustained-release and drug accumulation in the lungs. The electrostatic protein binding was investigated with micro-sized gel particles of carboxymethyl cellulose crosslinked with sodium trimetaphosphate [40]. Carriers prepared with various crosslinker concentrations were loaded with lysozyme as a model protein. The impact of ionic strength on the protein adsorption and swelling degree was investigated. An increase in ionic strength resulted in a decline in protein uptake and a reduced swelling degree. Authors suggested that the decrease in lysozyme uptake in the presence of a higher ion concentration was associated with the salt screening phenomenon resulting in reduced attraction between charged microgel and lysozyme molecules. The same mechanism was probably responsible for the mitigation of repulsion between polymer chains and subsequent decrease in swelling degree. The ion responsive properties of dextran-poly(acrylic acid) copolymer were evaluated in ibuprofen-loaded particles. It was observed in the release experiments that ion presence in the acceptor medium significantly decreased ibuprofen release rate, especially in an acidic environment.

Furthermore, an increase in ionic strength resulted in increased particle hydrodynamic diameters. According to the authors, ion introduction had a significant impact on hydrogen bonding, ionic interactions, and surface charge, which subsequently affected particle size and release rate of ibuprofen [41]. Eudragit RS/LS coated beads were examined to elucidate the mechanism of diltiazem release. The ionic strength of the release media in a particular range resulted in higher release rates and a shorter drug release lag time. This observation was linked to the mechanism based on the exchange of a counterion of quaternary ammonium groups present in the eudragit molecule. The initial increase in ionic strength promoted a faster exchange of chloride anions and thus a faster release and shorter lag time to a certain ionic strength value. A further increase in ionic strength had the reverse effect, explained as a result of increased osmotic pressure [42]. Multiresponsive cryogels based on methacrylates and acrylamide-derived polycations exhibited some ionic strength-dependent behavior. The interaction between the counterion and the charged polymer groups led to a decrease in electrostatic repulsion inside supramolecular gel structure and thus a reduced distance between neighboring polymer chains. This phenomenon was manifested as a reduced equilibrium water uptake and may potentially have an impact on the release of the incorporated pharmaceutically active ingredient [43]. A therapeutic system employing porous polyvinylidene fluoride membrane grafted with acrylic acid was proposed as an ionic strength sensitive barrier for drug delivery. Membrane containers filled with solid drug doses were investigated as potential DDSs for propranolol, caffeine, and sodium salicylate. Release experiments in pH 7.0 and varying salt concentration resulted in a diminished release rate of caffeine and propranolol in higher ionic strength. On the contrary, sodium salicylate showed a slight change in release profile when the media ionic strength was varied. It was suggested that ion interaction with polyacrylic chains might result in the rearrangement of polymeric architecture and thus affect membrane permeability for certain molecules [44]. Structurally differing systems based on metal-organic frameworks (MOFs) were described as potential ion responsive carriers. A single crystalline material based on zinc, adenine, and biphenyl dicarboxylic acid was described by Rosi et al. [45]. The pores present in the obtained structures enabled the incorporation of procainamide in its cationic form. A significant ion-dependence during drug release was observed in dissolution studies. The release profile observed in phosphate-buffered saline differed considerably from the profile obtained in the corresponding experiment performed with deionized water. According to the authors, the cation-dependent drug release mechanism was related to the ion substitution phenomenon.

2.4. Selective Ion-Sensitive Systems

Except for non-selective formulations, systems responding selectively to certain ions or valence in a particular concentration range may be distinguished (Table 2). Due to distinctive changes in ion concentration accompanying various pathological conditions, such behavior may be extremely beneficial in terms of therapy efficacy and safety.

Table 2. Summary of selective ion-responsive systems for pharmaceutical applications.

Ion-Sensitive Component	Incorporated Substance	Application	Ion-Induced Response	Ref.
MOFs	5-FU	Potential treatment of central nervous system diseases	Zn^{2+} dependent drug release	[46]
	5-FU	Potential treatment of central nervous system diseases	Zn^{2+} dependent drug release	[47]
	5-FU	n/a	Zn^{2+} and Ca^{2+} dependent drug release	[48]
Modified liposomal carriers	Dye	n/a	Ca^{2+} dependent drug release	[49]
	Chelating agent, fluorescein	Hg^{2+} neutralization and detection	Hg^{2+} dependent release	[50]
	Fluorescent dye	n/a	Cu^{2+} dependent release	[51]
D3F3 peptide	Doxorubicin	Possible prostate cancer treatment	Zn^{2+} dependent in situ hydrogel formation	[52]
Mesoporous silica nanoparticles modified with Pb^{2+}-activated DNAzyme	Fluorescein	Pb^{2+} detection	Pb^{2+} dependent release	[53]
Prodrug 1,2,4-trioxolane moiety	ML4118S	Plasmodium infection treatment	Fe^{2+} dependent activation	[54]

Table 2. *Cont.*

Ion-Sensitive Component	Incorporated Substance	Application	Ion-Induced Response	Ref.
Polynucleotide framework	AS1411 aptamer	Cancer treatment	K^+ and pH dependent release on cellular membrane	[55]
Pectin	n/a	n/a	Ca^{2+} dependent gelling	[58]
Polyacrylamide hydrogels	n/a	adhesive materials	Ion-dependent adhesion	[59]
PNI-co-CF3-PT0.2-co-DDDEEKC0.2	n/a	Biodevices and artificial nanochannels	Ca^{2+} concentration dependent channels	[60]
Cholesteric liquid crystalline polymer	n/a	Fast calcium level test	Color change in presence of Ca^{2+}	[56]
Single strain 30-nucleotide DNA absorbed on carbon nanotubes	n/a	Determination of Hg^{2+} concentration in biological systems	Hg^{2+} mediated shift in emission energy	[57]
(1,2-diaminocyclohexane) platinum (II)	Platinum derivatives	Antitumor activity	Cl^- induced intracellular activation of chemotherapeutic agent	[61]

2.5. MOFs

MOFs are structures composed of metal-oxo clusters connected by organic ligands characterized by high porosity. Due to a high pore volume and surface area, they may serve as potential carriers for a variety of molecules, including drugs.

Nanoporous UiO-66-NH_2 MOF was conjugated with quaternary ammonium salts to obtain a DDS sensitive to the presence of Zn^{2+} ions. Carboxylato-pillar [5] arene caps, with an electron-rich cavity, complexed positively charged quaternary ammonium groups on the MOF surface via a host-guest interaction. This capping mechanism prevented 5-fluorouracil (5-FU) release from the pores. 5-FU liberation occurred after the detachment of the capping agent, competitively complexed by introduced Zn^{2+} ions and the subsequent exposition of the incorporated cargo to the release media. The release rate of 5-FU was determined in different concentrations of Zn^{2+}; the increase in release rate was observed in higher Zn^{2+} concentrations. In physiological concentration of Zn^{2+}, only 5% of the incorporated dose was released, indicating a possibly low premature drug release after administration. In vitro cytotoxicity of the non-loaded system was negligible in low concentrations [46]. Another Zn^{2+} sensitive system was based on two topological types of indium MOFs. Carrier nanopores served as a 5-FU reservoir, and in the presence of Zn^{2+}, the drug was released in an ion-exchange mechanism. The release of different Zn^{2+} concentrations was evaluated, in the lack of Zn^{2+} ions, the release curve of 5-FU achieved a plateau on the level of 50–55% released drug. An increase in Zn^{2+} concentration resulted in higher plateau levels up to 90% [47]. Ion-responsive MOF-based DDSs may also serve as a diagnostic tool. One of the examples of theranostic carriers was obtained using a zirconium-based MOF deposited on the surface of Fe_3O_4 nanoparticles. Drug-loaded MOF was modified with 1-(6-bromohexyl) pyridine, which allowed the host-guest capping of MOF nanopores with carboxylate-pillar [6] arenes. The competitive complexing of capping agent with Zn^{2+} and Ca^{2+} in surrounding media resulted in complex detachment, and subsequent exposition of the drug-loaded pores. 5-FU release in media without Zn^{2+} or Ca^{2+} achieved a plateau below the level of 15% of the incorporated drug after 2 h. Approximately a two-fold plateau level was observed in the presence of divalent ions [48].

2.6. Liposomes

A few approaches were employed in order to modify liposomal bilayer structure with ion-responsive molecules to provide ion sensitivity. Michael D. Best et al. reported a Ca^{2+} activated release from the liposomal system obtained via the introduction of an ionic switch anchored in the phospholipid bilayer via lipophilic hydrocarbon chains. The tetracarboxylate chelating site of the sensor undergoes a conformational change after Ca^{2+} complexing, which results in liposome membrane disruption and encapsulated drug release. The triggering effect of Ca^{2+} was pronounced, whereas the presence of K^{+}, Na^{+}, and Zn^{2+} had no impact on release. Furthermore, a significant change in carrier hydrodynamic diameter was observed [49]. Ion responsive liposomes employed as theranostic systems, selective Hg^{2+} sensitive carriers for detection and neutralization of Hg^{2+} was reported. Liposomes with pegylated phosphatidylethanolamine embedded in their bilayer were loaded with fluorescein as the indicator and meso-2,3-dimercaptosuccinic acid as the chelating agent. The mechanism of Hg^{2+} mediated release is based on the interaction between Hg^{2+} and phosphatidylethanolamine headgroups, which results in molecule reorientation and destabilization of the liposomal membrane. Subsequently, indicator release occurs, and concurrently discharged chelating agent neutralizes Hg^{2+}. Further investigation confirmed high selectivity towards Hg^{2+} in comparison to other metal cations. An in vitro cell viability evaluation confirmed that the presence of liposomes containing a chelating agent increased cell viability in the presence of Hg^{2+} [50]. The Cu^{2+} responsive liposomal system was obtained by employing egg lecithin to form a carrier membrane enriched with an ion-sensitive switch. Employed Cu^{2+} sensitive agents were derivatives of 3,7-diazabicyclo[3.3.1]nonan-9-one obtained by substituting secondary amine groups with alkyl chains, which enabled incorporation in carrier membrane. Conformational changes induced by Cu^{2+} in incorporated switches led to the formation of membrane defects and a subsequent leak of liposome cargo. Fluorescent dye incorporated in obtained liposomes was released exclusively in the presence of Cu^{2+} [51].

2.7. Other Carriers

The in situ gelling ion-responsive peptide-based system that could be potentially employed in prostate cancer therapy was proposed. A novel D3F3 forky peptide was synthesized and employed as a Zn^{2+} responsive agent for in situ hydrogel formation in the targeted organ. The determined gelation-inducing Zn^{2+} concentration of D3F3 was lower than observed physiological levels. Exposition to other ions present in biological fluids in physiological concentrations did not trigger gel formation. The in vitro model showed that the peptide component of the system exhibited no toxicity, whereas the one loaded with doxorubicin had a higher efficacy than doxorubicin alone [52]. The mesoporous nanoparticle system was designed to respond selectively to Pb^{2+}. The carrier was based on epoxidated silica particles loaded with fluorescein, which was released specifically in the presence of Pb^{2+}. Silica particles were surface modified with a substrate DNA strand, which was subsequently hybridized with a biotinylated Pb^{2+} specific DNAzyme. The obtained pore "gate" was capped with avidin via the biotin component of the DNAzyme. In the presence of Pb^{2+}, the substrate underwent degradation as a result of enzyme activation and following pore uncapping occurred. The Pb^{2+} triggered fluorescein release, which was proportional to the ion concentration. The impact of other ions on the investigated system was negligible; thus, it was concluded that the system is Pb^{2+} selective [53]. The Fe^{2+} sensitive prodrug designed for Plasmodium infection treatment was developed based on microbial catabolism responsible for the formation of ferrous ions in a host organism. The drug molecules were conjugated with a 1,2,4-trioxolane ring serving as a sensing moiety. In the presence of Fe^{2+}, the 1,2,4-trioxolane ring underwent degradation accompanied by iron oxidation. Subsequently, the drug-sensor linker was eliminated, resulting in the activation of a drug molecule. The prodrug was characterized by a superior efficacy in the animal model, compared to a non-conjugated pharmaceutically active molecule [54]. A complex K^{+} and pH sensing system dedicated to drug liberation in a cancer microenvironment based on framework nucleic acid was evaluated by T. Li et al. The system was composed of tetrahedral DNA nanostructures combined with cholesterol molecules, as cell membrane anchoring agent, fluorophores as indicators, and a pharmacologically active AS1411 aptamer. In proper pH and in the presence of K^{+}, the system was subjected to supramolecular rearrangement, resulting in aptamer folding and subsequent aptamer release in the form of a G-quadruplex. Simultaneously, the spatial interaction between fluorescent probes became possible, resulting in a change in observed fluorescence [55]. The Ca^{2+} sensitive structure, based on a chiral imprinted cholesteric liquid crystalline polymer, was proposed as a possible system capable of ion serum level monitoring. The thin layer of the liquid crystalline phase was obtained via surface photopolymerization. Due to the presence of benzoic acid derivatives exhibiting an affinity to Ca^{2+}, the obtained system manifested a noticeable color change, selectively in the presence of these ions. System response was most pronounced in the concentration range observed in human plasma, and thus it was concluded that the obtained ion-sensitive sensor could be employed as a convenient test for serum calcium levels [56]. The phenomenon of a divalent ion-dependent change in emission spectra was reported for systems assembled via noncovalent interaction between single-strain 30-nucleotide DNA and carbon nanotubes sidewalls. The most pronounced impact on emission energy was observed in the presence of Hg^{2+}, which promoted a DNA transition from form B to conformation Z and a subsequent shift in band-gap fluorescence. The observed phenomenon was found to be fully reversible after ion removal. The potential application of heavy metal ion detection in body fluids was suggested. The most pronounced shift was observed in the presence of Hg^{2+}; nevertheless, the system exhibited minor affinity towards other ions, which may reduce detection sensitivity [57]. A nasal formulation supplemented with low methoxyl pectin, as an ion-responsive gelling agent, exhibited enhanced adhesiveness. Divalent ions present on the mucosal surface crosslinked the adjacent pectin chains via carboxyl groups resulting in an ordered network. As a result, dripping and throat flow were reduced. The described approach may be beneficial in the nasal administration route due to the prolonged exposure time and irritation alleviation [58]. A novel system composed of two hydrogel elements was proposed. The first gel exhibited an ion-induced adhesiveness towards the second one serving as a binding hydrogel component. The ion-responsive hydrogel was obtained via the grafting of β-cyclodextrin and 2,2′-bipyridyl moieties into polyacrylamide chains. The second binding hydrogel component was prepared by incorporation of N-tert-butyl acrylamide moieties capable of host-guest interaction with β-cyclodextrin. In the native state, 2,2′-bipyridyl moieties are complexed by β-cyclodextrin, and consequently, the adhesive properties were hindered. 2,2′-bipyridyl moieties competitively complexed cations, resulting in the liberation of β-cyclodextrins. The non-complexed cyclodextrins interacted with the N-tert-butyl acrylamide moieties of the binding hydrogel, and adhesion between the two components was observed. The type of cation impacted ion-responsive hydrogel properties leading to various results [59]. A complex system capable of mimicking biological Ca^{2+} was reported. To construct this nanostructure, a copolymer consisting of three components: N-isopropyl acrylamide, acrylamide-[4-(trifluoromethyl)phenyl]-2-thiourea, and DDDEEKC was immobilized on the surface of the nanoporous membrane. DDDEEKC is a Ca^{2+} responsive selective ion-binding heptapeptide. Acrylamide-[4-(trifluoromethyl)phenyl]-2-thiourea is bound to the DDDEEKC peptide via hydrogen bonds formed via thiourea moieties. In a sufficient concentration of Ca^{2+}, the hydrogen bonds underwent rearrangement leading to the transition of the copolymer from the globular to the coil form and thus leading to channel blocking. The system was described as highly selective and capable of operating in physiological ion concentrations [60]. A distinctive micellar system releasing cargo in reduced pH in the presence of chloride ions was designed to deliver a platinum-derived antitumor agent. Nanocarriers entered cancer cells via endocytosis, where (1,2-diaminocyclohexane) platinum aqua complexes were activated in the reduced pH and in the presence of nucleophilic Cl^- ions. The carrier exhibited enhanced antitumor activity and, due to drug delivery to the tumor cell cytoplasm, avoided drug-resistance mechanisms [61].

3. Summary

Ion-sensitive DDSs, with their capability to provide a variety of effects in a controlled manner, are a promising category of medical devices. The mechanisms of carrier response may be based on an ion-exchange phenomenon, non-covalent crosslinking, a salt screening effect, selective complexing, or enzyme activation. This variety of ion-induced phenomena translates into a number of pharmaceutical and pharmacological effects occurring after exposition to these stimuli. The most explored type of ion-responsive behavior is an ion-triggered release of small pharmaceutically active molecules and macromolecular species. Drugs are liberated from the described systems with different selectivity towards particular ions. Non-selective DDSs release cargo mainly via ion exchange and salt screening effect. A release in response to specific ions was achieved in porous MOFs, where pore size and charge were suited for particular ions. Other selective approaches were based on ion-specific molecular switches and macromolecules such as enzymes and polynucleotides. Conformational changes occurring in the presence of specific ions are responsible for the release-inducing phenomena. Apart from drug delivery purposes, the described approaches may be applicable in a DDS suspended in aqueous media to avoid drug leak during storage. In situ gelling matrices enabled increased residence time with an associated change in drug delivery rate. Most of these approaches were based on ionically-crosslinked polysaccharides for topical administration of antimicrobial and antiglaucoma drugs. Nevertheless, DDSs aimed at oral, intrauterine, dental, or intravenous administration were proposed. Polysaccharide-based DDSs were generally well tolerated and non-toxic, which makes them promising candidates for clinical applications. The ion-induced rearrangement of carrier structure also resulted in the emission of diagnostic signals, which in combination with drug release, may be highly beneficial in terms of controlled drug deposition. Especially in the case of chemotherapeutics with a narrow therapeutic index, such an approach may greatly improve treatment safety. Notwithstanding, significantly less attention was paid to the ion-sensitivity approach in comparison to pH-sensitivity, enhanced permeability effect, or thermosensitive DDS [62]. Most of the carriers described above operate in physiological millimolar ion concentrations. However, physiological fluctuations observed in extracellular fluids occurred in a narrow range and varied between individuals, which presumably is insufficient to induce a proper response [63]. On the contrary, more significant variations in ion concentration were observed in the intracellular environment of particular tissues, both in physiological and pathological conditions [64–66]. An approach combining active targeting and intracellular ion-dependent drug activation is considered advantageous and was employed in the design of a carrier of drug currently under clinical trials [60]. Due to the variety of available ion-sensitive switches, chelating agents, and ionic polymers, further advancements in the described category of DDSs are expected. In the case of topical formulations, a simple addition of an in situ gelling agent may be an extremely beneficial and safe way to increase therapeutic efficacy. Systemically administered carriers dedicated to more precise drug delivery may benefit from ion sensitivity, mainly by a reduction in premature drug release. However, the most beneficial outcomes are expected as a result of the combination of ion-sensitive properties with other smart drug delivery strategies. The ongoing development of multi-stimuli responsive smart carriers will presumably result in applicable solutions with an ion-sensitive component.

Author Contributions: Conceptualization, W.M. and T.U.; methodology, T.U. and M.R.; resources, W.M.; data curation, T.U., M.R.; writing—original draft preparation, M.R. and T.U.; writing—review and editing, W.M. and T.U.; visualization, T.U.; supervision, W.M.; funding acquisition, W.M. All authors have read and agreed to the published version of the manuscript.

Funding: This research was funded by Wroclaw Medical, University, grant number ST.D060.21.054.

Institutional Review Board Statement: Not applicable.

Informed Consent Statement: Not applicable.

Data Availability Statement: Not applicable.

Conflicts of Interest: The authors declare no conflict of interest.

References

1. Manzari, M.T.; Shamay, Y.; Kiguchi, H.; Rosen, N.; Scaltriti, M.; Heller, D.A. Targeted drug delivery strategies for precision medicines. *Nat. Rev. Mater.* **2021**, *6*, 351–370. [CrossRef]
2. James, H.P.; John, R.; Alex, A.; Anoop, K.R. Smart polymers for the controlled delivery of drugs—A concise overview. *Acta Pharm. Sin. B* **2014**, *4*, 120–127. [CrossRef] [PubMed]
3. Kotze, M.J.; van Velden, D.P.; van Rensburg, S.J.; Erasmus, R. Pathogenic mechanisms underlying iron deficiency and iron overload: New insights for clinical application. *eJIFCC* **2009**, *20*, 108–123. [PubMed]
4. Portbury, S.D.; Adlard, P.A. Zinc signal in brain diseases. *Int. J. Mol. Sci.* **2017**, *18*, 2506. [CrossRef] [PubMed]
5. Reid, I.R.; Gamble, G.D.; Bolland, M.J. Circulating calcium concentrations, vascular disease and mortality: A systematic review. *J. Intern. Med.* **2016**, *279*, 524–540. [CrossRef] [PubMed]
6. Vardhan, K.H.; Kumar, P.S.; Panda, R.C. A review on heavy metal pollution, toxicity and remedial measures: Current trends and future perspectives. *J. Mol. Liq.* **2019**, *290*, 111197. [CrossRef]
7. Zia, K.M.; Tabasum, S.; Khan, M.F.; Akram, N.; Akhter, N.; Noreen, A.; Zuber, M. Recent trends on gellan gum blends with natural and synthetic polymers: A review. *Int. J. Biol. Macromol.* **2018**, *109*, 1068–1087. [CrossRef] [PubMed]
8. Burke, W. The ionic composition of nasal fluid and its function. *Health (Irvine. Calif.)* **2014**, *6*, 720–728. [CrossRef]
9. Fijorek, K.; Püsküllüoğlu, M.; Tomaszewska, D.; Tomaszewski, R.; Glinka, A.; Polak, S. Serum potassium, sodium and calcium levels in healthy individuals—Literature review and data analysis. *Folia Med. Cracov.* **2014**, *54*, 53–70.
10. Ruiz-Ederra, J.; Levin, M.H.; Verkman, A.S. In situ fluorescence measurement of tear film [Na^+], [K^+], [Cl^-], and pH in mice shows marked hypertonicity in aquaporin-5 deficiency. *Investig. Ophthalmol. Vis. Sci.* **2009**, *50*, 2132–2138. [CrossRef]
11. Cao, S.L.; Ren, X.W.; Zhang, Q.Z.; Chen, E.; Xu, F.; Chen, J.; Liu, L.C.; Jiang, X.G. In situ gel based on gellan gum as new carrier for nasal administration of mometasone furoate. *Int. J. Pharm.* **2009**, *365*, 109–115. [CrossRef]
12. Belgamwar, V.S.; Chauk, D.S.; Mahajan, H.S.; Jain, S.A.; Gattani, S.G.; Surana, S.J. Formulation and evaluation of in situ gelling system of dimenhydrinate for nasal administration. *Pharm. Dev. Technol.* **2009**, *14*, 240–248. [CrossRef]
13. Wang, S.; Chen, P.; Zhang, L.; Yang, C.; Zhai, G. Formulation and evaluation of microemulsion-based in situ ion-sensitive gelling systems for intranasal administration of curcumin. *J. Drug Target.* **2012**, *20*, 831–840. [CrossRef]
14. Gänger, S.; Schindowski, K. Tailoring formulations for intranasal nose-to-brain delivery: A review on architecture, physico-chemical characteristics and mucociliary clearance of the nasal olfactory mucosa. *Pharmaceutics* **2018**, *10*, 116. [CrossRef]
15. Sun, Y.; Li, L.; Xie, H.; Wang, Y.; Gao, S.; Zhang, L.; Bo, F.; Yang, S.; Feng, A. Primary studies on construction and evaluation of ion-sensitive in situ gel loaded with paeonol-solid lipid nanoparticles for intranasal drug delivery. *Int. J. Nanomed.* **2020**, *15*, 3137–3160. [CrossRef]
16. Hao, J.; Zhao, J.; Zhang, S.; Tong, T.; Zhuang, Q.; Jin, K.; Chen, W.; Tang, H. Fabrication of an ionic-sensitive in situ gel loaded with resveratrol nanosuspensions intended for direct nose-to-brain delivery. *Colloids Surfaces B Biointerfaces* **2016**, *147*, 376–386. [CrossRef]
17. Rajput, A.P.; Butani, S.B. Fabrication of an ion-sensitive in situ gel loaded with nanostructured lipid carrier for nose to brain delivery of donepezil. *Asian J. Pharm.* **2018**, *12*, 6–11.
18. Sun, J.; Zhou, Z. A novel ocular delivery of brinzolamide based on gellan gum: In vitro and in vivo evaluation. *Drug Des. Devel. Ther.* **2018**, *12*, 383–389. [CrossRef]
19. Patel, P.; Patel, G. Formulation, ex-vivo and preclinical in-vivo studies of combined ph and ion-sensitive ocular sustained in situ hydrogel of timolol maleate for the treatment of glaucoma. *Biointerface Res. Appl. Chem.* **2021**, *11*, 8242–8265.
20. Yu, S.; Wang, Q.M.; Wang, X.; Liu, D.; Zhang, W.; Ye, T.; Yang, X.; Pan, W. Liposome incorporated ion sensitive in situ gels for opthalmic delivery of timolol maleate. *Int. J. Pharm.* **2015**, *480*, 128–136. [CrossRef]
21. Sultana, Y.; Aqil, M.; Ali, A. ion-activated, gelrite®-based in situ ophthalmic gels of pefloxacin mesylate: Comparison with conventional eye drops. *Drug Deliv. J. Deliv. Target. Ther. Agents* **2006**, *13*, 215–219. [CrossRef] [PubMed]
22. Nayak, N.S.; Srinivasa, U. Design and evaluation of ion activated in situ ophthalmic gel of moxifloxacin hydrochloride and ketorolac tromethamine combination using carboxy methylated tamarind kernel powder. *Saudi J. Med. Pharm. Sci.* **2017**, *3*, 1–8.
23. Janga, K.Y.; Tatke, A.; Balguri, S.P.; Lamichanne, S.P.; Ibrahim, M.M.; Maria, D.N.; Jablonski, M.M.; Majumdar, S. ion-sensitive in situ hydrogels of natamycin bilosomes for enhanced and prolonged ocular pharmacotherapy: In vitro permeability, cytotoxicity and in vivo evaluation. *Artif. Cells Nanomedicine Biotechnol.* **2018**, *46*, 1039–1050. [CrossRef] [PubMed]
24. Terreni, E.; Zucchetti, E.; Tampucci, S.; Burgalassi, S.; Monti, D.; Chetoni, P. Combination of nanomicellar technology and in situ gelling polymer as ocular drug delivery system (Odds) for cyclosporine-a. *Pharmaceutics* **2021**, *13*, 192. [CrossRef]
25. Göttel, B.; de Souza e Silva, J.M.; Santos de Oliveira, C.; Syrowatka, F.; Fiorentzis, M.; Viestenz, A.; Viestenz, A.; Mäder, K. Electrospun nanofibers—A promising solid in-situ gelling alternative for ocular drug delivery. *Eur. J. Pharm. Biopharm.* **2020**, *146*, 125–132. [CrossRef]
26. Zhang, X.; Pan, Y.; Li, S.; Xing, L.; Du, S.; Yuan, G.; Li, J.; Zhou, T.; Xiong, D.; Tan, H.; et al. Doubly crosslinked biodegradable hydrogels based on gellan gum and chitosan for drug delivery and wound dressing. *Int. J. Biol. Macromol.* **2020**, *164*, 2204–2214. [CrossRef]
27. Ng, J.Y.; Zhu, X.; Mukherjee, D.; Zhang, C.; Hong, S.; Kumar, Y.; Gokhale, R.; Ee, P.L.R. Pristine gellan gum–collagen interpenetrating network hydrogels as mechanically enhanced anti-inflammatory biologic wound dressings for burn wound therapy. *ACS Appl. Bio Mater.* **2021**, *4*, 1470–1482. [CrossRef]

28. Bellini, D.; Cencetti, C.; Meraner, J.; Stoppoloni, D.; D'Abusco, A.S.; Matricardi, P. An in situ gelling system for bone regeneration of osteochondral defects. *Eur. Polym. J.* **2015**, *72*, 642–650. [CrossRef]
29. Harish, N.; Prabhu, P.; Charyulu, R.; Gulzar, M.; Subrahmanyam, E.V. Formulation and evaluation of *in situ* gels containing clotrimazole for oral candidiasis. *Indian J. Pharm. Sci.* **2009**, *71*, 421. [CrossRef]
30. Abd Ellah, N.H.; Abouelmagd, S.A.; Abbas, A.M.; Shaaban, O.M.; Hassanein, K.M.A. Dual-responsive lidocaine in situ gel reduces pain of intrauterine device insertion. *Int. J. Pharm.* **2018**, *538*, 279–286. [CrossRef]
31. Hecht, H.; Srebnik, S. Structural characterization of sodium alginate and calcium alginate. *Biomacromolecules* **2016**, *17*, 2160–2167. [CrossRef] [PubMed]
32. Liu, Z.; Li, J.; Nie, S.; Liu, H.; Ding, P.; Pan, W. Study of an alginate/HPMC-based in situ gelling ophthalmic delivery system for gatifloxacin. *Int. J. Pharm.* **2006**, *315*, 12–17. [CrossRef] [PubMed]
33. Pandya, T.P.; Modasiya, M.K.; Patel, V.M. Sustained ophthalmic delivery of ofloxacin hydrochloride from an ion-activated in situ gelling system. *Der Pharm. Lett.* **2011**, *3*, 404–410.
34. Shelley, H.; Rodriguez-Galarza, R.M.; Duran, S.H.; Abarca, E.M.; Babu, R.J. In situ gel formulation for enhanced ocular delivery of nepafenac. *J. Pharm. Sci.* **2018**, *107*, 3089–3097. [CrossRef] [PubMed]
35. Kubo, W.; Miyazaki, S.; Attwood, D. Oral sustained delivery of paracetamol from in situ-gelling gellan and sodium alginate formulations. *Int. J. Pharm.* **2003**, *258*, 55–64. [CrossRef]
36. Hu, C.; Feng, H.; Zhu, C. Preparation and characterization of rifampicin-PLGA microspheres/sodium alginate in situ gel combination delivery system. *Colloids Surfaces B Biointerfaces* **2012**, *95*, 162–169. [CrossRef] [PubMed]
37. Hori, Y.; Winans, A.M.; Irvine, D.J. Modular injectable matrices based on alginate solution/microsphere mixtures that gel in situ and co-deliver immunomodulatory factors. *Acta Biomater.* **2009**, *5*, 969–982. [CrossRef]
38. Zhu, Q.; Wei, Y.; Li, C.; Mao, S. Inner layer-embedded contact lenses for ion-triggered controlled drug delivery. *Mater. Sci. Eng. C* **2018**, *93*, 36–48. [CrossRef]
39. Liu, H.; Zhu, J.; Bao, P.; Ding, Y.; Shen, Y.; Webster, T.J.; Xu, Y. Construction and in vivo/in vitro evaluation of a nanoporous ion-responsive targeted drug delivery system for recombinant human interferon α-2b delivery. *Int. J. Nanomedicine* **2019**, *14*, 5339–5353. [CrossRef]
40. Zhang, B.; Sun, B.; Li, X.; Yu, Y.; Tian, Y.; Xu, X.; Jin, Z. Synthesis of pH- and ionic strength-responsive microgels and their interactions with lysozyme. *Int. J. Biol. Macromol.* **2015**, *79*, 392–397. [CrossRef]
41. Zheng, Y.; Sun, J.; Jin, X.; Wu, X. Influence of ionic strength on the ph-sensitive in vitro ibuprofen release from dextran-poly(acrylic acid) copolymer. *Indian J. Pharm. Sci.* **2018**, *80*, 298–306. [CrossRef]
42. Bodmeier, R.; Guo, X.; Sarabia, R.E.; Skultety, P.F. The influence of buffer species and strength on diltiazem HCl release from beads coated with the aqueous cationic polymer dispersions, eudragit RS, RL 30D. *Pharm. Res.* **1996**, *13*, 52–56. [CrossRef]
43. Dragan, E.S.; Cocarta, A.I. Smart macroporous IPN hydrogels responsive to pH, temperature, and ionic strength: Synthesis, characterization, and evaluation of controlled release of drugs. *ACS Appl. Mater. Interfaces* **2016**, *8*, 12018–12030. [CrossRef]
44. Jarvinen, K.; Akerman, S.; Svarfvar, B.; Tarvainen, T.; Viinikka, P.; Paronen, P. Drug release from pH and ionic strength responsive poly(acrylic acid) grafted poly(vinylidenefluoride) membrane bags in vitro. *Pharm. Res.* **1998**, *15*, 802. [CrossRef] [PubMed]
45. An, J.; Geib, S.J.; Rosi, N.L. Cation-triggered drug release from a porous zinc-adeninate metal-organic framework. *J. Am. Chem. Soc.* **2009**, *131*, 8376–8377. [CrossRef] [PubMed]
46. Tan, L.L.; Li, H.; Zhou, Y.; Zhang, Y.; Feng, X.; Wang, B.; Yang, Y.W. Zn^{2+}-triggered drug release from biocompatible zirconium MOFs equipped with supramolecular gates. *Small* **2015**, *11*, 3807–3813. [CrossRef]
47. Du, X.; Fan, R.; Qiang, L.; Xing, K.; Ye, H.; Ran, X.; Song, Y.; Wang, P.; Yang, Y. Controlled Zn^{2+}-triggered drug release by preferred coordination of open active sites within functionalization indium metal organic frameworks. *ACS Appl. Mater. Interfaces* **2017**, *9*, 28939–28948. [CrossRef]
48. Wu, M.X.; Gao, J.; Wang, F.; Yang, J.; Song, N.; Jin, X.; Mi, P.; Tian, J.; Luo, J.; Liang, F.; et al. Multistimuli responsive core–shell nanoplatform constructed from Fe_3O_4@MOF equipped with pillar[6]arene nanovalves. *Small* **2018**, *14*, 1–6.
49. Lou, J.; Best, M.D. *Calcium-Responsive Liposomes: Toward ion-Mediated Targeted Drug Delivery*, 1st ed.; Elsevier Inc.: Houston, TX, USA, 2020; Volume 640, ISBN 9780128211533.
50. Yigit, M.V.; Mishra, A.; Tong, R.; Cheng, J.; Wong, G.C.L.; Lu, Y. Inorganic mercury detection and controlled release of chelating agents from ion-responsive liposomes. *Chem. Biol.* **2009**, *16*, 937–942. [CrossRef]
51. Veremeeva, P.N.; Lapteva, V.L.; Palyulin, V.A.; Sybachin, A.V.; Yaroslavov, A.A.; Zefirov, N.S. Bispidinone-based molecular switches for construction of stimulus-sensitive liposomal containers. *Tetrahedron* **2014**, *70*, 1408–1411. [CrossRef]
52. Tao, M.; Liu, J.; He, S.; Xu, K.; Zhong, W. In situ hydrogelation of forky peptides in prostate tissue for drug delivery. *Soft Matter* **2019**, *15*, 4200–4207. [CrossRef]
53. Song, W.; Li, J.; Li, Q.; Ding, W.; Yang, X. Avidin-biotin capped mesoporous silica nanoparticles as an ion-responsive release system to determine lead(II). *Anal. Biochem.* **2015**, *471*, 17–22. [CrossRef]
54. Deu, E.; Chen, I.T.; Lauterwasser, E.M.W.; Valderramos, J.; Li, H.; Edgington, L.E.; Renslo, A.R.; Bogyo, M. Ferrous iron-dependent drug delivery enables controlled and selective release of therapeutic agents in vivo. *Proc. Natl. Acad. Sci. USA* **2013**, *110*, 18244–18249. [CrossRef]
55. Peng, P.; Wang, Q.; Du, Y.; Wang, H.; Shi, L.; Li, T. Extracellular ion-responsive logic sensors utilizing DNA dimeric nanoassemblies on cell surface and application to boosting AS1411 internalization. *Anal. Chem.* **2020**, *92*, 9273–9280. [CrossRef]

56. Moirangthem, M.; Arts, R.; Merkx, M.; Schenning, A.P.H.J. An optical sensor based on a photonic polymer film to detect calcium in serum. *Adv. Funct. Mater.* **2016**, *26*, 1154–1160. [CrossRef]
57. Heller, D.H.; Jeng, E.S.; Yeung, T.-K.; Martinez, B.M.; Moll, A.E.; Gastala, J.B.; Strano, M.S. Optical detection of DNA conformational polymorphism on single-walled carbon nanotubes. *Science* **2006**, *311*, 508–511. [CrossRef]
58. Castile, J.; Cheng, Y.H.; Simmons, B.; Perelman, M.; Smith, A.; Watts, P. Development of in vitro models to demonstrate the ability of PecSys®, an in situ nasal gelling technology, to reduce nasal run-off and drip. *Drug Dev. Ind. Pharm.* **2013**, *39*, 816–824. [CrossRef]
59. Nakamura, T.; Takashima, Y.; Hashidzume, A.; Yamaguchi, H.; Harada, A. A metal-ion-responsive adhesive material via switching of molecular recognition properties. *Nat. Commun.* **2014**, *5*, 1–9. [CrossRef]
60. Li, Y.; Xiong, Y.; Wang, D.; Li, X.; Chen, Z.; Wang, C.; Qin, H.; Liu, J.; Chang, B.; Qing, G. Smart polymer-based calcium-ion self-regulated nanochannels by mimicking the biological Ca^{2+}-induced Ca^{2+} release process. *NPG Asia Mater.* **2019**, *11*, 1–13. [CrossRef]
61. Murakami, M.; Cabral, H.; Matsumoto, Y.; Wu, S.; Kano, M.R.; Yamori, T.; Nishiyama, N.; Kataoka, K. Improving drug potency and efficacy by nanocarrier-mediated subcellular targeting. *Sci. Transl. Med.* **2011**, *3*, 64ra2. [CrossRef]
62. Li, J.; Kataoka, K. Chemo-physical Strategies to Advance the in vivo functionality of targeted nanomedicine: The next generation. *J. Am. Chem. Soc.* **2021**, *143*, 538–559. [CrossRef] [PubMed]
63. Smith, G.L.; Eisner, D.A. Calcium buffering in the heart in health and disease. *Circulation* **2019**, *139*, 2358–2371. [CrossRef] [PubMed]
64. Litan, A.; Langhans, S.A. Cancer as a channelopathy: Ion channels and pumps in tumor development and progression. *Front. Cell. Neurosci.* **2015**, *9*, 1–11. [CrossRef]
65. Bagur, R.; Hajnóczky, G. Intracellular Ca^{2+} sensing: Role in calcium homeostasis and signaling. *Mol. Cell* **2017**, *66*, 1–18. [CrossRef] [PubMed]
66. Bafaro, E.; Liu, Y.; Xu, Y.; Dempski, R.E. The emerging role of zinc transporters in cellular homeostasis and cancer. *Signal Transduct. Target. Ther.* **2017**, *2*, 1–12. [CrossRef]

polymers

MDPI

Article

Enhancement of the Thermal Performance of the Paraffin-Based Microcapsules Intended for Textile Applications

Virginija Skurkyte-Papieviene [1,*], Ausra Abraitiene [1], Audrone Sankauskaite [1], Vitalija Rubeziene [2] and Julija Baltusnikaite-Guzaitiene [2]

1 Department of Textile Technologies, Center for Physical Sciences and Technology, 48485 Kaunas, Lithuania; ausra.abraitiene@ftmc.lt (A.A.); audrone.sankauskaite@ftmc.lt (A.S.)
2 Department of Textiles Physical-Chemical Testing, Center for Physical Sciences and Technology, 48485 Kaunas, Lithuania; vitalija.rubeziene@ftmc.lt (V.R.); julija.baltusnikaite@ftmc.lt (J.B.-G.)
* Correspondence: virginija.skurkyte@ftmc.lt

Abstract: Phase changing materials (PCMs) microcapsules MPCM32D, consisting of a polymeric melamine-formaldehyde (MF) resin shell surrounding a paraffin core (melting point: 30–32 °C), have been modified by introducing thermally conductive additives on their outer shell surface. As additives, multiwall carbon nanotubes (MWCNTs) and poly (3,4-ethylenedioxyoxythiophene) poly (styrene sulphonate) (PEDOT: PSS) were used in different parts by weight (1 wt.%, 5 wt.%, and 10 wt.%). The main aim of this modification—to enhance the thermal performance of the microencapsulated PCMs intended for textile applications. The morphologic analysis of the newly formed coating of MWCNTs or PEDOT: PSS microcapsules shell was observed by SEM. The heat storage and release capacity were evaluated by changing microcapsules MPCM32D shell modification. In order to evaluate the influence of the modified MF outer shell on the thermal properties of paraffin PCM, a thermal conductivity coefficient (λ) of these unmodified and shell-modified microcapsules was also measured by the comparative method. Based on the identified optimal parameters of the thermal performance of the tested PCM microcapsules, a 3D warp-knitted spacer fabric from PET was treated with a composition containing 5 wt.% MWCNTs or 5 wt.% PEDOT: PSS shell-modified microcapsules MPCM32D and acrylic resin binder. To assess the dynamic thermal behaviour of the treated fabric samples, an IR heating source and IR camera were used. The fabric with 5 wt.% MWCNTs or 5 wt.% PEDOT: PSS in shell-modified paraffin microcapsules MPCM32D revealed much faster heating and significantly slower cooling compared to the fabric treated with the unmodified ones. The thermal conductivity of the investigated fabric samples with modified microcapsules MPCM32D has been improved in comparison to the fabric samples with unmodified ones. That confirms the positive influence of using thermally conductive enhancing additives for the heat transfer rate within the textile sample containing these modified paraffin PCM microcapsules.

Keywords: paraffin PCM—melamine-formaldehyde microcapsules; outer shell; modification by MWCNTs and PEDOT: PSS; differential scanning calorimetry; dip coating; thermal conductivity and heat storage and release capacity; dynamic thermal behaviour

Citation: Skurkyte-Papieviene, V.; Abraitiene, A.; Sankauskaite, A.; Rubeziene, V.; Baltusnikaite-Guzaitiene, J. Enhancement of the Thermal Performance of the Paraffin-Based Microcapsules Intended for Textile Applications. *Polymers* **2021**, *13*, 1120. https://doi.org/10.3390/polym13071120

Academic Editor: M. Ali Aboudzadeh

Received: 24 February 2021
Accepted: 26 March 2021
Published: 1 April 2021

1. Introduction

Nowadays the textile industry is more dynamic than ever before and successfully combines together the long-standing traditions with rapid technological progress. These trends are reflected in the increased focus on phase-changing materials (PCMs) known as "smart materials" which can improve thermal insulation and thermal comfort of textiles, through the thermo-regulating effect of the PCMs. The active temperature-regulating ability of PCMs nowadays is widely applied in thermal management of such areas as buildings, electronics, medicine, automotive, textiles. Considering application in textiles, PCMs integrated into various fibres and fabrics provide temporary warmth or coolness effect, thereby diminishing thermal discomfort [1–5]. Among available PCMs, organic PCMs with a phase change temperature range of 18 °C to 35 °C are the most appropriate for textile application [5]. Paraffinic hydrocarbons in the solid-liquid phase in the organic materials group are the most preferred and practical materials for the production of microencapsulated PCMs (MPCMs) [6]. The thermal, physical, chemical, and mechanical properties of MPCMs are heavily dependent on the raw materials and synthesis processes during microencapsulation [7]. The usually applied encapsulation material for paraffin PCM is physically and chemically stable [8] melamine-formaldehyde (MF) resin [9–14]. The polyurea-formaldehyde resin [15], polystyrene [16], and poly (methylmethacrylate) derivatives [17] were investigated as well.

MPCMs can be introduced into the textile materials by several main finishing methods: coating [18–22], impregnation and exhaust [23,24], and embedding into the fibre during the polymer matrix spinning [25–27]. The problem of paraffin PCM is its low thermal conductivity, for example, paraffin has a thermal conductivity of 0.22 W/(m·K) [28] when compared with >3000 W/(m·K) for multiwall carbon nanotubes (MWCNTs) [29]. Moreover, microencapsulated PCMs have a polymeric shell, which not only prevents the content from leaking but also resists heat transition at the same time [30]. It was observed that the improvement of MPCMs thermal properties and especially the effective thermal conductivity depends on microcapsules structure [31]. One of the possibilities is to improve the thermal conductivity of the core materials.

Carbon-based nanostructures (nanofibers, nanoplatelets, and graphene flakes), carbon nanotubes, both metallic (Ag, Al, C/Cu, and Cu) and metal oxide (Al_2O_3, CuO, MgO, and TiO_2) nanoparticles and silver nanowires were investigated as the thermal conductivity promoters for materials of the PCMs [32–34]. However, it is difficult to ensure that the additives are uniformly distributed in the PCMs; moreover, the additives increase the weight of the PCMs and will decrease the latent heat storage capacity in general [35]. Another known opportunity is to enhance the thermal conductivity of MPCMs through the incorporation of nano-filler additives on polymeric shell materials. The thermal conductivity of the melamine urea-formaldehyde microcapsules containing paraffin PCM as a core has been significantly enhanced from 0.1944 W/(m·K) to 1.0540 W/(m·K) with a little influence to their enthalpy by a coating of 10 wt.% graphene sheets onto the polymeric shell [36]. The micro-PCM particles containing paraffin core with graphene/methanol modified melamine-formaldehyde hybrid shell have been successfully prepared and their microstructure and thermal performance were investigated as well [30,37]. Thermally conductive PCM paraffin-wax-embedded polymer microcapsules with graphene oxide (GO) platelet patched shell structure have been developed by researchers [38]. It was identified that the excellent thermal sensitivity of these microcapsules with GO platelet patched shell provides an efficient way to regulate thermal radiance according to the surrounding background, which made these microcapsule-embedded composites a promising material for active thermal camouflage and stealth applications. Electrically and thermally conductive paraffinic PCMs with melamine-formaldehyde shell microcapsules have been manufactured by coating with polypyrrole (PPy) and it is expected that these microcapsules will widen the application possibilities of PCMs in camouflage technology and electronic cooling [37].

Although different studies pertaining to the resistivity and conductivity properties of various fibre content fabrics modified by applying PPy [39] and PPy/carbon black composite [40] were analysed, a limited number of publications on the influence of the microencapsulated PCMs for the thermal resistance of textile materials were found. The researchers [41] found that the textile sample coated with polyethylene glycol (PEG) microcapsules during the testing with the Sweating Guarded-Hotplate apparatus has radiated 20% less heat than the untreated one.

This study is aimed to improve the thermal performance of paraffin PCM microcapsules for textile application by modifying their outer shell. For this purpose, the paraffin microcapsules MPCM32D with a transition temperature of 32.02 °C were modified using Layer-by-Layer (LbL) self-assembly technique to form multilayered thin coatings by electrostatic interaction among cationically charged MF resin shell and applied anionically charged thermal conductivity enhancing additives—MWCNTs or PEDOT: PSS. The presence of the layer of these additives on the outer shell of these modified PCM microcapsules was observed by a scanning electron microscope (SEM). The latent heat and thermal conductivity of the microcapsules MPCM32D modified with various concentrations of MWCNTs or PEDOT: PSS were measured, respectively, by the differential scanning calorimetry (DSC) technique and thermal conductivity determination device. The obtained results were evaluated by comparing them to the unmodified ones. In addition, a 3D warp-knitted spacer fabric from PET was dip-coated with shell-modified microcapsules MPCM32D that demonstrated optimal thermal characteristics, and its thermal performance—heat storage and release capacity, thermal conductivity, and dynamic thermal behaviour—was evaluated.

2. Materials and Methods

2.1. Materials

PCM microcapsules MPCM32D (composition: 17.4% melamine resin, 79.6% paraffin wax, water $\leq$ 3%) in a dry white powder were purchased from Microtek Laboratories Inc., Dayton, OH, USA. Multiwall carbon nanotubes NC7000 (average diameter: 9.5×10^{-9} m, average length: 1.5 μm) in form of 3 wt.% waterborne dispersion called Aquacyl AQ0302 were obtained from Nanocyl S.A., Sambreville, Belgium. Poly(3,4-ethylenedioxythiophene) poly(styrene sulfonate) (PEDOT: PSS) (1.3 wt.% dispersion in water, conductive grade), poly(diallyldimethylammonium chloride) (20 wt.% in water), sodium dodecylbenzenesulfonate (SDBS) and ethanol (95% denatured) were purchased from Sigma-Aldrich, Taufkirchen, Germany. Aqueous synthetic dispersion based on polyurethane (PU) Tubicoat MP and acrylic resin binder Itobinder PCM were obtained, respectively, from CHT Germany GmbH, Tübingen, DE and LJ Specialties Ltd., Chesterfield, Derbyshire UK.

2.1.1. Modification of the Outer Shell of PCM Microcapsules with the Thermally Conductive Additives

The MF resin shell of the powdered PCM microcapsules MPCM32D was modified with thermally conductive additives MWCNTs and PEDOT: PSS, respectively, using a layer-by-layer self-assembly method [42]. First, 10 g of these PCM microcapsules and 1.2 g of cationic surfactant poly(diallyldimethylammonium chloride) (20 wt.% aqueous solution) were dispersed in 500 mL of ethanol and stirred at room temperature for 10 min using a high-speed disperser (1000 rpm). In parallel, the different mass fractions (1 wt.%, 5 wt.%, and 10 wt.%), respectively, 3.3 g, 16.5 g, and 33 g of MWCNTs 3% waterborne dispersion Aquacyl AQ0302, or 7.5 g, 37.5 g and 75g of 1.3 wt.% PEDOT: PSS aqueous dispersions and 0.1 g of the SDBS (anionic surfactants) were added to 500 mL of deionized water and stirred with a high-speed disperser (5000 rpm) for another 10 min. Finally, the prepared suspensions of ionized microcapsules MPCM32D and different parts of the thermal conductivity enhancing additives calculated by mass weight, respectively, were mixed under intensive shaking at 20 $\pm$ 5 °C for 60 min (considering the fact that positively charged outer shell of the microcapsules could attract the negatively charged conductive additives due to the electrostatic interaction). Then the powder of microcapsules was filtered and washed several times with 40 °C deionized water. The MWCNTs shell-modified PCM microcapsules MPCM32D were dried at room temperature. In case of modification with PEDOT: PSS, these microcapsules were dried in an oven for 10 min at 100 °C temperature for PEDOT: PSS film formation.

2.1.2. Introduction of Shell-Modified PCM Microcapsules into Textile Materials

The unmodified and shell-modified PCM microcapsules analysed in this study were embedded in three-dimensional (3D) warp-knitted spacer fabric from PET and elastane, the technical data of which are presented in Table 1.

Table 1. Characteristics of 3D warp-knitted spacer fabric.

3D Warp-Knitted Spacer Fabric View—Warp-Wise	Type of Yarn and Linear Density, Tex	Content of Yarn, %	Mass per Unit Area, g/m^2	Thickness, mm
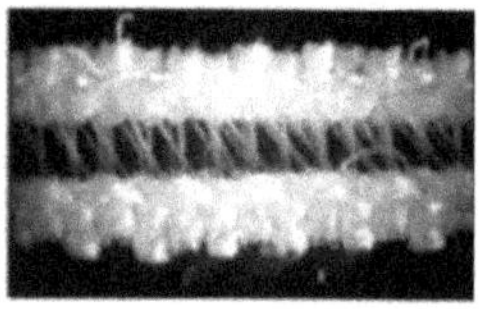	PET textured, 11.0	70		
	PET monofilament, 5.6	20	358	2.9
	Elastane, 7.8	10		

This knitted fabric was impregnated, respectively, with 5 wt.% MWCNTs or 5 wt.% PEDOT-PSS shell-modified PCM microcapsules MPCM32D, that demonstrated the best latent heat capacity and thermal conductivity characteristics, in composition with acrylic resin binder Itobinder PCM on a laboratory padder EVP-350 (Roaches International Ltd., West Yorkshire, UK) according to the recipe and process parameters presented in Table 2.

Table 2. Parameters of dip coating and drying-curing processes.

No.	Process	Auxiliaries		Parameters
		Microcapsules MPCM32D, g/L	Itobinder PCM, g/L	
1	Dip coating with microcapsules MPCM32D: Unmodified 5 wt.% MWCNTs shell-modified 5 wt.% PEDOT: PSS shell-modified	66	200	Wet pick-up: 80% Nip rolls: 2 bar
2	Drying—curing			Temperature: 120 °C Time: 6–8 min

2.2. *Methods of Investigation*

2.2.1. SEM Analysis

The surface morphology of the shell-modified PCM microcapsules MPCM32D and textile samples with these microcapsules was examined applying scanning electron microscopy (SEM) and using a Quanta 200 FEG device (FEI, Eindhoven, Netherlands) at low vacuum, 80 Pa, detector—LFD. All microscopic images were made under the same technical and technological conditions: electron beam heating voltage (10.00–30.00) kV, beam spot (3.0–5.0), magnification of 5000× and 10,000×, work distance (6.0–10.0) mm.

2.2.2. DSC Analysis

The thermal and phase transition characteristics of the unmodified and shell-modified PCM microcapsules, as well as of the textile fabric containing these microcapsules were determined by standard method EN 16806-1 [43], using DSC module (DSC Q10, TA Instruments, New Castle, DE, USA) equipment under a nitrogenous atmosphere. The mass of test specimens was about 10 mg. A lid was pressed on the crucible to ensure good contact between the specimen and the bottom of the crucible. The micro-encapsulated PCM samples were dried at 60 °C for 24 h to remove all water. The samples underwent a heating-cooling-heating cycle from −20 °C to +60 °C at a heating and cooling rate of 5 °C/min. Based on the recorded second heating cycle, enthalpy of fusion in J/g was determined by measuring the area under the peak to the baseline constructed, using software of DSC Q10. In the same way, the enthalpy of crystallization in J/g was determined based on the recorded cooling cycle.

The transition temperatures—peak melting and peak crystallization temperatures—extrapolated onset and end melting as well as crystallization temperatures, were also determined based on respectively the recorded second heating or based on the recorded cooling cycle, as defined in EN ISO 11357-3 [44].

2.2.3. Thermal Conductivity of PCM Microcapsules

Primarily, the unmodified and shell-modified PCM microcapsules have been incorporated into a polymeric matrix of polyurethane (PU) aqueous dispersion TUBICOAT MP and the homogeneous tablets in diameter of 55 mm and thickness of 3 mm were prepared. For this purpose, the 3 g of unmodified and shell-modified microcapsules MPCM32D and 3 g of TUBICOAT MP were intensively stirred until rigid consistency paste was achieved. Finally, the prepared mass was inserted into the round metal form and dried at 50 °C for about 8 h. For easier removal of the content from the form, a cellulose film was placed at the bottom. In the beginning, this film was overheated at a higher temperature than the drying temperature of the tablets to prevent the film from shrinking. Next, the tablets were stored in a desiccator for 24 h. The images of the formed tablets are shown in Figure 1.

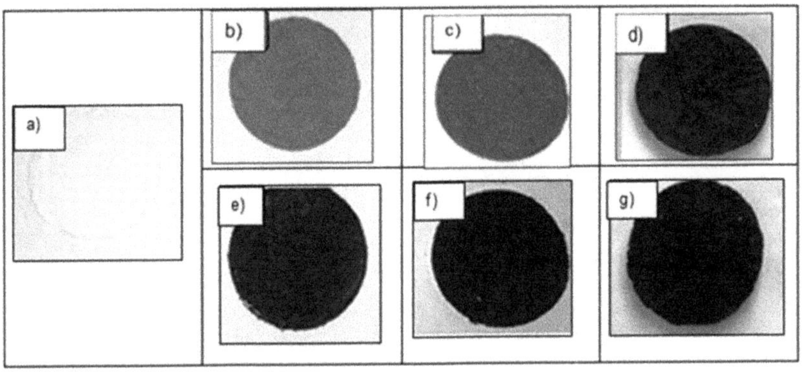

Figure 1. Images of unmodified microcapsules (**a**) and shell-modified MPCM32D microcapsules with different mass fraction: (**b**) poly (3,4-ethylenedioxyoxythiophene) poly (styrene sulphonate) (PEDOT: PSS) 1 wt.%, (**c**) PEDOT: PSS 5 wt.%, (**d**) PEDOT: PSS 10 wt.%, (**e**) multiwall carbon nanotubes (MWCNTs) 1 wt.%, (**f**) MWCNTs 5 wt.%, (**g**) MWCNTs 10 wt.% in a PU matrix.

The thermal conductivity measurements of the shell-modified microcapsules MPCM32D in a PU matrix were performed using an original thermal conductivity determination device [45] where the known thermal conductivity material as a reference was compared with the testing one. Next, these shell-modified PCM microcapsules with the best thermal performance were introduced into the textile.

2.2.4. Thermal Conductivity of Textile Materials Containing PCM Microcapsules

The thermal conductivity of the developed 3D warp-knitted spacer PET fabric containing microcapsules MPCM32D was evaluated based on the measurements of the reciprocal parameter—thermal resistance (R_{ct}). This parameter was measured under steady-state conditions using sweating guarded—hotplate M259B (SDL Atlas, Rock Hill, SC, USA) according to the standard method EN ISO 11092 [46]. The principle of this method is that the sample to be tested is placed on an electrically heated plate with conditioned air ducted flow across and parallel to its upper surface. For the determination of the thermal resistance, the heat flux through the test specimen is measured after steady-state conditions have been reached. Thermal resistance was calculated by the formula [46]:

$$R_{ct} = \frac{(T_m - T_a)\,A}{H - \Delta H_c} - R_{ct0}\left(\mathrm{m^2\,K/W}\right), \tag{1}$$

where T_m = 35 °C (temperature of the measuring unit); T_a = 20 °C (air temperature in the test enclosure); H—heating power supplied to the measuring unit, in watts; A—area of the measuring unit in square meters; R_{ct0}—the apparatus constant, in square meters Kelvin per Watt; $\Delta H_c = 0$ (correction term for heating power).

The coefficient of thermal conductivity λ, W/(m·K) was calculated according to the formula [47]:

$$\lambda = D/R_{ct}\ (\mathrm{W/(m\cdot K)}), \quad (2)$$

where: D—thickness of the sample (m); R_{ct}—thermal resistance (m^2 K/W).

2.2.5. Evaluation of Dynamic Thermal Behaviour of Textile Materials Containing PCM Microcapsules

The dynamic thermal behaviour of 3D warp-knitted spacer PET fabric dip-coated with unmodified and shell-modified microcapsules MPCM32D were analysed by using a thermal camera (spectral range: $\lambda = 7.5 \div 13$ μm) InfraCAM (FLIR SYSTEMS AB, Täby, Sweden) and an IR emitting lamp (250 W, 240 V, λ = 500–3000 nm, Ø = 125 mm) as the heat source (Figure 2). After 4 min as the IR lamp was switched on, the hottest location on the testing table surface was detected by using the InfraCAM. Thereafter, a polystyrene foam plate with a flatly laid spacer fabric was centred on the marked hottest location. The samples were heated for 4 min with an IR lamp placed about 50 cm above the knitted fabric to achieve a temperature of 4–5 °C higher than the melting point (32.02 °C) of paraffin PCM. Then the lamp was switched off, the knitted fabric was allowed to cool for another 4 min. The fabric samples were observed by the thermal camera every 15 s during the entire 8 min of the test. Five temperature measurements for each sample were performed and the imaginary temperature value was calculated. The variation coefficient was <5%.

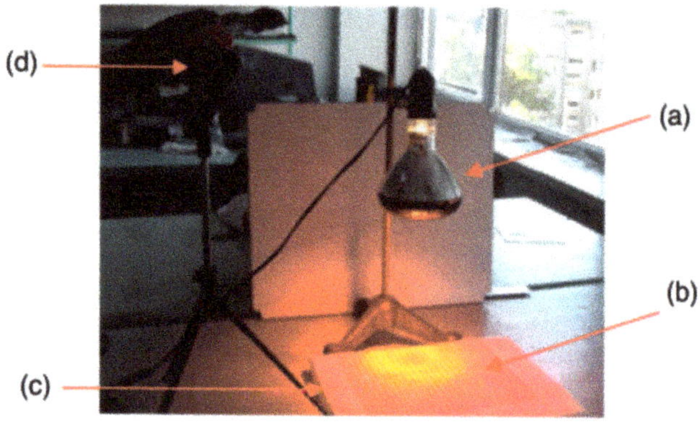

Figure 2. Stand for the analysis of fabric dynamic thermal behaviour: (**a**) infrared lamp; (**b**) 3D warp-knitted spacer PET fabric; (**c**) polystyrene foam plate; (**d**) thermal camera.

3. Results and Discussion

3.1. *Surface Morphology of The Shell-Modified PCM Microcapsules*

Before the microscopic analysis, the visual observation of the shell-modified microcapsules MPCM32D was performed. It was observed that after the modification with the thermally conductive additives MWCNTs or PEDOT: PSS, the initial white colour of these paraffin PCM microcapsules has changed to black or blue, respectively. The colour and its depth of these shell-modified microcapsules depended on the nature of the used additive and its concentration in the modification process (see Figure 3). It is especially expected that textile with the embedded blue coloured PEDOT: PSS shell-modified PCM microcapsules could be promising for leisure and protective clothing not only because of the enhanced thermal performance but also due to the expansion of colouristic possibilities to avoid the black colour.

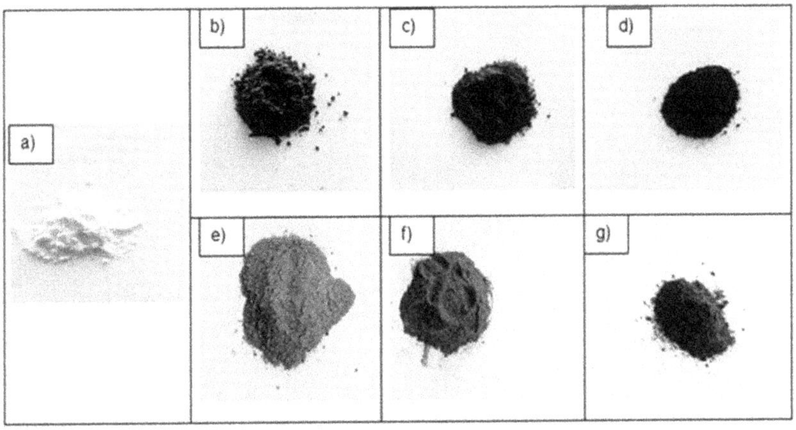

Figure 3. Pictures of microcapsules MPCM32D in powder form: unmodified (**a**) and shell-modified with different mass fraction: (**b**) MWCNTs 1 wt.%, (**c**) MWCNTs 5 wt.%, (**d**) MWCNTs 10 wt.%, (**e**) PEDOT: PSS 1 wt.%, (**f**) PEDOT: PSS 5 wt.%, (**g**) PEDOT: PSS 10 wt.%.

In order to estimate the surface morphology and the form of the shell-modified microcapsules MPCM32D, the SEM analysis was made. The micrographs of the modified, as well as unmodified microcapsules detected by SEM, are shown in Figure 4. A microgram (Figure 4a) shows an optical view of the unmodified MPCM32D microcapsules—they have a smooth and compact surface, and have a regular sphere shape. The images of shell-modified microcapsules presented in Figure 4 clearly show the newly formed coating of MWCNTs (Figure 4b,c) and PEDOT: PSS (Figure 4d,e), and this confirms the presence of these conductive additives on the outer shell of the PCM microcapsules.

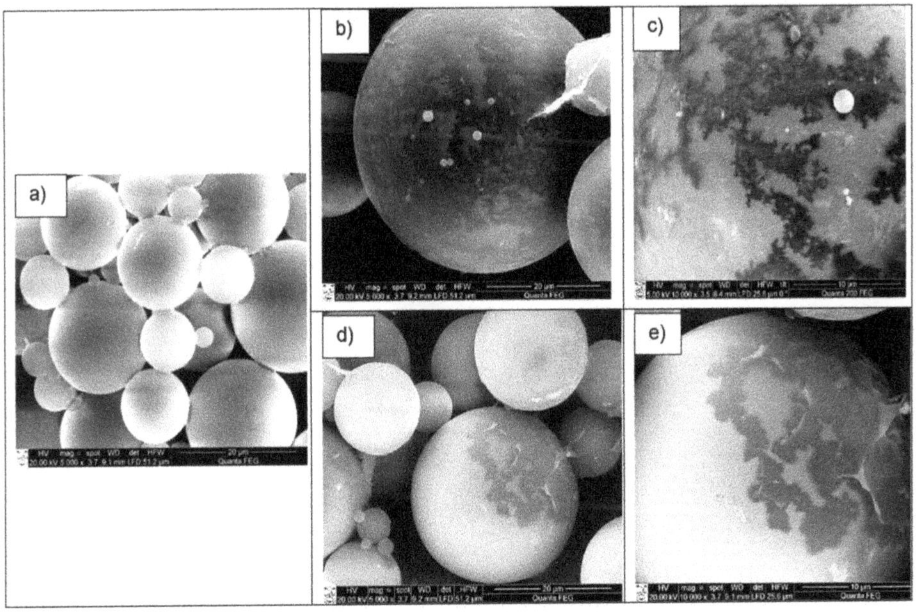

Figure 4. SEM micrograms of microcapsules MPCM32D: (**a**) unmodified, magnification 5000×; (**b**) with 5 wt.% MWCNTs shell-modified, magnification 5000×; (**c**) with 5 wt.% MWCNTs shell-modified, magnification 10,000×; (**d**) with 5 wt.% PEDOT: PSS shell-modified, magnification 5000×; (**e**) with 5 wt.% PEDOT: PSS shell-modified, magnification 10,000×.

3.2. Heat Storage and Release Capacity of the Shell-Modified PCM Microcapsules

The results of DSC analysis for unmodified and shell-modified microcapsules MPCM32D are presented in Figure 5 as the DSC curves—thermograms. The data extracted from DSC curves for heating and cooling processes of all tested paraffin PCM microcapsules are summarized in Table 3. The initial, unmodified microcapsules MPCM32D consisting of an MF resin shell surrounding a paraffin core, demonstrated the peak melting temperature of 32.02 °C (Figure 5a). This result falls within the melting temperature range (30–32 °C) specified by the manufacturer and corresponds to a typical paraffin two-peak DSC thermogram. The sharp main peak of the curve represents the solid-liquid phase change of the tested encapsulated paraffin PCM and the minor peak at the left side of the main peak corresponds to the solid-solid phase transition of paraffin. In this case, for unmodified microcapsules MPCM32D, the enthalpy of fusion (ΔH_f) of the solid-solid and the solid-liquid transition is 10.4 J/g and 106.3 J/g, respectively. After the modification, these two-phase change peaks remain (see Figure 5b,c) and this implies that modified microcapsules maintain good phase change behaviour. As enthalpies of fusion (ΔH_f) and crystallization (ΔH_c) during the solid-solid phase changing stage are significantly lower (for unmodified and shell-modified microcapsules) if compared to the solid-liquid stage, therefore for the further analysis only enthalpies of the solid-liquid stage were considered. The data calculated from DSC curves for shell-modified microcapsules MPCM32D with MWCNTs and PEDOT: PSS in different mass fractions (1 wt.%, 5 wt.%, and 10 wt.%) are presented in Table 3.

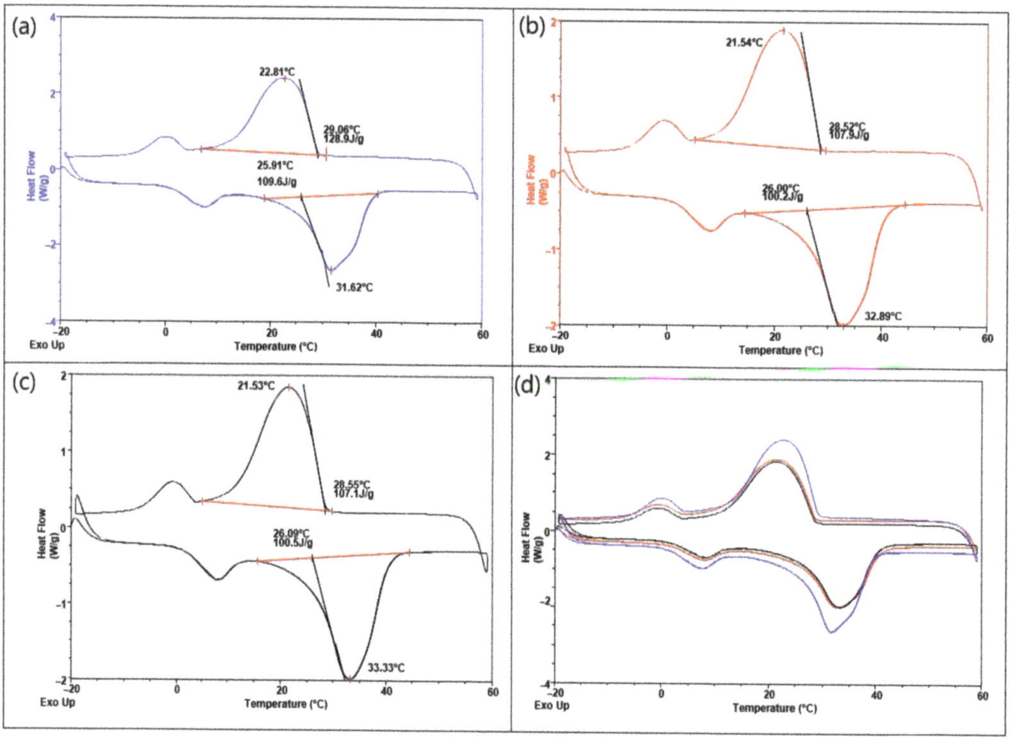

Figure 5. DSC thermograms of the microcapsules MPCM32D: unmodified (**a**), with 5 wt.% MWCNTs (**b**) and 5 wt.% PEDOT: PSS (**c**) shell-modified, and their summed thermogram (**d**).

Table 3. DSC results of the unmodified and shell-modified microcapsules MPCM32D.

Microcapsules MPCM32D	Melting				Crystallization			
	Peak Melting Temperature, Tp.m. °C	**Extrapolated Onset Melting Temperature, Tei.m.°C**	**Extrapolated End Melting Temperature, Tef.m. °C**	**Enthalpy of Fusion, ΔH_f, J/g**	**Peak Crystallization Temperature, Tp.c.°C**	**Extrapolated Onset Crystallization Temperature, Tei.c. °C**	**Extrapolated End Crystallization Temperature, Tef.c. ° C**	**Enthalpy of Crystallization, ΔHc, J/g**
Unmodified (initial)	32.02	13.19	40.09	106.3	21.44	30.61	5.49	113.0
Modified with PEDOT:PSS:								
1 wt.%	32.73	15.68	44.9	103.9	21.56	30.13	5.01	112.9
5 wt.%	33.33	15.56	45.06	100.5	21.53	30.25	3.95	108.5
10 wt.%	33.49	15.4	44.9	93.22	21.27	29.8	4.09	103.1
Modified with MWCNTs:								
1 wt.%	33.85	15.0	42.2	104.0	22.9	30.3	5.31	106.1
5 wt.%	32.89	14.49	45.42	100.2	21.54	29.42	4.42	108.7
10 wt.%	33.35	14.85	43.05	89.77	22.21	31.3	5.25	96.4

The melting peaks of the modified microcapsules MPCM32D with both thermally conductive additives, MWCNTs or PEDOT: PSS, shifted to a slightly higher temperature (max. by ~2 °C) and their crystallization peaks moved to a slightly lower temperature (max. by ~3 °C). Hence, it may be concluded that for the modified microcapsules MPCM32D, in both cases—using as additives either MWCNTs or PEDOT: PSS—phase transition temperatures are quite similar to those of the unmodified. The variations between these temperature values are rather small and are within the measurement uncertainty limits. It demonstrates that the incorporation of used outer shell modifiers does not affect the structure of microcapsules MPCM32D paraffin core. Moreover, some researchers [48–50] have proved that there was no chemical reaction between the paraffin and the CNTs or graphite in the preparation of their composites.

As seen in Table 3 and Figure 6, the modification of the MF resin outer shell of these microcapsules has a certain influence on their heat storage and release capacity.

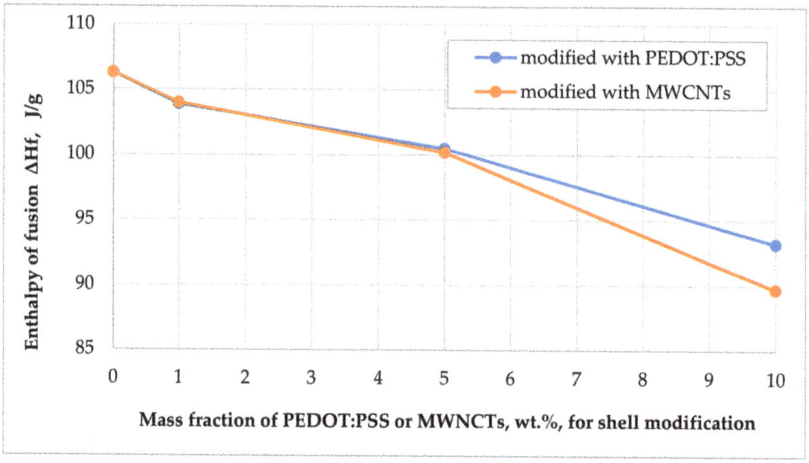

Figure 6. The influence of shell modifiers PEDOT: PSS and MWCNTs on fusion enthalpy of paraffin phase changing materials (PCMs) microcapsules.

These properties are the most important part of PCM application and define their thermoregulatory activity. The enthalpy of fusion (ΔH_f) of the 1 wt.% and 5 wt.% MWCNTs, as well as PEDOT: PSS, shell-modified paraffin microcapsules MPCM32D was found to be approximately the same as that of the unmodified ones. Only an increase in the amount of these thermally conductive additives until 10 wt.%, significantly reduced the heat storage capacity of the investigated PCM microcapsules. If compared with the unmodified ones, ΔH_f decreased 15.6% for MWCNTs and 12.3% in the case of PEDOT: PSS. A similar situation was observed with the enthalpy of crystallization (ΔH_c). This can be explained by the fact, that the addition of 10 wt.% or more of this thermal conductivity and heat transfer enhancing additives has quite considerably decreased the relative amount of paraffin PCM (or relative fraction of this PCM) in the microcapsules. Therefore, on the basis of this analysis, it could be stated, that the shell-modified microcapsules MPCM32D with the mass fraction of additives—MWCNTs or PEDOT: PSS—at about 5 wt.%, may be optimal for intended textile applications. The thermograms of 5 wt.% MWCNTs and 5 wt.% PEDOT: PSS shell-modified microcapsules MPCM32D, respectively, are shown in Figure 5b,c, and a thermogram of these three curves is summarized in Figure 5d.

3.3. Influence of Conductive Additives on the Thermal Conductivity of Modified PCM Microcapsules

In order to evaluate the influence of the used modifiers—MWCNTs or PEDOT: PSS—for the heat transfer efficiency of microcapsules MPCM32D, the thermal conductivity of the modified and unmodified microcapsules was measured. The thermal conductivity is considered as an important contributing factor to the total thermoregulatory capability of PCMs microcapsules, as it may delay or promote the thermal response to the heat storage and release [49,51]. With an aim to determine the thermal conductivity, the microcapsules MPCM32D have been incorporated into a polymeric matrix of polyurethane (PU) and obtained samples were tested. Therefore, the results presented in Figure 7 are not directly related to the particular tested microcapsules, but to the composites containing unmodified microcapsules MPCM32D and modified ones with different mass fractions (1 wt.%, 5 wt.%, and 10 wt.%) of MWCNTs or PEDOT: PSS. Thus, these results (Figure 7) allow for comparing the thermal conductivity of PCM microcapsules modified with different quantities of thermally conductive additives and to evaluate the effect. The thermal conductivity coefficient λ (W/(m K)) of each prepared microcapsules sample in the PU matrix is expressed in Figure 7. The effect of different mass fractions of both modifiers (MWCNTs and PEDOT: PSS) on the thermal conductivity of tested samples is also provided. It is evident that the thermal conductivity of the tested samples showed a linear increase with the increment of MWCNTs or PEDOT: PSS mass fraction by modifying the outer shell of the paraffin PCM microcapsules. Furthermore, it is obvious that the nature of the used additives has influenced these properties. The samples of microcapsules MPCM32D modified with MWCNTs demonstrated a higher enhancement of thermal conductivity compared to the ones modified with PEDOT: PSS. This is due to the higher internal conductivity of MWCNTs. After the comparison of the thermal conductivity values of the samples containing unmodified microcapsules MPCM32D (λ = 0.105 W/(m·K)) with the samples modified with 10 wt.% MWCNTs (λ = 0.264 W/(m·K) a remarkable improvement (2.5 times or 151%) is seen. Meanwhile, the thermal conductivity of microcapsules samples which were shell-modified with 10 wt.% PEDOT: PSS revealed less improvement (1.8 times or 79%).

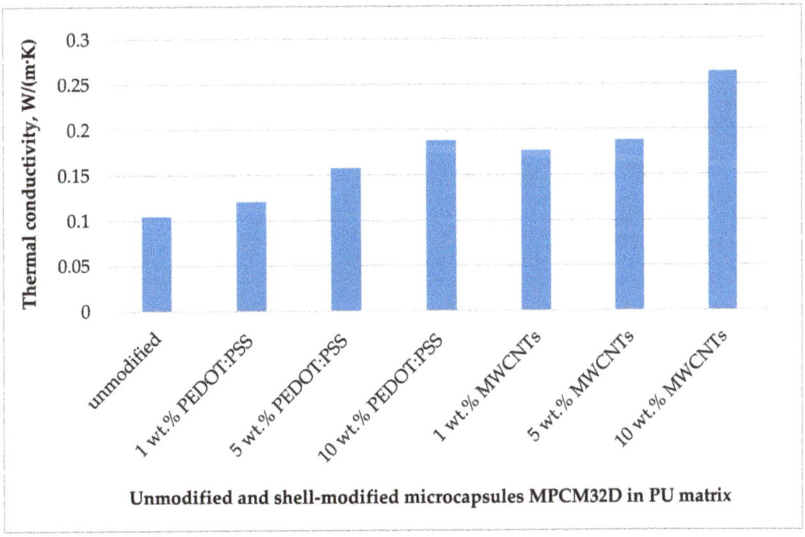

Figure 7. Thermal conductivity of unmodified and shell-modified PCM microcapsules MPCM32D in a polyurethane matrix.

3.4. Thermal Performance of Knitted Fabrics Containing Modified PCM Microcapsules

Based on the gained results of the phase change characteristics and the thermal conductivity of the shell-modified microcapsules MPCM32D, it was determined that mass fraction of MWCNTs or PEDOT: PSS, not exceeding 5 wt.%, may be the optimal one for textile applications, as the higher amount of these conductive additives reduces the heat storage and release capacity of paraffin PCM. Consequently, the 3D warp-knitted spacer PET fabric (Table 1) was dip-coated with the Itobinder PCM and 5 wt.% MWCNTs or 5 wt.% PEDOT: PSS shell-modified microcapsules MPCM32D according to the recipe and parameters presented in Table 2. For comparison, samples of knitted fabrics—untreated, dip-coated with unmodified microcapsules MPCM32D, and treated only with Itobinder PCM were used and their codes are given in Table 4.

Table 4. Thermal properties of 3D warp-knitted spacer PET fabric untreated and dip-coated with microcapsules MPCM32D.

Sample Code	Method of Textile Sample Treatment	Thermal Resistance, R_{ct} ($m^2 \cdot K/W$)	Thermal Conductivity Coefficient, λ ($W/(m \cdot K)$)
1	Untreated (initial)	0.068	0.043
2	Dip-coated with Itofinish PCM	0.069	0.042
3	Dip-coated with Itofinish PCM and unmodified microcapsules MPCM32D	0.073	0.039
4	Dip-coated with Itofinish PCM and 5 wt.% PEDOT:PSS shell-modified microcapsules MPCM32D	0.061	0.047
5	Dip-coated with Itofinish PCM and 5 wt.% MWCNTs shell-modified microcapsules MPCM32D	0.060	0.048

* The thickness D (m) of all samples is the same, D = 0.0029 m.

For the evaluation of the thermal performance of these samples, three testing methods were applied—DSC analysis (except Samples 1 and 2), determination of thermal resistance (R_{ct}) under steady-state conditions using sweating guarded-hotplate, and monitoring of dynamic thermal behaviour, during the temperature changes, based on infrared thermography. The DSC results of Samples 4 and 5 presented in Table 5, revealed that the fabric with 5 wt.% MWCNTs or 5 wt.% PEDOT: PSS shell-modified microcapsules MPCM32D feature quite similar enthalpies of fusion (ΔH_f) and crystallization (ΔH_c) when compared to the fabric containing the unmodified microcapsules (Sample 3). Hence, it can be concluded that the heat storage and release capacities for 3D warp-knitted spacer PET fabric impregnated with 5 wt.% MWCNTs or PEDOT: PSS shell-modified paraffin microcapsules MPCM32D remain the same as in the case of unmodified microcapsules.

Table 5. DSC analysis results of 3D warp-knitted PET fabric with unmodified and shell-modified microcapsules MPCM32D.

Sample Code	Knitted Fabric with Microcapsules MPCM32D	Melting		Crystallization	
		Peak Melting Temperature, Tp.m, °C	Enthalpy of Fusion, ΔH_f, J/g	Peak Crystallization Temperature, Tp.c, °C	Enthalpy of Crystallization, ΔH_c, J/g
3	Unmodified	31.66	21.78	24.78	24.64
4	5 wt.% PEDOT:PSS shell-modified	31.42	22.71	24.28	23.19
5	5 wt.% MWCNTs shell-modified	31.42	24.22	24.65	26.31

The DSC analysis allows obtaining the basic thermal properties of the textile samples treated with PCM microcapsules, mostly related to their thermoregulatory capacity. However, to assess the thermal performance of the tested textile samples considering the thermal response rate, two more methods were used. To assess the influence of the microcapsules modification on the thermal conductivity of textile samples, the thermal resistance R_{ct}, ($m^2 \cdot K/W$) was measured. Based on the obtained results the thermal conductivity coefficient λ, ($W/(m \cdot K)$) was calculated. The received values are presented in Table 4. The structure of the knitted fabric and the layer of polymeric binder used for the incorporation of PCM microcapsules have influenced the heat transportation to these microcapsules and the temperature values. This might be because of the differences in the thermal conductivity of individual components of the tested material and their distribution. The tested textile samples consist of PET/elastane yarns, an acrylic resin binder, and PCM microcapsules, therefore the thermal conductivity was determined separately for knitted fabric: untreated (initial) (Sample 1), dip-coated only with Itofinish PCM (Sample 2), dip-coated with unmodified (Sample 3) and with 5 wt.% PEDOT: PSS and MWCNTs shell-modified (Samples 4 and 5) microcapsules MPCM32D, respectively. As it is seen from the results presented in Table 4, the thermal conductivity of Sample 2 is slightly lower than that of Sample 1. It suggests the conclusion that the acrylic resin binder is a barrier, so it can delay heat access and its emission. Furthermore, the thermal conductivity of the fabric sample with unmodified microcapsules (Sample 3) decreased even more, as one more barrier—MF resin shell of microcapsules—appeared. However, after modification of the outer shell of the microcapsules MPCM32D with 5 wt.% of used add-on's, the thermal conductivity of investigated knitted fabric has improved in comparison to the fabric samples with unmodified ones: ~12% in case of MWCNTs modifier and ~ 9% for PEDOT: PSS. That confirms the positive influence of the used thermal conductivity enhancing additives for the heat transfer rate within the textile sample containing these modified paraffin PCM microcapsules.

To obtain supplementary information regarding the heat transfer capabilities of modified PCM microcapsules the dynamic thermal behaviour of fabric samples containing a composition of the unmodified and shell-modified microcapsules MPCM32D and acrylic resin binder Itofinish PCM, during the temperature changes was investigated. The results of the analysis (Figure 8) showed that Samples 4 and 5 containing shell-modified PCM microcapsules heated up faster than the samples containing unmodified (Sample 3) microcapsules whose behaviour was practically indistinguishable from that of untreated fabric sample's (Sample 1). The results of these measurements showed that the formed coating of the MWCNTs and PEDOT: PSS on the outer shell of paraffin microcapsule MPCM32D accelerated their melting process and increased the heat transfer rate of the material, which ensured a rapid thermal response to the ambient temperature. The results of these measurements revealed that coating of the outer shell of the paraffin microcapsules MPCM32D with the thermal conductivity enhancing additives MWCNTs and PEDOT: PSS accelerates their melting process and increases the heat transfer rate of the fabric, which ensures a rapid thermal response to the ambient temperature.

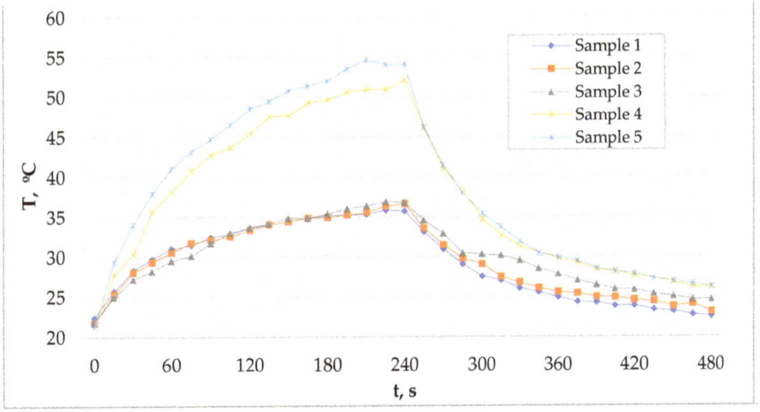

Figure 8. Comparison of dynamic thermal behaviour of the fabric samples containing unmodified and shell-modified PCM microcapsules during the temperature changes.

4. Conclusions

Shell-modification of PCM microcapsules MPCM32D with MWCNTs and PEDOT: PSS additives resulted in enhancement of their thermal conductivity. The SEM micrographs have shown a clearly visible change in surface topography of modified microcapsules due to the presence of modifiers used on their outer shell surface. This fact, as well as essentially unaltered phase transition temperatures of modified microcapsules, confirms that both conductive additives—either MWCNTs or PEDOT: PSS—are adhered only to the outer shell of PCM microcapsules.

It was determined the optimal amount of conductive additive in the modified microcapsules, which is about 5 wt.% for both modifiers. The higher amount of applied additives (10 wt.% and more) despite significant improvement of thermal conductivity, rather reduces the heat storage and release capacity of investigated PCM microcapsules.

The investigation of the thermal performance of textile samples confirmed the positive influence of these paraffinic microcapsules' shell-modification. For samples treated with shell-modified microcapsules MPCM32D in comparison with samples treated with unmodified ones: the thermal conductivity increased ~12% in the case of MWCNTs modifier, and ~9% in the case of PEDOT: PSS; the heat transfer rate through the material increased, ensuring more rapid thermal response to the ambient temperature. DSC analysis results proved that these modified microcapsules incorporated into textile maintain good heat storage and release capacity similar to unmodified microcapsules MPCM32D. Besides, it should be noted that investigations on the durability of fabrics treated with composites of modified microcapsules and acrylic resin binder will be the subject of our subsequent study.

Author Contributions: Conceptualization, A.A., A.S. and V.S.-P.; methodology, V.S-P., A.S. and V.R.; investigation, V.S.-P. and A.S.; writing—original draft preparation, V.S.-P., A.S. and V.R.; writing—review and editing, A.S., V.R. and J.B.-G.; supervision, A.A. All authors have read and agreed to the published version of the manuscript.

Funding: This research received no external funding.

Institutional Review Board Statement: Not applicable.

Informed Consent Statement: Not applicable.

Data Availability Statement: The data presented in this study are available on request from the corresponding author.

Conflicts of Interest: The authors declare no conflict of interest.

References

1. Peng, G.; Dou, G.; Hu, Y.; Sun, Y.; Chen, Z. Phase change material (PCM) microcapsules for thermal energy storage. *Adv. Polym. Technol.* **2020**. [CrossRef]
2. Mondal, S. Phase change materials for smart textiles—An overview. *Appl. Therm. Eng.* **2008**, *28*, 1536–1550. [CrossRef]
3. Keyan, K.; Ramachandran, T.; Shamugasundaram, O.L.; Balasubramaniam, M.; Ragavendra, T. Microencapsulation of PCMs in textiles: A review. *J. Text. Appar. Technol. Manag.* **2012**, *7*, 1–10.
4. Prajapati, D.G.; Kandasubramanian, B. A review on polymeric-based phase change material for thermo-regulating fabric application. *Polym. Rev.* **2020**, *60*, 389–419. [CrossRef]
5. Sarier, N.; Onder, E. Organic phase change materials and their textile applications: An overview. *Thermochim. Acta* **2012**, *540*, 7–60. [CrossRef]
6. Yeşilyurt, M.K.; Nadaroglu, H.; Çomakli, Ö. Materials and Selection Thereof for Encapsulated Phase Change Materials for Heat Transfer Applications. *Int. J. Innov. Res. Rev.* **2019**, *3*, 16–22.
7. Jamekhorshid, A.; Sadrameli, S.M.; Farid, M. A review of microencapsulation methods of phase change materials (PCMs) as a thermal energy storage (TES) medium. *Renew. Sust. Energ. Rev.* **2014**, *31*, 531–542. [CrossRef]
8. Alkan, C.; Sarı, A.; Karaipekli, A. Preparation, thermal properties and thermal reliability of microencapsulated n-eicosane as novel phase change material for thermal energy storage. *Energy Convers. Manag.* **2011**, *52*, 687–692. [CrossRef]
9. Mohaddes, F.; Islam, S.; Shanks, R.; Fergusson, M.; Wang, L.; Padhye, R. Modification and evaluation of thermal properties of melamine-formaldehyde/n-eicosane microcapsules for thermo-regulation applications. *Appl. Therm. Eng.* **2014**, *71*, 11–15. [CrossRef]
10. Zhang, X.X.; Fan, Y.F.; Tao, X.M.; Yick, K.L. Fabrication and properties of microcapsules and nanocapsules containing *n*-octadecane. *Mater. Chem. Phys.* **2004**, *88*, 300–307. [CrossRef]
11. Fan, Y.F.; Zhang, X.X.; Wu, S.Z.; Wang, X.C. Thermal stability and permeability of microencapsulated *n*-octadecane and cyclohexane. *Thermochim. Acta* **2005**, *429*, 25–29. [CrossRef]
12. Chen, Z.; Wang, J.; Yu, F.; Zhang, Z.; Gao, X. Preparation and properties of graphene oxide-modified poly (melamine-formaldehyde) microcapsules containing phase change material n-dodecanol for thermal energy storage. *J. Mater. Chem. A* **2015**, *3*, 11624–11630. [CrossRef]

13. Huang, R.; Li, W.; Wang, J.; Zhang, X. Effects of oil-soluble etherified melamine-formaldehyde prepolymers on in situ microencapsulation and macroencapsulation of n-dodecanol. *New J. Chem.* **2017**, *41*, 9424–9437. [CrossRef]
14. Su, J.; Wang, L.; Ren, L. Fabrication and thermal properties of microPCMs: Used melamine-formaldehyde resin as shell material. *J. Appl. Polym. Sci.* **2006**, *101*, 1522–1528.
15. Su, J.F.; Wang, L.X.; Ren, L. Synthesis of polyurethane microPCMs containing n-octadecane by interfacial polycondensation: Influence of styrene-maleic anhydride as a surfactant. *Colloids Surf. A Physicochem. Eng. Asp.* **2007**, *299*, 268–275. [CrossRef]
16. Sánchez, L.; Sánchez, P.; de Lucas, A.; Carmona, M.; Rodríguez, J.F. Microencapsulation of PCMs with a polystyrene shell. *Colloid. Polym. Sci.* **2007**, *285*, 1377–1385. [CrossRef]
17. Alay Aksoy, S.; Alkan, C.; Tözüm, M.S.; Demirbağ, S.; Altun Anayurt, R.; Ulcay, Y. Preparation and textile application of poly (methyl methacrylate-co-methacrylic acid)/n-octadecane and n-eicosane microcapsules. *J. Text. Inst.* **2017**, *108*, 30–41. [CrossRef]
18. Kim, J.; Cho, G. Thermal storage/release, durability, and temperature sensing properties of thermostatic fabrics treated with octadecane-containing microcapsules. *Text. Res. J.* **2002**, *72*, 1093–1098. [CrossRef]
19. Koo, K.; Choe, J.; Park, Y. The application of PCMMcs and SiC by commercially direct dual-complex coating on textile polymer. *Appl. Surf. Sci.* **2009**, *255*, 8313–8318. [CrossRef]
20. Koo, K.; Park, Y.; Choe, J.; Kim, E. The application of microencapsulated phase-change materials to nylon fabric using direct dual coating method. *J. Appl. Polym. Sci.* **2008**, *108*, 2337–2344. [CrossRef]
21. Sanchez, P.; Sanchez-Fernandez, M.V.; Romero, A.; Rodriguez, J.F.; Sanchez-Silva, L. Development of thermo-regulating textiles using paraffin wax microcapsules. *Thermochim. Acta* **2010**, *498*, 16–21. [CrossRef]
22. Sánchez-Silva, L.; Rodríguez, J.F.; Romero, A.; Sánchez, P. Preparation of coated thermo-regulating textiles using Rubitherm-RT31 microcapsules. *J. Appl. Polym. Sci.* **2012**, *124*, 4809–4818. [CrossRef]
23. Alay, S.; Alkan, C.; Göde, F. Synthesis and characterization of poly (methyl methacrylate)/n-hexadecane microcapsules using different cross-linkers and their application to some fabrics. *Thermochim. Acta* **2011**, *518*, 1–8. [CrossRef]
24. Alay, S.; Göde, F.; Alkan, C. Synthesis and thermal properties of poly (n-butyl acrylate)/n-hexadecane microcapsules using different cross-linkers and their application to textile fabrics. *J. Appl. Polym. Sci.* **2011**, *120*, 2821–2829. [CrossRef]
25. Gao, X.Y.; Han, N.; Zhang, X.X.; Yu, W.Y. Melt-processable acrylonitrile–methyl acrylate copolymers and melt-spun fibers containing MicroPCMs. *J. Mater. Sci.* **2009**, *44*, 5877–5884. [CrossRef]
26. Zhang, X.X.; Wang, X.C.; Tao, X.M.; Yick, K.L. Energy storage polymer/MicroPCMs blended chips and thermo-regulated fibers. *J. Mater. Sci.* **2005**, *40*, 3729–3734. [CrossRef]
27. Zhang, X.X.; Wang, X.C.; Tao, X.M.; Yick, K.L. Structures and properties of wet spun thermo-regulated polyacrylonitrile-vinylidene chloride fibers. *Text. Res. J.* **2006**, *76*, 351–359. [CrossRef]
28. Pan, N.; Sun, G. *Functional Textiles for Improved Performance, Protection and Health*; Elsevier: Cambridge, UK, 2011; pp. 1–552.
29. Paul, R. *Functional Finishes for Textiles: Improving Comfort, Performance and Protection*; Elsevier: Cambridge, UK, 2015; pp. 1–629.
30. Wang, X.; Guo, Y.; Su, J.; Zhang, X.; Han, N.; Wang, X. Microstructure and thermal reliability of microcapsules containing phase change material with self-assembled graphene/organic nano-hybrid shells. *Nanomaterials* **2018**, *8*, 364. [CrossRef]
31. Su, J.F.; Wang, S.B.; Zhang, Y.Y.; Huang, Z. Physicochemical properties and mechanical characters of methanol-modified melamine-formaldehyde (MMF) shell microPCMs containing paraffin. *Colloid Polym. Sci.* **2011**, *289*, 111–119. [CrossRef]
32. Liu, L.; Su, D.; Tang, Y.; Fang, G. Thermal conductivity enhancement of phase change materials for thermal energy storage: A review. *Renew. Sustain. Energy Rev.* **2016**, *62*, 305–317. [CrossRef]
33. Khodadadi, J.M.; Fan, L.; Babaei, H. Thermal conductivity enhancement of nanostructure-based colloidal suspensions utilized as phase change materials for thermal energy storage: A review. *Renew. Sustain. Energy Rev.* **2013**, *24*, 418–444. [CrossRef]
34. Fan, L.; Khodadadi, J.M. Thermal conductivity enhancement of phase change materials for thermal energy storage: A review. *Renew. Substain. Energy Rev.* **2011**, *15*, 24–46. [CrossRef]
35. Lin, Y.; Jia, Y.; Alva, G.; Fang, G. Review on thermal conductivity enhancement, thermal properties and applications of phase change materials in thermal energy storage. *Renew. Substain. Energy Rev.* **2018**, *82*, 2730–2742. [CrossRef]
36. Qiao, Z.; Mao, J. Multifunctional poly (melamine-urea-formaldehyde)/graphene microcapsules with low infrared emissivity and high thermal conductivity. *Mater. Sci. Eng. B* **2017**, *226*, 86–93. [CrossRef]
37. Liu, J.; Chen, Z.; Liu, Y.; Liu, Z.; Ren, Y.; Xue, Y.; Zhang, Q. Preparation of a PCM microcapsule with a graphene oxide platelet-patched shell and its thermal camouflage applications. *Ind. Eng. Chem. Res.* **2019**, *58*, 19090–19099. [CrossRef]
38. Yun, H.R.; Li, C.L.; Zhang, X.X. Fabrication and characterization of conductive microcapsule containing phase change material. *e-Polymers* **2019**, *19*, 519–526. [CrossRef]
39. Maity, S. Optimization of processing parameters of in-situ polymerization of pyrrole on woollen textile to improve its thermal conductivity. *Prog. Org. Coat.* **2017**, *107*, 48–53. [CrossRef]
40. Villanueva, R.; Ganta, D.; Guzman, C. Mechanical, in-situ electrical and thermal properties of wearable conductive textile yarn coated with polypyrrole/carbon black composite. *Mater. Res. Express* **2018**, *6*, 016307. [CrossRef]
41. Ghosh, S.; Bhatkhande, P. Encapsulation of PCM for thermo-regulating fabric application. *Int. J. Org. Chem.* **2012**, *2*, 366–370. [CrossRef]
42. Decher, G.H.J.D.; Hong, J.D.; Schmitt, J. Buildup of ultrathin multilayer films by a self-assembly process: III. Consecutively alternating adsorption of anionic and cationic polyelectrolytes on charged surfaces. *Thin Solid Film.* **1992**, *210*, 831–835. [CrossRef]

43. EN 16806-1: 2016, Textiles and Textile Products—Textiles Containing Phase Change Materials (PCM)—Part 1: Determination of the Heat Storage and Release Capacity. Available online: https://standards.cen.eu/dyn/www/f?p=204:110:0::::FSP_PROJECT,FSP_ORG_ID:41048,6229&cs=1D7ABBC6B2A40DDC22B23ACBD972BC93C (accessed on 31 March 2021).
44. EN ISO 11357-3: 2018, Plastics—Differential Scanning Calorimetry (DSC)—Part 3: Determination of Temperature and Enthalpy of Melting and Crystallization. Available online: https://standards.cen.eu/dyn/www/f?p=204:110:0::::FSP_PROJECT,FSP_ORG_ID:63684,6230&cs=1BA699BFF2F14B41B465D6C6991F0DEC8 (accessed on 31 March 2021).
45. Jančauskas, A.; Buinevičiūtė, M. Effect of different amounts of alumina and aluminium oxide on thermal conductivity of autopolymerized denture base resin. In *Šilumos Energetika Ir Technologijos—2015: Konferencijos Pranešimų Medžiaga*; Kaunas LEI 2015; Kaunas University of Technology: Kaunas, Lithuania, 2015; pp. 168–173, ISSN 2335-2485.
46. ISO 11092:2014; *Textiles—Physiological Effects—Measurement of Thermal and Water-Vapour Resistance under Steady-State Conditions (Sweating Guarded-Hotplate Test)*; ISO: Geneva, Switzerland, 2014.
47. Stanković, S.B.; Popović, D.; Poparić, G.B. Thermal properties of textile fabrics made of natural and regenerated cellulose fibers. *Polym. Test.* **2008**, *27*, 41–48. [CrossRef]
48. Rao, Z.H.; Zhang, G.Q. Thermal Properties of Paraffin Wax-based Composites Containing Graphite. *Energy Sources Part A Recovery Util. Environ. Eff.* **2011**, *33*, 587–593. [CrossRef]
49. Cui, W.; Xia, Y.; Zhang, H.; Xu, F.; Zou, Y.; Xiang, C.; Sun, L. Microencapsulation of phase change materials with carbon nanotubes reinforced shell for enhancement of thermal conductivity. In *IOP Conference Series: Materials Science and Engineering*; IOP Publishing: Bristol, England, 2017; Volume 182, p. 012015.
50. Qiao, Z.; Mao, J. Enhanced thermal properties with graphene oxide in the urea-formaldehyde microcapsules containing paraffin PCMs. *J. Microencapsul.* **2017**, *34*, 1–9. [CrossRef]
51. Vakhshouri, A.R. Paraffin as Phase Change Material. Paraffin—An Overview. *IntechOpen* **2020**. [CrossRef]

polymers

MDPI

Article

Design of New Polyacrylate Microcapsules to Modify the Water-Soluble Active Substances Release

Valentina Sabatini [1,2], Laura Pellicano [1], Hermes Farina [1,2,3], Eleonora Pargoletti [1,2], Luisa Annunziata [2,3], Marco A. Ortenzi [1,2,3], Alessandro Stori [4] and Giuseppe Cappelletti [1,2,3,*]

1 Dipartimento di Chimica, Università degli Studi di Milano, Via Golgi 19, 20133 Milan, Italy; vsabatini@outlook.com (V.S.); laura.pellicano@studenti.unimi.it (L.P.); hermes.farina@unimi.it (H.F.); eleonora.pargoletti@unimi.it (E.P.); marco.ortenzi@unimi.it (M.A.O.)

2 Consorzio Interuniversitario per la Scienza e Tecnologia dei Materiali (INSTM), Via Giusti 9, 50121 Firenze, Italy; luisa.annunziata@unimi.it

3 CRC Materiali Polimerici "LaMPo", Dipartimento di Chimica, Università degli Studi di Milano, Via Golgi 19, 20133 Milano, Italy

4 AMVIC srl, Piazza Santo Stefano 6, 20122 Milano, Italy; alessandro.stori@icloud.com

* Correspondence: giuseppe.cappelletti@unimi.it; Tel.: +39-0250314228

Abstract: Despite the poor photochemical stability of capsules walls, polyacrylate is one of the most successful polymers for microencapsulation. To improve polyacrylate performance, the combined use of different acrylate-based polymers could be exploited. Herein butyl methacrylate (BUMA)-based lattices were obtained via free radical polymerization in water by adding (i) methacrylic acid (MA)/methyl methacrylate (MMA) and (ii) methacrylamide (MAC) respectively, as an aqueous phase in Pickering emulsions, thanks to both the excellent polymer shells' stability and the high encapsulation efficiency. A series of BUMA_MA_MMA terpolymers with complex macromolecular structures and BUMA_MAC linear copolymers were synthesized and used as dispersing media of an active material. Rate and yield of encapsulation, active substance adsorption onto the polymer wall, capsule morphology, shelf-life and controlled release were investigated. The effectiveness of the prepared BUMA-based microcapsules was demonstrated: BUMA-based terpolymers together with the modified ones (BUMA_MAC) led to slow (within ca. 60 h) and fast (in around 10 h) releasing microcapsules, respectively.

Keywords: polyacrylate; water-based latex; Pickering emulsion; microencapsulation; controlled release

Citation: Sabatini, V.; Pellicano, L.; Farina, H.; Pargoletti, E.; Annunziata, L.; Ortenzi, M.A.; Stori, A.; Cappelletti, G. Design of New Polyacrylate Microcapsules to Modify the Water-Soluble Active Substances Release. *Polymers* **2021**, *13*, 809. https://doi.org/10.3390/polym13050809

Academic Editor: M. Ali Aboudzadeh

Received: 9 February 2021
Accepted: 2 March 2021
Published: 6 March 2021

1. Introduction

In the last decade, microencapsulation technique of effective liquid and solid agents has attracted wide interest from several industrial sectors, such as the food chemistry, building, health and beauty industries [1–4]. The advantages of such a procedure are manifold: for example, it can be used to protect active materials, such as probiotics and drugs, during their passage through animals and human bodies [5] or in the construction industry, for the protection of change phase and thermal energy storage materials [6]. Moreover, encapsulation allows several fertilizers and agricultural handlings to be released over time [7]. The key to the success of every microencapsulation process is the formation of a resistant and continuous protective envelope for the entrapped active substance and keep it safe from the surrounding environment. Methods and materials adopted for the encapsulation process contribute to define microcapsules properties and kinetic release. In particular, the methods commonly available for the formulation of microcapsules can be divided into physical and chemical approaches [8]. Both these methods confer some useful properties depending on the desired characteristics of the microcapsules produced. These include microcapsules' shape and size, shells thickness and mechanical resistance [9]. Furthermore, the selection of the most suitable industrial production technique depends on multiple process parameters, such as the physical state of the host material (liquid or solid, hydrophilic or hydrophobic, and so on), the inorganic or organic nature, the miscibility as well as the chemical compatibility between envelope and active agent [10–13]. The most reported physical techniques are pan coating [14] and spray drying [15] due to their easy preparation technology that favors fast and cheap lab-to-industrial scale productions [16]. However, particles of high granulometry (often with average sizes larger than 1 mm) and not uniformly coated microcapsules have been reported as part of such processes, respectively [17]. On the other side, chemical microencapsulation techniques are strictly based on chemical interactions between the different materials used. These chemical interactions result in polymerization of several substances that form the wall of microcapsules. The most common types of chemical technique used are the interfacial polymerization [18] and in situ polymerization [19]. Although chemical methods for microencapsulation are very popular thanks to the possibility to tailor the final properties during the synthesis of protective materials and for this purpose have been used widely in many industrial sectors, they have several drawbacks. Indeed, to maintain some operative conditions, strong solvents and acids are required [20]. Moreover, polymerization catalysts are usually toxic [21]: for example, chemical scavengers can be used to reduce the amount of free unreacted harmful chemicals, but this increases production costs [22]. Moreover, both chemical methods are known to be very successful in producing microcapsules with diameters lower than 100 microns. The smaller their size the larger the volume fraction that would be necessary to deliver the desirable results and thus, the poor margin of tolerance of the obtainable microcapsules with different dimensions raise skepticism over the use of chemical methods to produce them [22].

In this scenario, the combination of physical and chemical methods is an emerging approach, thanks to the hopeful benefits deriving from each of these techniques and the minimization of the already cited drawbacks. The physicochemical methods are mainly based on the formation of walls from preformed polymers, and among the most used methods, ionic gelation [23], complex coacervation [24] and Pickering emulsion [25] can be found. Ionic gelation produces generally very porous and permeable hydrogels [26], and this can be beneficial in some applications, such as in flavors delivery [27] but in others, such as building, can be problematic [28]. Complex coacervation can produce a wide range of microcapsules dimension, from 10 to 1000 microns, and utilizes environmentally friendly materials, but sometimes the yield of production is very poor depending on the type of emulsifier chosen [29]. Moreover, in this plethora of methods, Pickering emulsion is the process during which, for example in a water-in-oil emulsion, micro-sized droplets of water are dispersed in a continuous oil phase and polymer nanoparticles are located at the interface. Depending on their surface tension, these nanoparticles can stabilize the emulsion in the colloidal sense [30]. Nagayama et al. [31] were the first to demonstrate that Pickering emulsions can be transformed into microcapsules when the interfacial nanoparticles are either linked or fused together to form a continuous wall. The greatest advantage of such kind of procedure is that it allows process adjustments during the wall formation step. This in-process intervention cannot be done for the other techniques discussed above. On the other side, the choice of microcapsules "building" materials draws the morphological properties of the resulting protective shells, such as permeability and release performances. For instance, biopolymers such as alginates and polysaccharides favor the development of microcapsules with tailored surface features [32,33]. Moreover, the use of biopolymers is useful for applications involving the biodegradable nature of the protective shells, for example in the case of zootechnic probiotics [34]. However, the use of synthetic polymers as encapsulating materials is of particular interest in the field of porous and permeable microcapsules preparation [35] and, in this latter case, polyacrylates are the main protagonists [36], due to their easy and cheap preparation using both solvent and water-based dispersing media [37–39], the possibility to use ad hoc functionalized monomers to obtain smart shells [40] and, last but not least, the biocompatible and biodegradable nature of some acrylic monomers, such as acrylamide [41]. Thus, the fine modulation of pores dimension and distribution are the key properties of such enveloping materials [42].

Nowadays, in the field of microcapsules preparation processes, one of the main issues to overcome is to obtain a stable product during storage with consistent encapsulation and release efficiency. The latter demand is quite difficult to reach because the common materials used for lattices formulation, e.g., alginates [43], polyacrylates [44], pectin/casein blends [45,46] and gelatines [47], are very sensitive to processing and environmental factors, such as temperature, solar irradiation, humidity and pH variation [48], and this results in poor encapsulation efficacy, reduced microcapsules shelf-life and uncontrolled release kinetics. Vincent et al. [49] reported the preparation of butyl methacrylate–methacrylic acid (BUMA_MA) latex for the microencapsulation and release process of different microorganisms as pesticide agents. Despite the comparable pesticide behaviour of their latex with common chemical products, they described that during emulsion process upon adding ethanol as co-solvent, the latex particles within the aqueous droplets are colloidally unstable leading to coalescence and resulting in a poor microcapsules formation. This behaviour should be due to the high affinity of MA hydroxyl groups with water (latex dispersing medium), that obstructs the formation of polymer particles at the interphase between water and oil phases [50].

To the best of the authors' knowledge, few studies in the literature about microcapsules formation from Pickering emulsion procedure, tailoring different polyacrylate-based polymers, have been reported so far. Hence, in this work, we prepared novel polyacrylic lattices (i.e., a series of BUMA-based terpolymers with complex macromolecular structure and BUMA_methacrylamide (MAC) linear copolymers) as protective materials for the microencapsulation of an active agent, by water-based free radical polymerization procedure during Pickering emulsion. These microcapsules were designed to: (i) be formulated exploiting environmentally friendly materials to achieve more biodegradable polymeric matter; (ii) preserve the active material from the surrounding environment; (iii) tune the effective agent release; and (iv) be re-dispersed in water to form a free-flowing non-viscous aqueous formulation. Therefore, a deep characterization of the prepared polymeric latex, via ^{1}H nuclear magnetic resonance (NMR), Fourier-transform infrared and thermal analyses, and of microcapsules through dynamic light scattering and scanning electron microscopy studies, was conducted and is discussed here as well as the corresponding kinetics of active material release.

2. Materials and Methods

2.1. Materials

N-butyl methacrylate (BUMA, 99%), methacrylic acid (MA, 99%), methyl methacrylate (MMA, 99%), pentaerythritol triacrylate (T3, technical grade), methacrylamide (MAC, 98%), sodium persulfate ($Na_2S_2O_8$, ≥99.9%), sodium dodecyl sulfate (SDS, ≥99.9%), trifluoroacetic acid-d ($C_2DF_3O_2$, 99.5 atom % D), sunflower oil, sodium chloride (NaCl, ≥99.0%), iron (II, III) oxide nanopowder (Fe_3O_4, 50–100 nm particle size), ethanol (96%), methyl orange (MO), KNO_3 10^{-2} M and distilled water Chromasolv® (≥99.9%) were supplied by Sigma Aldrich (Milan, Italy) and used without further purification.

2.2. Synthesis of Butyl Methacrylate–Methacrylic Acid (BUMA_MA), BUMA_MA_Methyl Methacrylate (MMA), BUMA_MA_MMA_Pentaerythritol Triacrylate (T3) and BUMA_Methacrylamide (MAC) Copolymers

BUMA_MA copolymer (used as reference), two terpolymers, i.e., BUMA_MA_MMA and BUMA_MA_MMA_T3, and BUMA_MAC copolymers were synthesized via free radical polymerization. Table 1 and Table S1 show the molar ratio and the exact amount of the monomers used for the syntheses, respectively.

In a typical polymerization procedure, a 250 cm^3 three-necked round bottom flask was equipped with a reflux condenser having a nitrogen inlet adapter, an internal thermometer adapter and a mechanical stirrer. The flask was flushed with nitrogen, charged with 100 cm^3 of distilled water, 0.5 g of sodium dodecyl sulfate (SDS) used as latex stabilizer [49], $Na_2S_2O_8$ (mol $Na_2S_2O_8$ = 1% mol $mol^{-1}{}_{\Sigma \text{acrylate monomers}}$) and acrylate monomers, according to the desired polymer. The polymerization mixture was put in an oil bath, mechanically stirred at around 280–300 rpm, heated for 24 h at 69 °C and then gradually cooled down to room temperature. The reaction yields, determined gravimetrically, were around 100% for all the investigated systems.

Table 1. Molar ratio of reagents adopted for the synthesis of butyl methacrylate (BUMA)-based polymers. MA = methacrylic acid; BUMA = n-butyl methacrylate; MMA = methyl methacrylate; T3 = pentaerythritol triacrylate; MAC = methacrylamide.

Sample	MA/BUMA (mol mol^{-1})	MMA/BUMA (mol mol^{-1})	T3/BUMA (mol mol^{-1})	MAC/BUMA (mol mol^{-1})
BUMA_MA	0.83	-	-	-
BUMA_MA_MMA	0.25	0.75	-	-
BUMA_MA_MMA_T3	0.25	0.75	0.005	-
BUMA_MAC_25	-	-	-	0.33
BUMA_MAC_50	-	-	-	1.00
BUMA_MAC_75	-	-	-	3.00

2.3. Polymer Latex Characterization: Fourier Transform Infrared (FT-IR), Nuclear Magnetic Resonance (^{1}H NMR), Dynamic Light Scattering (DLS) and Differential Scanning Calorimetry (DSC) Analyses

For spectroscopic and thermal characterizations, an aliquot of the lattices prepared was previously dried under N_2 at 50 °C for 24 h.

Fourier transform-infrared (FT-IR) spectra were obtained on a Spectrum 100 spectrophotometer (Perkin Elmer) in attenuated total reflection (ATR) mode using a resolution of 4.0 and 256 scans, in a range of wavenumbers between 4000 and 450 cm^{-1}. A single-bounce diamond crystal was used with an incidence angle of 45°. ^{1}H NMR spectra were collected at 25 °C with a BRUKER 400 MHz spectrometer. All samples were prepared dissolving 8–10 mg of polymer in 1 cm^3 of $C_2DF_3O_2$.

Dynamic light scattering (DLS) analyses were carried out to study the lattices particles size distribution. Samples were analysed after a 1:100 dilution in KNO_3 10^{-2} M to adjust the ionic strength (this step does not affect the correctness of the results obtained, as already reported in the literature) [51,52], using a Malvern Zetasizer NANO ZS (at 25 °C; Malvern Panalytical, Malvern). Measurements were performed on two different aliquots, for 30 scans each.

Differential scanning calorimetry (DSC) analyses were conducted using a Mettler Toledo DSC1; the analyses were conducted weighting 5–10 mg of each sample in a standard 40 µL aluminium pan, using the following temperature cycles:

1. heating from 0 °C to 100 °C at 20 °C min^{-1};
2. 2 min isotherm at 100 °C;
3. cooling from 100 °C to 0 °C at 20 °C min^{-1};
4. 2 min isotherm at 0 °C;
5. heating from 0 °C to 150 °C at 20 °C min^{-1}.

using an empty 40 µL aluminium pan as reference.

2.4. Preparation Process for Methyl Orange (MO) Encapsulation within Polyacrylates Microcapsules and Scanning Electron Microscopy (SEM) Characterization

In a 300 cm^3 glass jacket reactor 88 cm^3 of sunflower oil, 224 mg of Fe_3O_4 and 721 mg of NaCl were added. The mixture was processed with a mechanical stirrer (around 380–400 rpm) for 15 min at room temperature. In a separate tube, 1 cm^3 of a methyl orange (MO) solution with a concentration of 600 ppm was added to 16 cm^3 of the as-synthesized latex (see Scheme 1). Then, 6 cm^3 of ethanol and the solution of latex/methyl orange were added dropwise and the mixture was mechanically stirred (around 250–300 rpm) for 20–30 min at room temperature. The mixture was decanted for 2 h in order to separate the yellowish microcapsules from the supernatant oil phase. The oil phase was removed, and the resulting microcapsules were re-dispersed in water. The process was repeated until oil was no longer observable.

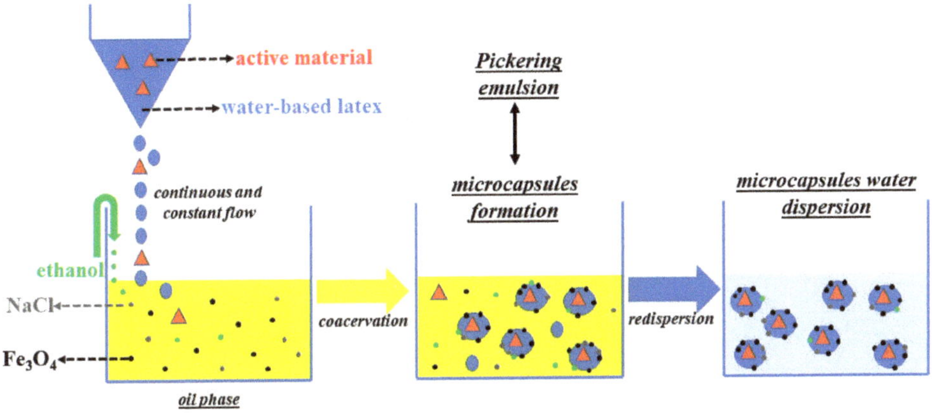

Scheme 1. Schematic representation of methyl orange (MO) encapsulation within polyacrylate microcapsules.

An aliquot of the microcapsules obtained was dried for 2 h at room temperature on a blotting paper in order to perform scanning electron microscopy (SEM) images. This analysis was carried out using a SEM Hitachi TM-1000 (Tokio, Japan).

2.5. MO Kinetic Release

To deeply investigate the encapsulation and release capacities, methyl orange was chosen to make easier the correct evaluation through ultraviolet/visible (UV/Vis) spectroscopy. Specifically, once the MO-encapsulated microcapsules were obtained, 1 g of them were dispersed into 100 cm^3 of distilled water (pH 5.5) under magnetic stirring (ω = 50–80 rpm). The absorbance spectra (recorded in the range between 650 and 350 nm by using SHIMADZU UV 2600 spectrophotometer (Kyoto, Japan)) of each sampling (1.5 cm^3) were acquired at regular time intervals for a total of 120 h, in order to monitor the release of MO over time. Moreover, to compute the exact MO amount, a calibration curve for MO in water was obtained at pH 5.5 (see Figure S1), following the absorbance value at the maximum wavelength of 465 nm and obtaining a molar extinction coefficient (ε) equal to (25,320 $\pm$ 70) $dm^3\ mol^{-1}\ cm^{-1}$. The encapsulation efficiency percentage (EE%) was computed according to [53].

3. Results and Discussion

3.1. Synthesis and Characterization of BUMA-Based Lattices

Several approaches have been already reported in literature to improve microcapsules formation either using hydroxyl functionalized monomers, such as MA neutralized by $CaCl_2$ or NaOH, or by adding emulsification agents [54]. However, the difficulty in reaching a complete salt-based protection of hydroxyl groups and the possible interference between process additives and active substances, result in lattices that are not stable with the formation of flocculation by-products [55]. Herein, to improve the performances of polyacrylic lattices, a novel BUMA-based terpolymer adding MMA as comonomer to BUMA and MA was synthesized. In this way, a decrease in the affinity between the polymer latex and the aqueous dispersing media, thanks to the presence of apolar -CH_3 side groups instead of hydroxyl functionalities, was attained. Specifically, two new kinds of BUMA-based polymers were obtained via free radical polymerization in water by adding (i) MA/MMA as comonomers plus T3, as branching agent (adopting MA/BUMA molar ratio of 0.25% mol/mol for both BUMA_MA_MMA samples) and (ii) MAC, improving both the polymer shells' stability and the encapsulation efficiency. As a reference, BUMA_MA latex (having MA/BUMA molar ratio of 0.83 mol mol^{-1}) was prepared as already described in literature [49].

Figure 1a shows the comparison between BUMA_MA, BUMA_MA_MMA and BUMA_MA_MMA_T3 FT-IR spectra. Notably, it is possible to appreciate several peaks related to: the bending of C–H aliphatic bonds at ~3100–2800 cm^{-1} (*δ-CH*), the stretching of carbonyl ester groups -C=O in the range between ~1750 cm^{-1} and ~1600 cm^{-1} (*υ-C=O*), and the characteristic absorption band for the symmetric stretching vibration of -C–O conjugated to carbonyl ester groups (1350–1100 cm^{-1}) ($υ_s$*–C–O*) [56]. In the case of MMA-based samples, the stretching of –CH_3 groups conjugated to carbonyl functionalities is also detectable in the range ~3100–2800 cm^{-1} (*υ–*CH_3) [57]. As expected, the BUMA_MA sample, having a relatively high content of methacrylic acid, shows the presence of the broad band related to –OH stretching above 3000 cm^{-1} and a double C=O peak due to the simultaneous presence of ester and acid moieties along the macromolecular chain.

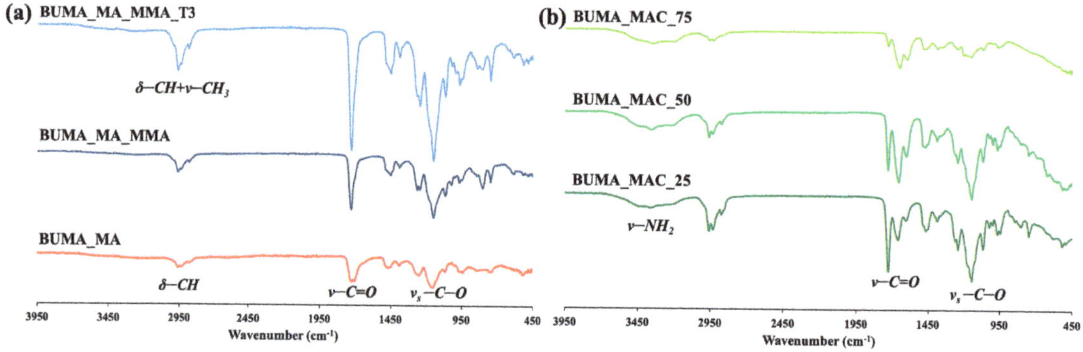

Figure 1. Fourier transform infrared (FT-IR) spectra of (**a**) BUMA_MA and BUMA-based terpolymers and (**b**) BUMA_MAC copolymers.

Since T3 is present in very small quantities, no significant differences can be observed between BUMA_MA_MMA and BUMA_MA_MMA_T3 samples. As discussed by Mallo et al. [58], the use of branching agents can be a way to obtain non-linear polymer lattices, which are insoluble in water in the form of three-dimensional networks but, at the same time tend to swell. As such, the polymeric shell can form at the interphase between water and oil phases during the emulsion process. Thus, to further reduce the affinity between latex and water, a novel terpolymer with complex macromolecular structure was developed (namely BUMA_MA_MMA_T3) by adding 0.5% mol/mol of pentaerythritol triacrylate, T3, as branching agent. Such a low concentration of T3, i.e., from 0.05% to 1.0% mol mol^{-1}, is mandatory to avoid the flocculation of parts of polymer from water media, as also confirmed by literature data [58]. Figure 2a shows the BUMA_MA_MMA_T3 ^{1}H NMR spectrum, where the presence of the branching agent cannot be highlighted due to its exiguous amount. For clarity, the reference and BUMA_MA_MMA ^{1}H NMR spectra have been compared to the one relative to BUMA_MA_MMA_T3 sample (Figure S2).

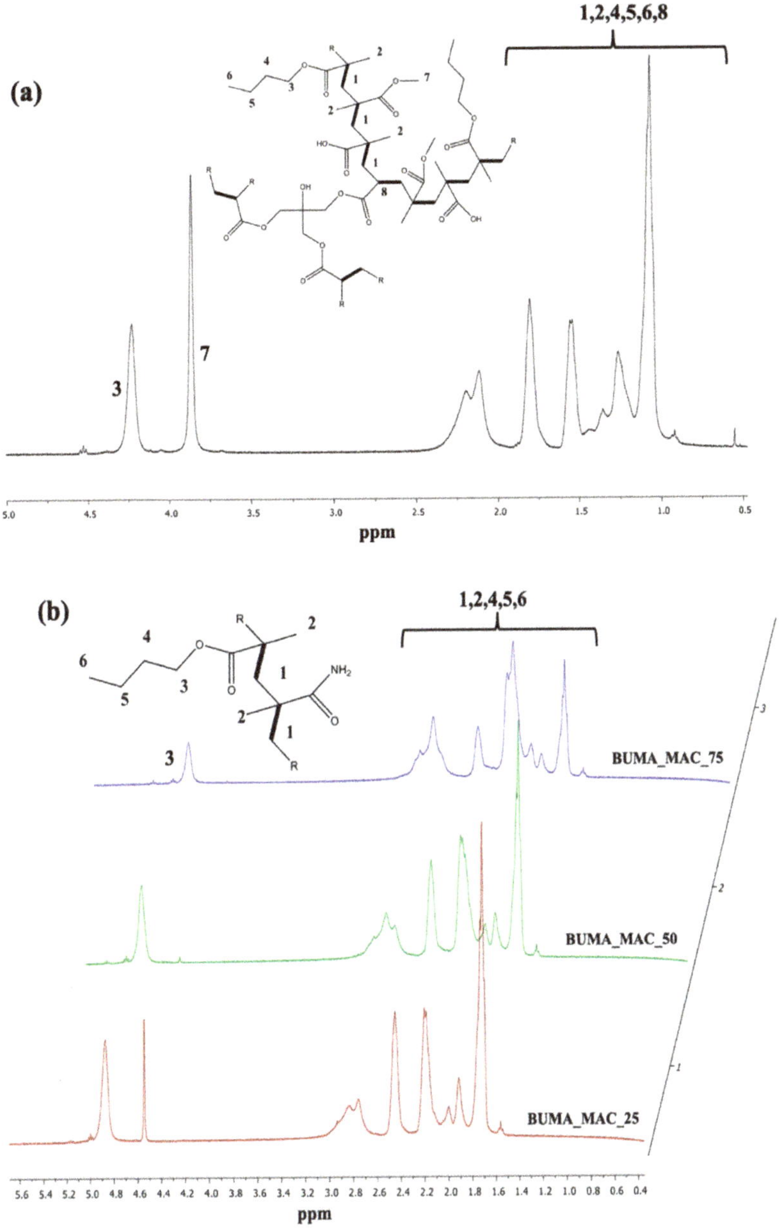

Figure 2. Nuclear magnetic resonance (^{1}H NMR) spectra of (**a**) BUMA_MA_MMA_T3 and (**b**) BUMA_MAC copolymers.

Furthermore, since our goals are to increase the microcapsules stability and to modulate the active substance (herein MO) encapsulation and its release, other kinds of polyacrylate-based material were investigated. As reported by Stranimaier et al. [59], amide functional monomers are useful for the preparation of waterborne lattices due to their apolar nature that favors: (i) the formation of polymer shells at water/oil interphase, the improved (ii) photochemical resistance and (iii) stable appearance of the corresponding microcapsules. Moreover, alongside these promising features, acrylamide-based materials are susceptible to different biodegradation processes [60]. Hence, methacrylamide (MAC) was selected as comonomer in combination with BUMA in order to synthesize new lattices with improved photochemical stability and a potential biodegradable behavior. As summarized in Table 1, a series of BUMA_MAC copolymers were prepared exploiting different molar ratios between the two monomers adopted. Figure 1b reports the comparison between BUMA_MAC_25, BUMA_MAC_50 and BUMA_MAC_75. The presence of -NH_2 moieties is clearly visible due to the broad signal at 3600–3200 cm^{-1} related to the stretching of NH_2 group, that decreases as the quantity of MAC gets lower in the copolymers. The same holds true for the double C=O peak in the 1750–1600 cm^{-1} region, with the peak at higher frequencies relative to the ester moiety of BUMA and the one at lower frequencies of the amide moiety of MAC. Moreover, the comparison among the ^1H NMR spectra reported in Figure 2b highlights the decrease of NMR peak "**3**" relative to BUMA height and width as the amount of MAC gets higher.

To unravel the stability of the as-prepared lattices, DLS measurements and DSC analyses were performed. At first, Table 2 (2nd column) and Figure S3 show the data obtained relative to DLS measurements, indicating the average particles diameter for each copolymer. By comparing the same polymers family, we can assess that BUMA_MA_MMA_T3 latex seems to be the most performing one, since the polymer particles have a single average dimension of about 80 ± 20 nm, lower than the BUMA_MA (180 ± 70 nm) and BUMA_MA_MMA (two populations: 85 ± 15 and 250 ± 60 nm) ones.

Table 2. Molar ratio of reagents adopted for the synthesis of BUMA-based polymers.

Sample	<d^{DLS}> (nm)	Fox T_g (°C)	Real T_g (°C)
BUMA_MA	180 ± 70	67.1	n.d.
BUMA_MA_MMA	(85 ± 15); (250 ± 60)	55.5	78.2
BUMA_MA_MMA_T3	80 ± 20	55.5	75.2
BUMA_MAC_25	(37 ± 7); (170 ± 30)	40.8	46.1
BUMA_MAC_50	(25 ± 4); (80 ± 10)	71.1	51.0
BUMA_MAC_75	12 ± 3	120.3	n.d.

It is worth noting that in the latter case, the addition of only an apolar monomer, i.e., MMA, is not enough to ensure the formation of homogenous and stable latex, since polymeric aggregates of ca. 250 nm can be formed. Conversely, concerning the BUMA_MAC family, BUMA_MAC_75 latex seems to be the most stable one and DLS measurements reveal the presence of particles with a very small hydrodynamic diameter of around 12 ± 3 nm.

Lastly, during Pickering emulsions a latex can be transformed into microcapsules when the interfacial nanoparticles are either linked or fused together to form a continuous wall. This step, usually called "locking", can be achieved via several approaches, such as chemical cross-linking, in situ polymerization, thermal softening and solvent-instability of the particles [61]. However, when nanoparticles are polymeric, a continuous wall can thus be formed around the aqueous droplet, leading to the formation of a microcapsule. The glass transition temperature, T_g, of a polymer is a key value for the occurrence of the locking phenomena, since it corresponds to the temperature at which a polymer switches from hard and glassy form to soft and viscous one. In particular, T_g values higher than room temperature favour the fusion of the latex polymer particles in a continuous polymeric shell and, therefore, promote high microcapsules yield [62].

Herein, according to Table 1, the molar ratio between monomers was established via the Fox equation (Equation (1)) [63] in order to ensure, during the Pickering emulsion process, the formation of stable and consistent polymer walls and therefore favour the locking process:

$$1/T_g = w_1/T_{g,1} + w_n/T_{g,n} \quad (1)$$

where T_g corresponds to the theoretical glass transition temperature value of the resultant copolymer or terpolymer; w_1 and w_n are the weight fractions of each component; $T_{g,1}$ and $T_{g,n}$ are the glass transition temperatures of BUMA, MA, MMA and MAC homopolymers,

293–501–378–486 K, respectively. Glass transitions were determined during the first heating scan (see Figure S4). Notably, from the comparison of data reported in the 3rd and 4th columns of Table 2, the T_g values of our samples are not in good agreement with theoretical values and in some cases cannot be detected; this might be due to the presence of sodium salts in the copolymers containing MA, and to the water uptake of $-CONH_2$ moieties in copolymers containing MAC. Nevertheless, T_g is always higher than room temperature.

3.2. Pickering Emulsions and MO Kinetic Release of BUMA_MA_MMA_T3 and BUMA_MAC_75 Lattices

Once the lattices of BUMA_MA, BUMA_MA_MMA, BUMA_MA_MMA_T3 and BUMA_MAC copolymers were prepared, several Pickering emulsions, according to the procedure summarized in Scheme 1, were formulated. NaCl and iron oxide nanoparticles were added in the oil phase respectively to stabilize the emulsion against Ostwald ripening in the coacervation step and to serve as light absorbing material in the final formulation [62]. Upon mixing the water-based latex and the oil phase, a Pickering water-in-oil emulsion was obtained. The latex nanoparticles have a surface energy which is intermediate between oil and water, thus they position themselves at the interface between water and oil. With the addition of ethanol as co-solvent [64], the latex particles form a continuous wall of polymer around the aqueous droplets, that results in the formation of insoluble microcapsules (MC). Remarkably, the fabrication of these capsules does not require any organic solvent and is entirely performed at room temperature. According to Vincent et al. [49], after the addition of ethanol, the latex particles of BUMA_MA reference collapsed and coalesced, resulting in poor microcapsule formation yield. Despite the addition of MMA monomer, BUMA_MA_MMA latex also showed the same behaviour, i.e., an unsatisfactory conversion of latex particles in polymer capsules. On the other hand, it is worth noting both the successful process of microcapsules formation and the MO encapsulation for BUMA_MA_MMA_T3 latex, thanks to the combined used of apolar and branching monomers that allow the formation of a cross-linked latex. Lastly, the series of BUMA_MAC copolymers were also tested in the Pickering emulsion process but only the sample with the higher amount of MAC, namely BUMA_MAC_75, resulted in promising microcapsules formation. The behaviour of BUMA_MAC_25 and BUMA_MAC_50 samples is probably due to the low amount of MAC monomer that was not enough to promote the formation of polymer shells at the water/oil interphase [65].

Hence, considering the optimal microcapsules formation obtained with our novel BUMA_MA_MMA_T3 and BUMA_MAC_75 lattices, both microscopy analyses and kinetic tests were carried out to investigate these systems in depth. In particular, the trend in the DLS lattices particles size was also corroborated by the aspect and dimension of the microcapsules, observed by SEM. Indeed, as regards, BUMA_MA_MMA_T3-based microcapsules (MC(BUMA_MA_MMA_T3)), they show a rough surface with an average diameter of 220–270 μm, as clearly visible in Figure 3b. This may suggest the formation of a continuous polymer wall probably characterized by the presence of pores large enough to allow the passage of the active substance contained in the MC core, thus favouring its release [66,67].

Figure 3. Scanning electron microscopy (SEM) micrographs displaying the actual capsules size (highlighted with red arrows) relative to: (**a**) MC(BUMA_MA), (**b**) MC(BUMA_MA_MMA_T3) and (**c**) MC(BUMA_MAC_75) (inset: high magnification image).

On the other hand, the reference BUMA_MA latex gave rise to microcapsules with both an average diameter comparable with the one of MC(BUMA_MA_MMA_T3) and, even, smaller (ca. 100 μm; Figure 3a), but characterized by a very smooth surface. By contrast, MC(BUMA_MAC_75) exhibits a very small average diameter of about 30–55 μm (Figure 3c), with a quite rough external surface. Hence, we can infer that the difference in both microcapsules dimensions and surface porosity can suggest a different release behavior of the same active substance, i.e., MO. For this reason, kinetic tests were performed by using the best performing samples (namely, MC(BUMA_MA_MMA_T3) and MC(BUMA_MAC_75)) following the MO release over time by UV/Vis spectroscopy.

Figure 4a clearly shows the released dye concentration up to 120 h, alongside with the corresponding spectroscopic data (Figure 4b,c) and photos displaying the actual presence of MO in the aqueous medium surrounding the microcapsules at the end of the kinetics (Figure 4d,e). In particular, with the bigger MC(BUMA_MA_MMA_T3) system, a partial release was achieved only after 60 h (Figure 4a,b; encapsulation efficiency percentage, EE% of 46%), thus revealing a slower kinetic with respect to the one relative to MC(BUMA_MAC_75). Indeed, for the latter, the curve trend at the beginning is very sharp leading to an almost full MO release already after 10 h (EE% ca. 86%; see Figure 4a,c). However, a possible MO adsorption together with a direct encapsulation cannot be excluded. Actually, the different release behavior of the aforementioned microcapsules could be rather complex depending on (i) the adsorption/encapsulation features; (ii) the chemical affinity of the active substance towards the polymer wall (having different polarity and macromolecular structure); (iii) the porosity of the microcapsules and the relative size. Notably, in the present case, it seems that mainly the porosity and the size of the microcapsules play a pivotal role in affecting the release features, being MC(BUMA_MAC_75) three-fold smaller than MC(BUMA_MA_MMA_T3).

Hence we successfully demonstrated that, by tailoring the polymeric materials used in the Pickering emulsions, it is possible to synthesize microcapsules with ad hoc morphological and surface features to be applied in different application fields, according to the desired releasing rate, and in particular, BUMA_MA_MMA_T3-based microcapsules appear to be the most effective for the slow release process of MO; on the other hand, BUMA_MAC shells are performing as fast-releasing microcapsules.

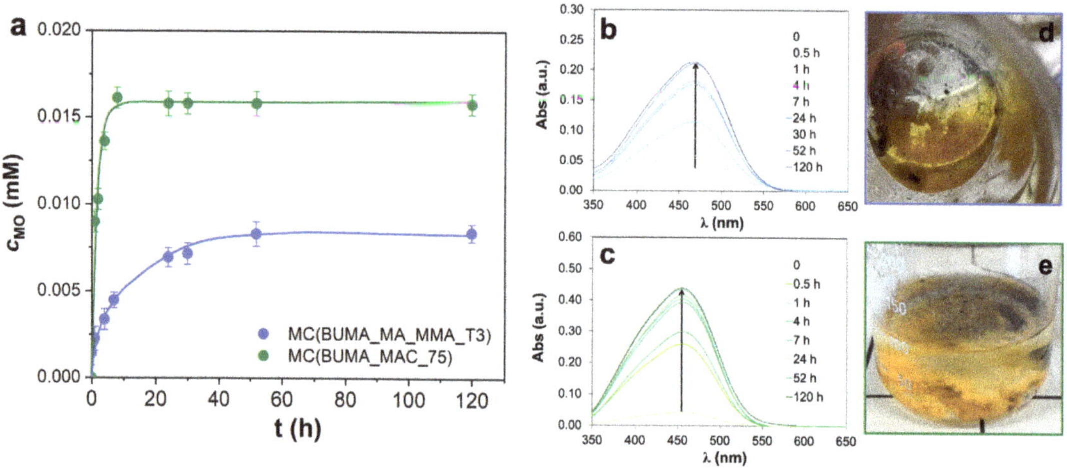

Figure 4. (**a**) Released MO concentration curves relative to MC(BUMA_MA_MMA_T3) and MC(BUMA_MAC_75) systems. Error bars computed on three different replicates. (**b**,**c**) Corresponding kinetics followed through ultraviolet/visible (UV/Vis) analysis, along with the relative (**d**,**e**) photos taken at the end of the test (120 h).

4. Conclusions

In the field of microencapsulation, the need to develop new polymer lattices, characterized by improved shelf-life and high encapsulation-release efficiency, is always pressing due to the wide plethora of industrial applications involved. In this context, the use of polyacrylates, bearing apolar, cross-linkable and potential biodegradable monomers along the polymeric chains, could be a way to create tailor-made microcapsules with permeable surfaces and controlled kinetics release.

Herein, butyl methacrylate (BUMA), methacrylic acid (MA) and methyl methacrylate (MMA) latex was synthesized via free radical polymerization in a water media using just 0.5% mol mol^{-1} of a branching monomer, pentaerythritol triacrylate (T3). Furthermore, a series of BUMA_methacrylamide (MAC) copolymers were prepared, varying the molar percentage of MAC monomer (25–50–75% mol mol^{-1}). At first, the macromolecular structures, thermal features and hydrodynamic volumes were investigated before and after the Pickering emulsion, i.e., the microencapsulation process used here; moreover, the morphology of the microcapsules obtained was assessed via scanning electron microscopy, demonstrating the formation of polymer shells with average diameters of 270 and 55 μm in the case of BUMA_MA_MMA_T3 and BUMA_MAC_75, respectively.

The microcapsules obtained were studied by kinetic release, focusing on rate of release and physical resistance of the microcapsules dispersed in water. Both polymeric lattices showed very promising features as protective envelopes. Notably, BUMA_MA_MMA_T3 latex seems to be the most suitable one in the case of slow release, due to its ability to form bigger microcapsules and thus protect for longer times the active agents entrapped from the surrounding environment. Conversely, for BUMA_MAC_75-based microcapsules fast kinetic release was observed. Hence, to the best of the authors' knowledge the synthesis of such kinds of lattice and their use in Pickering emulsion for microencapsulation process of effective agents has not been reported so far. Specifically, the present research has led to an in depth comprehension of microcapsules formation and kinetic release of BUMA-based lattices that can be successfully adopted as protective polymer shells for the controlled delivery of active materials.

Supplementary Materials: The following are available online at https://www.mdpi.com/2073-4360/13/5/809/s1, Table S1: amounts of the reagents adopted for the synthesis of BUMA-based polymers; Figure S1: (a) methyl orange UV/Vis spectra at different molecule concentrations, (b) relative calibration plot at wavelength fixed at 465 nm; Figure S2: ^1H NMR spectrum of (a) BUMA_MA and (b) BUMA_MA_MMA polymers; Figure S3: Dynamic light scattering data relative to BUMA_MA, BUMA_MA_MMA, BUMA_MA_MMA_T3 and BUMA_MAC_75 systems; Figure S4: DSC curves relative to (a) BUMA_MA, (b) BUMA_MA_MMA, (c) BUMA_MA_MMA_T3, (d) BUMA_MAC_25, (e) BUMA_MAC_50 and (f) BUMA_MAC_75.

Author Contributions: Investigation and experimental tests: V.S., L.P., E.P., L.A. and G.C.; Writing-Original Draft Preparation: V.S., E.P. and G.C.; Writing-Review and Editing: V.S., M.A.O. and G.C.; Supervision: H.F., A.S. and G.C. All authors have read and agreed to the published version of the manuscript.

Funding: This research received no external funding.

Institutional Review Board Statement: Not applicable.

Informed Consent Statement: Not applicable.

Data Availability Statement: Data are contained within the article or supplementary material.

Conflicts of Interest: The authors declare no conflict of interest.

References

1. Carvalho, I.T.; Estevinho, B.N.; Santos, L. Application of microencapsulated essential oils in cosmetic and personal healthcare products—A review. *Int. J. Cosmet. Sci.* **2016**, *38*, 109–119. [CrossRef] [PubMed]
2. Kissel, T.; Maretschek, S.; Packhauser, C.; Schnieders, J.; Seidel, N. *Microencapsulation. Methods and Industrial Applications*; CRC Press: Boca Raton, FL, USA; Taylor and Francis: New York, NY, USA, 2006.

3. Siler-Marinkovic, S.; Bezbradica, D.; Skundric, P. Microencapsulation in the textile industry. *Chem. Ind. Chem. Eng. Q.* **2006**, *12*, 58–62. [CrossRef]
4. Balassa, L.L.; Fanger, G.O.; Wurzburg, O.B. Microencapsulation in the food industry. *CRC Crit. Rev. Food Technol.* **1971**, *2*, 245–265. [CrossRef]
5. Yeo, Y.; Baek, N.; Park, K. Microencapsulation methods for delivery of protein drugs. *Biotechnol. Bioprocess Eng.* **2001**, *6*, 213–230. [CrossRef]
6. Giro-Paloma, J.; Martínez, M.; Cabeza, L.F.; Fernández, A.I. Types, Methods, Techniques, and Applications for Microencapsulated Phase Change Materials (MPCM): A Review. *Renew. Sustain. Energy Rev.* **2016**, *53*, 1059–1075. [CrossRef]
7. Scott, W.E.; Mcconnell, D.G. Microencapsulation Process. U.S. Patent 4956129A, 1981.
8. Yow, H.N.; Routh, A.F. Formation of liquid core-polymer shell microcapsules. *Soft Matter* **2006**, *2*, 940–949. [CrossRef] [PubMed]
9. Lee, D.; Rubner, M.F.; Cohen, R.E. Formation of nanoparticle-loaded microcapsules based on hydrogen-bonded multilayers. *Chem. Mater.* **2005**, *17*, 1099–1105. [CrossRef]
10. Becker, A.L.; Zelikin, A.N.; Johnston, A.P.R.; Caruso, F. Tuning the formation and degradation of layer-by-layer assembled polymer hydrogel microcapsules. *Langmuir* **2009**, *25*, 14079–14085. [CrossRef] [PubMed]
11. Jiang, Z.; Liu, H.; He, H.; Ribbe, A.E.; Thayumanavan, S. Blended Assemblies of Amphiphilic Random and Block Copolymers for Tunable Encapsulation and Release of Hydrophobic Guest Molecules. *Macromolecules* **2020**, *53*, 2713–2723. [CrossRef]
12. Xu, Z.; Zhang, J.; Pan, T.; Li, H.; Huo, F.; Zheng, B.; Zhang, W. Encapsulation of Hydrophobic Guests within Metal–Organic Framework Capsules for Regulating Host–Guest Interaction. *Chem. Mater.* **2020**, *32*, 3553–3560. [CrossRef]
13. Takahashi, R.; Miwa, S.; Rössel, C.; Fujii, S.; Lee, J.H.; Schacher, F.H.; Sakurai, K. Polymersome formation induced by encapsulation of water-insoluble molecules within ABC triblock terpolymers. *Polym. Chem.* **2020**, *11*, 3446–3452. [CrossRef]
14. Brophy, M.R.; Deasy, P.B. Influence of coating and core modifications on the in vitro release of methylene blue from ethylcellulose microcapsules produced by pan coating procedure. *J. Pharm. Pharmacol.* **1981**, *33*, 495–499. [CrossRef]
15. Gharsallaoui, A.; Roudaut, G.; Chambin, O.; Voilley, A.; Saurel, R. Applications of spray-drying in microencapsulation of food ingredients: An overview. *Food Res. Int.* **2007**, *40*, 1107–1121. [CrossRef]
16. Soottitantawat, A.; Yoshii, H.; Furuta, T.; Ohkawara, M.; Linko, P. Microencapsulation by spray drying: Influence of emulsion size on the retention of volatile compounds. *J. Food Sci.* **2003**, *68*, 2256–2262. [CrossRef]
17. Anwar, S.H.; Kunz, B. The influence of drying methods on the stabilization of fish oil microcapsules: Comparison of spray granulation, spray drying, and freeze drying. *J. Food Eng.* **2011**, *105*, 367–378. [CrossRef]
18. Tsuda, N.; Ohtsubo, T.; Fuji, M. Preparation of self-bursting microcapsules by interfacial polymerization. *Adv. Powder Technol.* **2012**, *23*, 724–730. [CrossRef]
19. Fan, C.; Zhou, X. Influence of operating conditions on the surface morphology of microcapsules prepared by in situ polymerization. *Colloids Surfaces A Physicochem. Eng. Asp.* **2010**, *363*, 49–55. [CrossRef]
20. Kobašlija, M.; McQuade, D.T. Polyurea microcapsules from oil-in-oil emulsions via interfacial polymerization. *Macromolecules* **2006**, *39*, 6371–6375. [CrossRef]
21. Jinglei, Y.; Keller, M.W.; Moore, J.S.; White, S.R.; Sottos, N.R. Microencapsulation of isocyanates for self-healing polymers. *Macromolecules* **2008**, *41*, 9650–9655.
22. Brandau, T. Preparation of monodisperse controlled release microcapsules. *Int. J. Pharm.* **2002**, *242*, 179–184. [CrossRef]
23. Prajapati, S.K.; Tripathi, P.; Ubaidulla, U.; Anand, V. Design and development of gliclazide mucoadhesive microcapsules: In vitro and In vivo evaluation. *AAPS PharmSciTech* **2008**, *9*, 224–230. [CrossRef]
24. Dong, Z.; Ma, Y.; Hayat, K.; Jia, C.; Xia, S.; Zhang, X. Morphology and release profile of microcapsules encapsulating peppermint oil by complex coacervation. *J. Food Eng.* **2011**, *104*, 455–460. [CrossRef]
25. Gao, Q.; Wang, C.; Liu, H.; Wang, C.; Liu, X.; Tong, Z. Suspension polymerization based on inverse Pickering emulsion droplets for thermo-sensitive hybrid microcapsules with tunable supracolloidal structures. *Polymer* **2009**, *50*, 2587–2594. [CrossRef]
26. Chatterjee, S.; Salaün, F.; Campagne, C.; Vaupre, S.; Beirão, A.; El-Achari, A. Synthesis and characterization of chitosan droplet particles by ionic gelation and phase coacervation. *Polym. Bull.* **2014**, *71*, 1001–1013. [CrossRef]
27. Martins, I.M.; Rodrigues, S.N.; Barreiro, M.F.; Rodrigues, A.E. Release of thyme oil from polylactide microcapsules. *Ind. Eng. Chem. Res.* **2011**, *50*, 13752–13761. [CrossRef]
28. Rossier-Miranda, F.J.; Schroën, C.G.P.H.; Boom, R.M. Colloidosomes: Versatile microcapsules in perspective. *Colloids Surfaces A Physicochem. Eng. Asp.* **2009**, *343*, 43–49. [CrossRef]
29. Butstraen, C.; Salaün, F. Preparation of microcapsules by complex coacervation of gum Arabic and chitosan. *Carbohydr. Polym.* **2014**, *99*, 608–616. [CrossRef]
30. Li, J.; Stöver, H.D.H. Pickering emulsion templated layer-by-layer assembly for making microcapsules. *Langmuir* **2010**, *26*, 15554–15560. [CrossRef] [PubMed]
31. Velev, O.D.; Furusawa, K.; Nagayama, K. Assembly of latex particles by using emulsion droplets as templates. 2. Ball-like and composite aggregates. *Langmuir* **1996**, *12*, 2385–2391. [CrossRef]
32. Wang, W.; Liu, X.; Xie, Y.; Zhang, H.; Yu, W.; Xiong, Y.; Xie, W.; Ma, X. Microencapsulation using natural polysaccharides for drug delivery and cell implantation. *J. Mater. Chem.* **2006**, *16*, 3252–3267. [CrossRef]
33. Lee, K.Y.; Park, W.O.N.H.O.; Ha, W.A.N.S. Polyelectrolyte complexes of sodium alginate with chitosan or its derivatives for microcapsules. *J. Appl. Polym. Sci.* **1997**, 425–432. [CrossRef]

34. Czarnecki-Maulden, G.L. Effect of dietary modulation of intestinal microbiota on reproduction and early growth. *Theriogenology* **2008**, *70*, 286–290. [CrossRef]
35. Lensen, D.; Vriezema, D.M.; van Hest, J.C.M. Polymeric microcapsules for synthetic applications. *Macromol. Biosci.* **2008**, *8*, 991–1005. [CrossRef]
36. Arshady, R. Microspheres and microcapsules, a survey of manufacturing techniques Part II: Coacervation. *Polym. Eng. Sci.* **1990**, *30*, 905–914. [CrossRef]
37. Sabatini, V.; Cattò, C.; Cappelletti, G.; Cappitelli, F.; Antenucci, S.; Farina, H.; Ortenzi, M.A.; Camazzola, S.; Di Silvestro, G. Protective features, durability and biodegration study of acrylic and methacrylic fluorinated polymer coatings for marble protection. *Prog. Org. Coat.* **2018**, *114*, 47–57. [CrossRef]
38. Sabatini, V.; Farina, H.; Montarsolo, A.; Pargoletti, E.; Ortenzi, M.A.; Cappelletti, G. Fluorinated Polyacrylic Resins for the Protection of Cultural Heritages: The Effect of Fluorine on Hydrophobic Properties and Photochemical Stability. *Chem. Lett.* **2018**, *47*, 280–283. [CrossRef]
39. Sabatini, V.; Pargoletti, E.; Comite, V.; Ortenzi, M.A.; Fermo, P.; Gulotta, D.; Cappelletti, G. Towards Novel Fluorinated Methacrylic Coatings for Cultural Heritage: A Combined Polymers and Surfaces Chemistry Study. *Polymers* **2019**, *11*, 1190. [CrossRef]
40. Shchukin, D.G. Container-based multifunctional self-healing polymer coatings. *Polym. Chem.* **2013**, *4*, 4871–4877. [CrossRef]
41. Gin, H.; Dupuy, B.; Caix, J.; Baquey, C.H.; Ducassou, D. In vitro diffusion in polyacrylamide embedded agarose microbeads. *J. Microencapsul.* **1990**, *7*, 17–23. [CrossRef] [PubMed]
42. Dupuy, B.; Cadic, C.; Gin, H.; Baquey, C.; Duty, B.; Ducassou, D. Microencapsulation of isolated pituitary cells by polyacrylamide microlatex coagulation on agarose beads. *Biomaterials* **1991**, *12*, 493–496. [CrossRef]
43. Polk, A.; Amsden, B.; De Yao, K.; Peng, T.; Goosen, M.F.A. Controlled release of albumin from chitosan—Alginate microcapsules. *J. Pharm. Sci.* **1994**, *83*, 178–185. [CrossRef]
44. Benita, S.; Hoffman, A.; Donbrow, M. Microencapsulation of paracetamol using polyacrylate resins (Eudragit Retard), kinetics of drug release and evaluation of kinetic model. *J. Pharm. Pharmacol.* **1985**, *37*, 391–395. [CrossRef] [PubMed]
45. Basuli, U.; Chattopadhyay, S.; Nah, C.; Chaki, T.K. Electrical Properties and Electromagnetic Interference Shielding Effectiveness of Multiwalled Carbon nanotubes—Reinforced EMA nanocomposites. *Polym. Compos.* **2012**. [CrossRef]
46. Santinho, A.J.P.; Ueta, J.M.; Freitas, O.; Pereira, N.L. Physicochemical characterization and enzymatic degradation of casein microcapsules prepared by aqueous coacervation. *J. Microencapsul.* **2002**, *19*, 549–558. [CrossRef] [PubMed]
47. McMahon, W.A.; Lew, C.W.; Branly, K.L. Controlled Release Microcapsules. U.S. Patent 5466460A, 1995.
48. Kydonieus, A.F. *Controlled Release Technologies: Methods, Theory, and Applications*; CRC Press: Boca Raton, FL, USA; Taylor and Francis: New York, NY, USA, 2019.
49. Bashir, O.; Claverie, J.P.; Lemoyne, P.; Vincent, C. Controlled-release of Bacillus thurigiensis formulations encapsulated in lightresistant colloidosomal microcapsules for the management of lepidopteran pests of Brassica crops. *PeerJ* **2016**, *2016*, 1–9.
50. Andersson Trojer, M.; Wendel, A.; Holmberg, K.; Nydén, M. The effect of pH on charge, swelling and desorption of the dispersant poly(methacrylic acid) from poly(methyl methacrylate) microcapsules. *J. Colloid Interface Sci.* **2012**, *375*, 213–215. [CrossRef]
51. Cambiella, A.; Benito, J.M.; Pazos, C.; Coca, J.; Ratoi, M.; Spikes, H.A. The effect of emulsifier concentration on the lubricating properties of oil-in-water emulsions. *Tribol. Lett.* **2006**, *22*, 53–65. [CrossRef]
52. Olietti, A.; Pargoletti, E.; Diona, A.; Cappelletti, G. A novel optimized mold release oil-in-water emulsion for polyurethane foams production. *J. Mol. Liq.* **2018**, *261*, 199–207. [CrossRef]
53. Budinčić, J.M.; Petrović, L.; Đekić, L.; Fraj, J.; Bučko, S.; Katona, J.; Spasojević, L. Study of vitamin E microencapsulation and controlled release from chitosan/sodium lauryl ether sulfate microcapsules. *Carbohydr. Polym.* **2021**, *251*, 116988. [CrossRef]
54. Nonami, K. NII-Electronic Library Service. *Chem. Pharm. Bull.* **2002**, *57*, 364–370.
55. Hayashi, M.; Furomoto, M. Microcapsules and Processes for Producing the Same. US20040195711A1, 2007.
56. Chiantore, O.; Trossarelli, L.; Lazzari, M. Photooxidative degradation of acrylic and methacrylic polymers. *Polymer* **2000**, *41*, 1657–1668. [CrossRef]
57. Dong, J.; Ozaki, Y. FTIR and FT-raman studies of partially miscible poly(methyl methacrylate)/poly(4-vinylphenol) blends in solid states. *Macromolecules* **1997**, *30*, 286–292. [CrossRef]
58. Braun, O.; Mallo, P. Novel Inverse Latex with a Low Content of Monomer Comprising a Strong Acid Functional Group and Use in the Manufacture of Topical Compositions. U.S. Patent No. 11/631,404, 15 November 2007.
59. Huybrechts, J. Coatings Comprising Self-Stabilized Lattices Prepared in a Aqueous Carrier. U.S. Patent No. 5,936,026, 10 August 1999.
60. Sathesh Prabu, C.; Thatheyus, A.J. Biodegradation of acrylamide employing free and immobilized cells of Pseudomonas aeruginosa. *Int. Biodeterior. Biodegrad.* **2007**, *60*, 69–73. [CrossRef]
61. Yin, D.; Zhang, Q.; Yin, C.; Zhao, X.; Zhang, H. Hollow microspheres with covalent-bonded colloidal and polymeric shell by Pickering emulsion polymerization. *Polym. Adv. Technol.* **2012**, *23*, 273–277. [CrossRef]
62. Destribats, M.; Lapeyre, V.; Wolfs, M.; Sellier, E.; Leal-Calderon, F.; Ravaine, V.; Schmitt, V. Soft microgels as Pickering emulsion stabilisers: Role of particle deformability. *Soft Matter* **2011**, *7*, 7689–7698. [CrossRef]
63. Fox, T.G.; Loshaek, S. Influence of molecular weight and degree of crosslinking on the specific volume and glass temperature of polymers. *J. Polym. Sci.* **1955**, *15*, 371–390. [CrossRef]
64. He, Y. Preparation and modification of ZnO microspheres using a Pickering emulsion as template. *Mater. Lett.* **2005**, *59*, 114–117. [CrossRef]

65. Ekkehard, J.; Dieter, B.; Werner, B.; Peter, N. Microcapsule Preparations and Detergents and Cleaning Agents Containing Microcapsules. U.S. Patent No. 6,951,836, 4 October 2005.
66. Blaiszik, B.J.; Caruso, M.M.; McIlroy, D.A.; Moore, J.S.; White, S.R.; Sottos, N.R. Microcapsules filled with reactive solutions for self-healing materials. *Polymer* **2009**, *50*, 990–997. [CrossRef]
67. Zhao, Y.; Zhang, W.; Liao, L.P.; Wang, S.J.; Li, W.J. Self-healing coatings containing microcapsule. *Appl. Surf. Sci.* **2012**, *258*, 1915–1918. [CrossRef]

polymers

MDPI

Communication

In-Situ One-Step Direct Loading of Agents in Poly(acrylic acid) Coating Deposited by Aerosol-Assisted Open-Air Plasma

Gabriel Morand [1,2], Pascale Chevallier [1], Cédric Guyon [2], Michael Tatoulian [2] and Diego Mantovani [1,*]

1 Laboratory for Biomaterials and Bioengineering (CRC-I), Department of Min-Met-Mat Engineering and the CHU de Québec Research Center, Laval University, PLT-1745G, 2325 Rue de l'Université, Québec, QC G1V 0A6, Canada; gabriel.morand.1@ulaval.ca (G.M.); pascale.chevallier@crchudequebec.ulaval.ca (P.C.)

2 Laboratoire Procédés, Plasmas, Microsystèmes (2PM), Institut de Recherche de Chimie Paris (IRCP-UMR 8247), Chimie ParisTech-PSL, PSL Research University, 11 Rue Pierre et Marie Curie, F-75005 Paris, France; cedric.guyon@chimieparistech.psl.eu (C.G.); michael.tatoulian@chimieparistech.psl.eu (M.T.)

* Correspondence: diego.mantovani@gmn.ulaval.ca; Tel.: +1-418-656-2131

Abstract: In biomaterials and biotechnology, coatings loaded with bioactive agents are used to trigger biological responses by acting as drug release platforms and modulating surface properties. In this work, direct deposition of poly(acrylic acid) coatings containing various agents, such as dyes, fluorescent molecules, was achieved by aerosol-assisted open-air plasma. Using an original precursors injection strategy, an acrylic acid aerosol was loaded with an aqueous aerosol and deposited on silicon wafers. Results clearly showed that agents dissolved in the aqueous aerosol were successfully entrapped in the final coating. The effect of aerosols concentration, flow rate, and treatment time, on the coating morphology and the amount of entrapped agents, was also investigated. It was demonstrated that this process has the potential to entrap a tunable amount of any sensible water-soluble agent without altering its activity. To the best of our knowledge, this is the first time that the loading of an aqueous aerosol in coatings deposited by plasma from a liquid aerosol precursor is reported. This innovative approach complements plasma deposition of coatings loaded with bioactive agents from aqueous aerosols with the use of non-volatile liquid precursors.

Keywords: aerosol-assisted deposition; open-air plasma; entrapment in coating; acrylic acid

Citation: Morand, G.; Chevallier, P.; Guyon, C.; Tatoulian, M.; Mantovani, D. In-Situ One-Step Direct Loading of Agents in Poly(acrylic acid) Coating Deposited by Aerosol-Assisted Open-Air Plasma. *Polymers* **2021**, *13*, 1931. https://doi.org/10.3390/polym13121931

Academic Editor: Mohammad Ali Aboudzadeh

Received: 29 April 2021
Accepted: 8 June 2021
Published: 10 June 2021

1. Introduction

Coatings containing bioactive agents constitute one of the most successful value-added strategies for engineering material surfaces while preserving bulk properties. They are of special interest in the biomedical field, where specific interactions between biomaterials and the biological surrounding environment need to be promoted and triggered. In fact, these coatings can modulate surface properties while acting as drug delivery systems [1,2], for example, to confer antibacterial properties or cell-growth control [3–5]. In general, coatings loaded with bioactive agents are obtained by "wet" chemistry methods such as dip- or spray-coating [4,6]. However, such processes involve time-consuming steps and require large amounts of solvent thus increasing their cost and hindering their wide-scale production. Finally, they generally exhibit poor adhesion properties, which constitutes a major drawback for medical devices and implants [7].

In this context, atmospheric pressure plasma deposition (APPD) has emerged as an appealing approach for the manufacturing of coatings because it efficiently allows depositing a large variety of coatings from a limited amount of chemicals, with an even enhanced adhesion on a number of substrates. Furthermore, this process can be easily included in an open-air industrial production line, which makes it rapid and cost-efficient, if compared to low-pressure plasma [8,9]. However, when bioactive agents need to be loaded into the coating, their successful deposition induces non-trivial challenges. In fact, the sensitive bioactive agents (such as organic molecules, for example) are expected to be entrapped, homogeneously distributed, and unaltered by the reactive plasma environment.

In order to prevent bioactive agent denaturation, the coating deposition step is carried out under conditions known as soft-plasma polymerization, achieved by using dielectric-barrier discharge (DBD) sources [10,11]. In such an approach, the overall plasma reactivity is decreased to a point where it does not reasonably alter the structure of the molecules. However, soft-plasma polymerization is limited to the use of specific polymerizable precursors (PPs), essentially alkenes or siloxanes [12]. These PPs are injected in the discharge as a gas, as a vapor from the evaporation of volatile components, or as a liquid from nebulization of non-volatile components in the so-called aerosol-assisted APPD (AA-APPD). Since bioactive agents of greatest interest, such as proteins, antibiotics, or nanoparticles, are mostly non-volatile, they need to be dispersed into a solution and nebulized into the discharge.

Two strategies are generally adopted to disperse the bioactive agents into the solutions and inject them into the discharge. First, the bioactive agents are dissolved into the PP solutions, prior to nebulization [13,14]. However, this approach is restrictive as it requires the bioactive agents to be soluble, and stable in the PP solution [15]. To overcome this issue, a second strategy consists of dissolving the bioactive agents into water, which is then nebulized and injected in the discharge, while the PP is introduced separately as gas [16–24], or as vapor [25]. The mechanism associated with this approach has already been reported [24], and presents the main advantage of being appropriate for water-soluble agents [15]. Furthermore, this approach successfully allows the entrapment of several agents, including enzymes [17], proteins [16,18–20], and antibiotics [21,23]. Moreover, Lo Porto et al. evidenced that the bioactive agent release could be controlled by changing process parameters [23]. However, mainly gaseous precursors have been investigated, e.g., acetylene and ethylene gas or evaporated hexamethyldisiloxane [15], which strongly limit the choice of PP. Indeed, PPs in a liquid state such as acrylic acid [26], lactic acid [14,27], caprolactone-based molecules [10], methacrylate anhydride [11], ethylene glycol-based molecules [27,28], constitute promising candidates for the plasma-deposition of polymeric drug carriers and have never been investigated, to our best of knowledge [15].

Therefore, in this work, a one-step deposition of coatings containing agents from a non-volatile liquid precursor, herein acrylic acid, using open-air AA-APPD was investigated. To do so, the liquid PP is simultaneously nebulized with water aerosol containing agents. This innovative strategy consists of the in-flight loading of water droplets inside hydrophilic acrylic acid ones, as displayed in Figure 1, and is based on the original solvent exchange method used in the field of aerosols [29,30]. The impacts of different parameters such as aerosols concentration, flow rate, and treatment time on the coating morphology and the amount of entrapped agents were evaluated, and discussed in accordance with the loading mechanism.

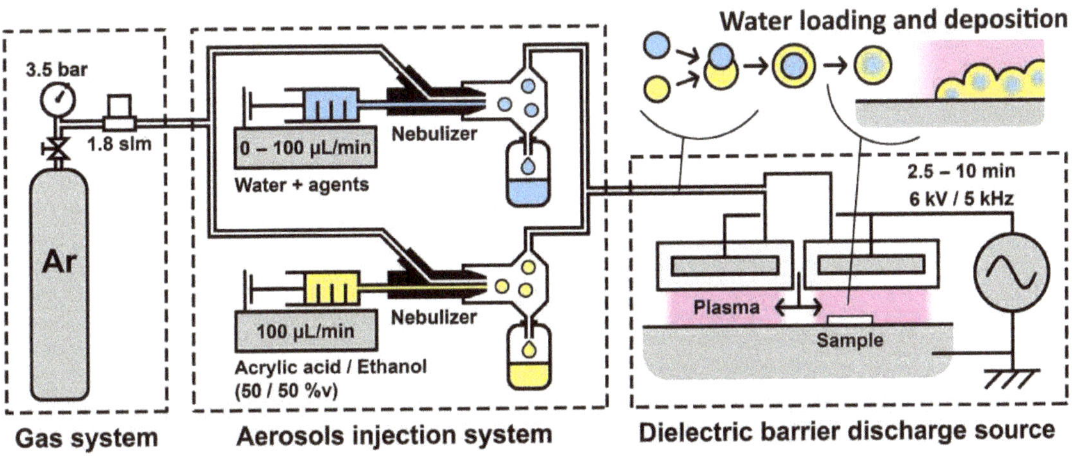

Figure 1. Scheme of the experimental set-up of the open-air aerosol-assisted atmospheric pressure plasma deposition; gas system feeds the aerosols injection system, aerosols are nebulized and mixed in the Ar flow, aerosols are injected into the dielectric barrier discharge to be deposited on silicon wafer samples. Scheme of the in-flight loading of the water droplets into the precursor aerosol, as depicted by the solvent exchange method, and deposition mechanism.

2. Materials and Methods

AA-APPD was performed with an AlmaPLUS reactor system (AlmaPlasma, Bologna, Italy). It consists of a gas system, a homemade aerosol injection system, and an open-air under-hood planar DBD source (Figure 1). The gas system was used to feed the aerosol injection system with a 1.8 standard liter per minute flow rate of Ar 99.999% (Messer Canada Inc., Mississauga, ON, Canada) by setting the pressure to 3.5 bar. The aerosol injection system consisted of two parallel nebulizers PEEK Mira Mist (Burgener, Mississauga, ON, Canada) allowing loading the Ar flow with a precursor aerosol and a water aerosol simultaneously. Nebulizers were fed by syringe pumps, which allowed controlling the aerosols flow rate. Both aerosols were mixed through 20 cm of gas line and a mixing chamber and then injected into the DBD. The DBD source consists of two high-tension electrodes placed above a grounded metallic table used as a sample holder. The inter-electrodes gap was set to 2 mm. High-tension electrodes are surrounded by 4.5 mm of dielectric high-density polyethylene. A 6 kV tension with a 5 kHz frequency was applied on the high-tension electrodes to produce the discharge. A 1 mm gap between the two high-tension electrodes was used to fill the DBD with the Ar flow loaded with the aerosols. AA-APPD was performed on 1×1 cm^2 clean substrates cut from (100)-oriented silicon wafers with a thickness of 280 µm. The effect of deposition time was studied between 2.5 min and 10 min (Table 1). A 50/50% v mixture of acrylic acid 99% (Sigma-Aldrich, Hamilton, ON, Canada) and ethanol anhydrous (Commercial Alcohols, Brampton, ON, Canada) was used as precursor solution. Ethanol addition is expected to increase water aerosol loading into the precursor aerosol [31–33]. Ultrapure water from a Purelab Flex (Elga Veolia, Woodridge, IL, USA) was used as a water solution. The precursor flow rate was fixed to 100 µL/min while the effect of the water flow rate was studied between 0 µL/min and 100 µL/min (Table 1).

The morphology of the coatings was characterized with a scanning electron microscope (SEM) FEI Quanta 250 (FEI Company Inc., Thermo-Fisher Scientific, OR, USA) using a 7.5 kV acceleration voltage in secondary electron mode, at an observation distance of 10 mm. Prior to analyses, samples were coated with a thin gold-palladium film to obtain scanning electron images with improved quality. Coated samples were fractured in order to observe the cross-section of the coatings and measure their thickness. The coatings composition was investigated by an Attenuated Total Reflectance Fourier-Transform Infrared

(ATR-FTIR) using a Cary 660 spectrometer (Agilent Technologies, Santa-Clara, CA, USA). In order to characterize the water aerosol deposition and its entrapment in the coating, the water solution was loaded with various tracers. Coatings were then deposited under condition C (Table 1) and characterized using three different characterization methods depending on the tracer. First, the water solution was loaded with commercial red food colorant (1 drop/mL, Club House, London, ON, Canada) and the resulting coating was observed with an optical microscope Olympus BX41M (Olympus America Corp., Center Valley, PA, USA) in bright field mode. Then, the water solution was loaded with $CuSO_4$ (1 mmol/mL, Sigma-Aldrich, Hamilton, ON, Canada) and the resulting coating was observed with an energy dispersive X-ray spectrometer (EDX) EDAX (Ametek Material Analysis, Mahwah, NJ, USA) coupled with SEM. Finally, the water solution was loaded with fluorescent Lucifer Yellow CH, Lithium Salt (LY, 428 nm_{ex}/536 nm_{em}, 0.5 mg/mL, Thermo-Fisher Scientific, Waltham, MA, USA), and the resulting coating was observed with a confocal microscope LSM800 Axio Observer 7 (Carl Zeiss, Jena, Germany) using the z-stack mode. A total of 28 slices were recorded over a thickness of 7.29 μm using a 488 nm laser source and a 450–700 nm range of detection. Ultimately, quantification of deposited LY was performed. After overnight aging under ambient conditions, coatings were immersed in 500 μL of ultrapure water for 60 min, until complete dispersion and dissolution. The immersion solution was collected and placed in 96 multi-well plates and the fluorescence was recorded by means of a SpectraMax i3x Multi-Mode Plate Reader (Molecular Devices, San Jose, CA, USA). Quantification errors were extracted from uncertainties on the calibration curve plotted using 20 measurements from 0 μg/mL to 0.6 μg/mL in ultrapure water, and from the standard deviation calculated from five different coated samples. The various conditions under which AA-APPD was performed are reported in Table 1. The impacts of water flow rate (Table 1, conditions A, B, C, and D), deposition time (Table 1, conditions E, C, and F), and LY concentration (Table 1, conditions G, C, and H) were studied.

Table 1. Different conditions used during the aerosol-assisted atmospheric pressure plasma deposition by varying the water flow rate injected in the nebulizer, the deposition time and the Lucifer Yellow concentration in the water for the deposited agent quantification study.

Condition	Water Flow Rate (μL/min)	Deposition Time (min)	LY Concentration (μg/mL) [1]
A	0	5	/
B	25	5	500
C	50	5	500
D	100	5	500
E	50	2.5	500
F	50	10	500
G	50	5	250
H	50	5	1000

[1] Abbreviation: LY, Lucifer Yellow.

3. Results and Discussion

3.1. *Poly(acrylic acid) Coating Deposition*

Coatings were deposited under the conditions A to F listed in Table 1. Images of the surface and cross-section of coatings were obtained from SEM and are shown in Figure 2. The as-deposited coatings are composed of agglomerated particles, whose sizes and size distributions vary depending on the deposition conditions, as shown in Figure 2.

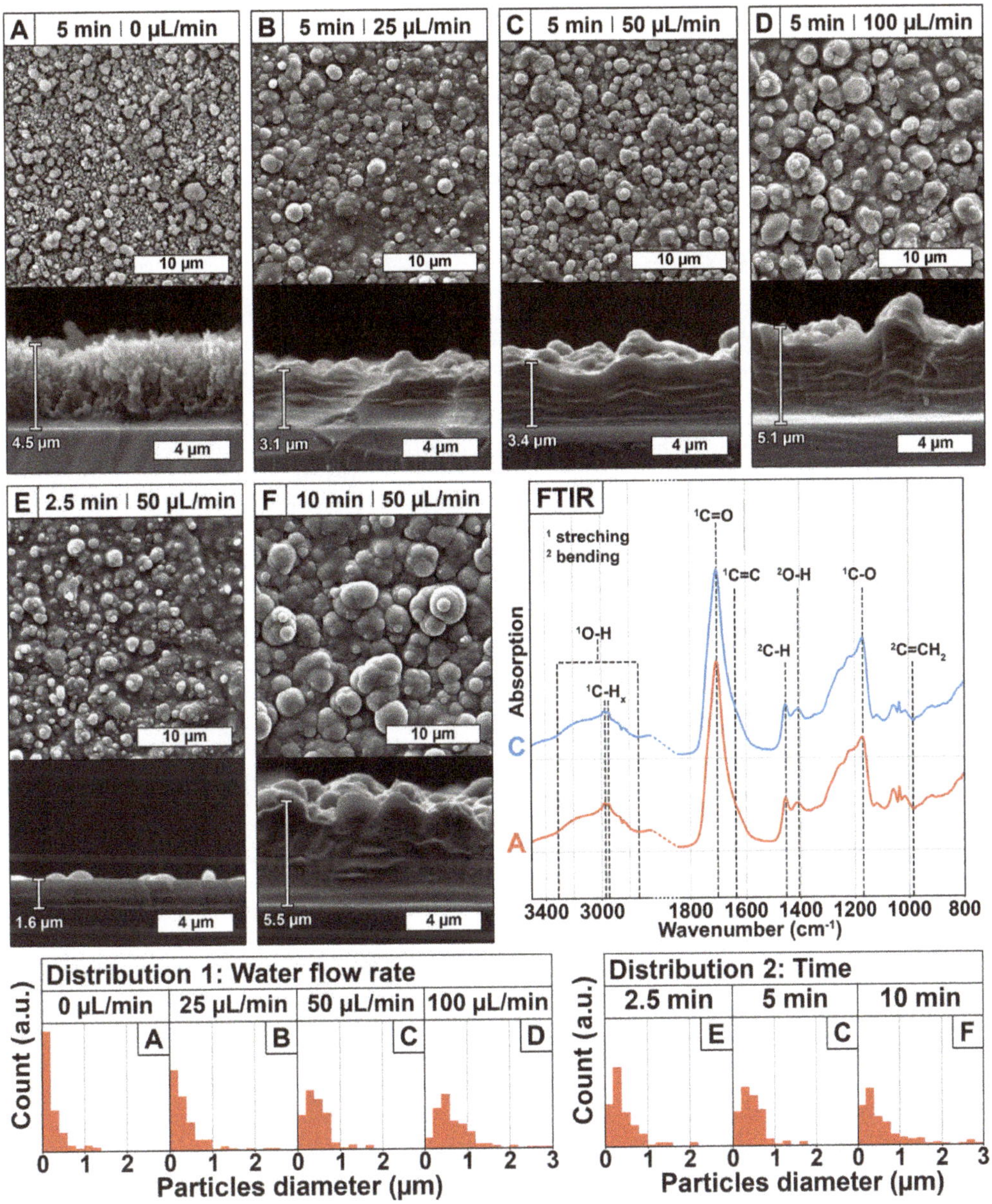

Figure 2. Scanning electron microscopy images of coatings surface morphology (×5000) and cross-section (×10,000) deposited with different water flow rate: 0 µL/min (**A**), 25 µL/min (**B**), 50 µL/min (**C**), and 100 µL/min (**D**); and for different time: 2.5 min (**E**), 5 min (**C**), and 10 min (**F**). Influence of plasma parameters, water flow (Distribution 1) and deposition time (Distribution 2), on the distribution of particles diameter, done on 200 particles. Fourier-transform infrared spectra of coatings chemistry deposited without water flow rate (FTIR, A, red spectrum), and with 50 µL/min water flow rate (FTIR, C, blue spectrum), and identification of major bonds signal.

The coating deposition from acrylic acid, as a non-volatile precursor, was performed without water flow (condition A) and is further used as control. This coating appears to be porous, with a thickness of ~4.5 µm and composed of particles whose diameters are mostly smaller than 200 nm (Figure 2A). The addition of a flow rate of 25 μL/min of water during the plasma deposition process (condition B), leads to a different coating morphology and thickness, as clearly observed in Figure 2B. Indeed, the coating appears to be continuous whereas it was porous in condition A and its thickness decreases to ~3.1 μm from ~4.5 μm without water. This coating is also composed of particles that exhibit higher diameters than the ones previously obtained without water, as seen with the widening of the distribution until 1 μm, the decrease of the proportion of particles below 200 nm, and the apparition of particles larger than 2 μm (Figure 2B). This increase in the dimension of the particles is correlated to the broadening of the size of the droplets in the aerosol. In fact, due to the increased probability of in-flight collisions induced by the injection of the water droplets, the coalescence of the droplets, and in consequence their size, is increased. Moreover, in the presence of water, the shape of the particles appears more defined and more spherical, which is explained by increased surface tension of the droplets due to their coalescence. In fact, the water surface tension is higher than both acrylic acid and ethanol (71 mN/m against 28 mN/m and 21 mN/m, respectively at 30 °C according to suppliers). Regarding the coatings differences in terms of morphology and thickness in presence of water (Figure 2B) or not (Figure 2A), they can be related to the low vapor pressure of water compared to the one of ethanol (2.3 kPa against 5.8 kPa at 20 °C according to suppliers). In fact, a part of water remains liquid inside the particles after deposition, and this residual water then induces the solvation of the surrounding poly(acrylic acid) chains, their reorganization, and finally the loss of the porous structure, thus leading to a continuous and thinner coating. All these observations corroborate the mechanism of in-flight water loading in the precursor droplets, as depicted in Figure 1, and its successful deposition.

Then, the impact of some parameters as water flow rate and deposition time on poly(acrylic acid) coatings morphology and thickness was investigated. The water flow rate was increased progressively from 25 μL/min (Figure 2B) to 50 μL/min (Figure 2C) and 100 μL/min (Figure 2D), while keeping the deposition time constant (5 min). Whatever the water flow rate, the coatings remained continuous and included particles. Nonetheless, SEM analyses clearly evidence that incrementing the water flow rate from 25 μL/min to 100 μL/min increases the thickness of the coating from ~3.1 μm to ~5.1 μm, respectively. In the same manner, particle sizes are impacted, as their diameters increase, as seen by a progressive decrease of the proportion of particles below 200 nm and the apparition of particles larger than 3 μm (Figure 2—Distribution 1). Both effects were correlated to a higher number of water droplets in the gas flow due to the higher water flow, which increases the probability of in-flight collision between droplets, thus enhancing the coalescence of the droplets. Hence, the diameter of droplets increases, and so does acrylic acid deposition. Therefore, increasing the water flow rate increases the deposition rate.

Regarding the impact of treatment time on coatings thickness, while keeping the water flow rate constant at 50 μL/min, SEM images clearly evidence that increasing deposition time increases the thickness of the coatings: from ~1.6 μm for 2.5 min (Figure 2E) to ~3.1 μm for 5 min (Figure 2C), until ~5.5 μm for 10 min (Figure 2F). From these measurements, it appears that the coating thickness increases linearly with deposition time ($R^2 = 0.995$), leading to a deposition rate value of ~0.6 μm/min (from 2.5 min to 10 min). Moreover, longer deposition time leads to an increase of the diameter of particles, as evidenced in SEM images as well as a change in the particles diameter distributions (Figure 2—Distribution 2): a decrease of the proportion of particles between 200 and 400 nm and an increase of the proportion of larger particles, up to 4 μm (Figure 2F). According to previous results, this is correlated to the enhanced coalescence of the droplets, meaning that the latter increases with deposition time.

After having demonstrated the impact of water presence, water flow rate, as well as deposition time on coatings morphology and thickness, and the composition of the coatings, were evaluated by FTIR analyses (Figure 2, FTIR). The spectra of coatings deposited without and with water, conditions A and C, respectively, are similar, meaning that the presence of water does not influence the plasma-induced process of acrylic acid polymerization. Moreover, these spectra display the characteristic bands of poly(acrylic acid) obtained by standard polymerization ("wet" chemistry), without any noticeable difference [34]. These main characteristic bands are the following: carboxylic acid O-H stretching at 2800 cm^{-1} to 3300 cm^{-1}, C-H stretching at 2970 cm^{-1} and 2950 cm^{-1}, C=O stretching at 1703 cm^{-1}, H-C-H scissoring at 1452 cm^{-1}, carboxylic acid O-H bending at 1406 cm^{-1}, and carboxylic acid C-O bending at 1169 cm^{-1} [34]. Moreover, it should be emphasized that no residual traces of bands from the initial double bond carbon of the acrylic acid, C=C stretching at 1637 cm^{-1} and $C{=}CH_2$ deformation out of the plane at 984 cm^{-1}, are evidenced [34]. This observation allows us to conclude that the plasma polymerization occurs via the vinyl group and is indeed a soft-plasma polymerization process as expected with AA-APPD [12].

3.2. Agents Deposition and Entrapment in the Coating

The feasibility of depositing poly(acrylic acid) coatings using an original precursors injection strategy, liquid PP simultaneously nebulized with a water aerosol, while keeping its chemical composition, was demonstrated. In order to determine if the innovative strategy proposed herein has the potential to entrap agents in the coating, while keeping its molecular integrity, various tracers were dissolved in the water solution and deposited under condition C (Table 1). For this purpose, different agents were tested, from simpler to more complex ones: red food colorant, copper sulfate and Lucifer Yellow (fluorescent molecule). The resulting coatings are shown in Figure 3.

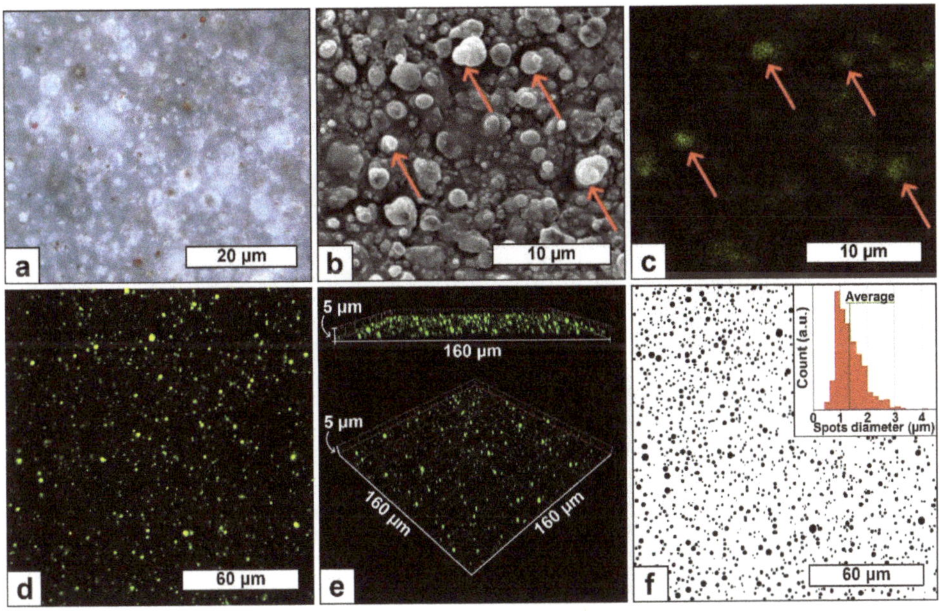

Figure 3. Characterization of different agents in coatings deposited with condition C (Water aerosol flow rate of 50 μL/min and deposition time of 5 min); bright field optical microscopy image (×100) of coating deposited with water loaded with 1 drop/mL of red colorant (**a**), scanning electron microscopy image of the surface morphology (×5000) (**b**) and corresponding energy-dispersive X-ray spectroscopy mapping of Cu (**c**) of coating deposited with water loaded with 1 mol/L of $CuSO_4$, confocal microscopy image (×50, 428 nm_{ex}/536 nm_{em}) (**d**) and corresponding three-dimensional representations (**e**) of coating deposited with water loaded with 0.5 mg/mL of Lucifer Yellow, LY, and binarized image of the fluorescent spots of the confocal microscopy image "d" (**f**) and extracted spots diameter distribution on 1603 spots (f, graph, top-right).

When loaded with red food colorant, red spots can be visualized in the image obtained by optical microscopy (Figure 3a). Their presence validates that the tracer, initially dissolved in the water solution, is indeed deposited. However, due to optical microscopy limitations, this result does not allow us to determine whether the red colorant is in fact entrapped in the coating or just sprayed. To assess the coating morphology while visualizing the loaded agent, $CuSO_4$ was used. In fact, the resulting coating morphology and the Cu distribution within the coating can be analyzed simultaneously by SEM and by EDX mapping (Figure 3b,c, respectively). Foremost, the presence of an agent in the water solution does not seem to affect the coating morphology nor has an effect on the diameter of particles (Figure 3b) when compared to the coating without the agent dissolved in the water solution and deposited in the same condition (Figure 2C). Thanks to EDX mapping, the presence of Cu in the coating is evidenced. The Cu distribution appears homogeneous all over the coating, with some Cu-richer areas pointed out by red arrows in Figure 3c. By comparison with the corresponding SEM image (Figure 3b), the Cu-rich areas are corresponding to specific particles (Figure 3b, red arrows), suggesting that the Cu is mainly localized into these specific deposited particles. This result corroborates that the Cu is in the acrylic acid droplets during their polymerization and deposition, meaning that the Cu droplets are indeed loaded in the precursor aerosol, as foreseen. Therefore, the in-situ one-step direct loading of agents in poly(acrylic acid) coating deposited by aerosol-assisted open-air plasma is successful. However, this result just demonstrates that Cu is loaded in the last coating micrometer (SEM depth analysis) but not within all the coating thickness. Furthermore, the statement that the loaded agent will keep its molecular integrity has yet not been proven.

To do so, a sensitive fluorescent molecule, LY, was used, and the as-obtained coating was characterized by confocal microscopy. Moreover, this technique permits 3-dimensional imaging, which allows observation of the agent within all the coating thicknesses (Figure 3d,e). A homogeneous distribution of green and fluorescent spots, associated with the presence of LY, is observed in the 2D image (Figure 3d), meaning that the plasma process has not altered this sensitive molecule. In addition, 3D confocal microscopy images (Figure 3e) show that these fluorescent spots are well distributed within all the coating thicknesses. The round shape of fluorescent spots, whose diameters ranged between 800 nm–2 μm (Figure 3f—average spot size of 1.37 μm), corroborates that the LY is entrapped in acrylic acid droplets. If this value is compared with the particles diameter distribution extracted from the SEM image (Figure 2C), it corroborates the presence of LY in the largest droplets, as expected by the loading mechanism by coalescence.

3.3. Deposited Agent Quantification

This innovative approach was developed with the aim to efficiently load agents in poly(acrylic acid) coating thanks to AA-APPD, and to control its concentration for application as drug-release systems. That said, the first step was to quantify the amount of LY deposited and entrapped in the coatings depending on the deposition conditions, as reported in Table 1. Briefly, the influence of water flow rate (Figure 4a), deposition time (Figure 4b) and initial LY concentration (Figure 4c) on the LY loaded concentration inside the coating, in $\mu g/cm^2$, was evaluated.

By increasing the water flow rate, the amount of deposited LY rises from 7 $ng/cm^2 \pm 1$ ng/cm^2 for 25 μL/min (Figure 4a, B), to 81 $ng/cm^2 \pm 17$ ng/cm^2 for 50 μL/min (Figure 4a, C), up to 155 $ng/cm^2 \pm 78$ ng/cm^2 for 100 μL/min (Figure 4a, D). This result can be correlated to previous SEM images showing that increasing the water flow rate leads to higher particle diameters due to the enlargement of the size of the droplets formed in the aerosol. However, there is no linear correlation between the water flow rate and the deposited LY concentration (Figure 4a). Indeed, at 25 μL/min (Figure 4a, B), the LY concentration is quite low compared to 50 μL/min and 100 μL/min ones, as if a minimal water flow rate is needed to induce an effective loading of the agent during the coalescence of the droplets. Regarding the effect of deposition time, increasing it increases the amount of deposited LY from 43 ng/cm^2 ± 35 ng/cm^2 for 2.5 min (Figure 4b, E), to 81 ng/cm^2 ± 17 ng/cm^2 for 5 min (Figure 4b, C), and to 152 ng/cm^2 ± 58 ng/cm^2 for 10 min (Figure 4b, F). When correlated to SEM images, this LY concentration increase with longer deposition time can be expected, as the diameter of the deposited particles is larger, and the coating thickness is higher. Furthermore, the deposited LY concentration is linear with the coating deposition time (R^2 = 0.998) in agreement with the constant deposition rate, as mentioned before (based on thickness evaluation from SEM images—Figure 2C,E,F). Finally, it can be observed that increasing water LY concentration increases the amount of deposited LY from 20 ng/cm^2 ± 4 ng/cm^2 for 0.25 mg/mL (Figure 4c, G), to 81 ng/cm^2 ± 17 ng/cm^2 for 0.50 mg/mL (Figure 4c, C), and to 149 ng/cm^2 ± 37 ng/cm^2 for 1.00 mg/mL (Figure 4c, H). Therefore, changing the initial LY concentration is one manner to control the amount of loaded agents without affecting others process parameters. This easy and promising approach will allow tuning released agent concentrations depending on the targeted drug delivery applications.

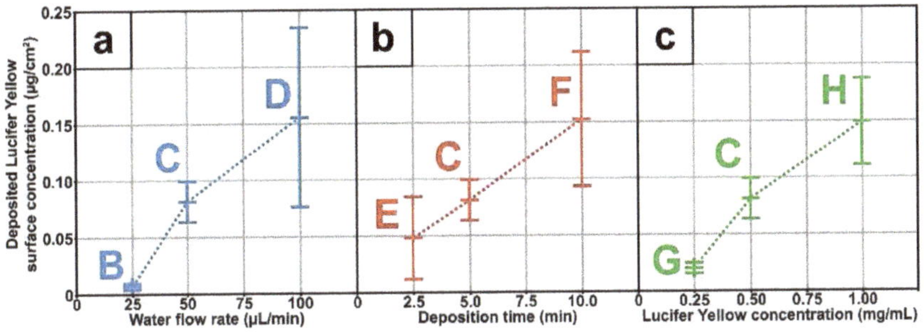

Figure 4. Quantification of Lucifer Yellow in coatings depos ited with different water flow rate (**a**): 25 µL/min (B), 50 µL/min (C), and 100 µL/min (D); for different time (**b**): 2.5 min (E), 5 min (C), and 10 min (F); and for different Lucifer Yellow concentration in the water aerosol (**c**): 0.25 mg/mL (G), 0.5 mg/mL (C), and 1 mg/mL (H).

3.4. Loading and Deposition Mechanism

The parameters studied herein showed that it is possible to tune the coating morphology, thickness and concentration of the deposited agent. Furthermore, the results provide evidence that agent deposition is correlated to the amount of water inside the precursor aerosol. Indeed, an increase in water flow rate, i.e., t higher water concentration in the aerosol led to bigger poly(acrylic acid) particles (SEM images—Figure 2) and higher loaded agent concentration (Figure 4a). These observations mean that water is loaded in the precursor droplets prior to PP polymerization, as depicted in Figure 1. Moreover, this coalescence process allows maintaining the integrity of the agent of interest, as seen with the use of a fluorescent-sensitive molecule (Figure 3d–f). In addition, the spots size of the different tracers, investigated herein either red colorant (Figure 3a), $CuSO_4$ (Figure 3c), and LY (Figure 3d), shows similar diameters, i.e., 2 µm and smaller. Moreover, coatings morphology seemed to be unmodified in presence of the agent (Figures 2C and 3b). This suggests that the presence of the agent has little to no impact on the loading and deposition mechanism, meaning that this AA-APPD process would allow entrapping various water-soluble agents in poly(acrylic acid) coatings.

Finally, this study demonstrates that the coalescence of water and precursor droplets is a critical step for successful agent loading and deposition. Such loading arises from the decrease in surface tension of the droplets when water is loaded in the precursor droplets [29,30]. However, this mechanism suggests that entrapment will be only possible with hydrophilic precursor solutions. This would limit the choice of PPs to hydrophilic precursors. Nevertheless, ethanol addition into the precursor solution represents a promising method in order to increase its hydrophilic behavior and thus stimulates coalescence of water droplets with the aerosol of precursor, even though it is hydrophobic [31–33].

4. Conclusions

The developed one-step AA-APPD process, from a non-volatile liquid precursor, allowed depositing various water-soluble agents loaded in a poly(acrylic acid) coating. Fluorescent agent deposition showed that this process maintained agent integrity during deposition. Process parameters, including deposition time, water flow, and water agent concentration, allowed us to fine-tune both contents in loaded agents and coating morphology/thickness. A mechanism based on the coalescence of water and precursor droplets was proposed to explain the as-obtained coatings. Thanks to the versatility of this innovative procedure, the AA-APPD developed herein from a non-volatile liquid precursor would be transferable to any other liquid precursor of interest. By controlling the coating morphology and the loaded agent concentration, it could therefore be expected to control the agent release, making this approach appealing for designing specific drug-release systems.

Author Contributions: G.M. conceived, designed and performed the experiments; G.M. and P.C. collected, and analyzed data and wrote the original manuscript; P.C. and C.G. supervised the work and provided resources; C.G. and D.M. reviewed and approved the final version of the manuscript; M.T. and D.M. managed the project, monitored its advancement, and acquired funding. All authors have read and agreed to the published version of the manuscript.

Funding: This work was partially funded by the Natural Sciences and Engineering Research Council of Canada (NSERC), the Canada Research Chairs program and the *Agence Nationale de la Recherche* (ANR). This work has received the support of *Institut Pierre-Gilles de Gennes* and the *Institut Carnot IPGG Microfluidique*.

Institutional Review Board Statement: Not applicable.

Informed Consent Statement: Not applicable.

Data Availability Statement: The data presented in this study are available on request from the corresponding author.

Acknowledgments: Authors thank Francesco Copes and Valentina Mariscotti Cumino, from Laval University Laboratory for Biomaterials and Bioengineering for help and guidance in reviewing this manuscript.

Conflicts of Interest: The authors declare no conflict of interest.

References

1. Zelikin, A.N. Drug releasing polymer thin films: New era of surface-mediated drug delivery. *ACS Nano* **2010**, *4*, 2494–2509. [CrossRef]
2. Visan, A.I.; Popescu-Pelin, G.; Socol, G. Degradation Behavior of Polymers Used as Coating Materials for Drug Delivery—A Basic Review. *Polymers* **2021**, *13*, 1272. [CrossRef]
3. Cloutier, M.; Mantovani, D.; Rosei, F. Antibacterial Coatings: Challenges, Perspectives, and Opportunities. *Trends Biotechnol.* **2015**, *33*, 637–652. [CrossRef] [PubMed]
4. Livingston, M.; Tan, A. Coating techniques and release kinetics of drug-eluting stents. *J. Med. Devices Trans. ASME* **2016**, *10*, 1–8. [CrossRef]
5. Wang, Z.; Wang, Z.; Lu, W.W.; Zhen, W.; Yang, D.; Peng, S. Novel biomaterial strategies for controlled growth factor delivery for biomedical applications. *NPG Asia Mater.* **2017**, *9*, e435-17. [CrossRef]
6. John, A.A.; Subramanian, A.P.; Vellayappan, M.V.; Balaji, A.; Jaganathan, S.K.; Mohandas, H.; Paramalinggam, T.; Supriyanto, E.; Yusof, M. Review: Physico-chemical modification as a versatile strategy for the biocompatibility enhancement of biomaterials. *RSC Adv.* **2015**, *5*, 39232–39244. [CrossRef]
7. Wiemer, M.; Butz, T.; Schmidt, W.; Schmitz, K.P.; Horstkotte, D.; Langer, C. Scanning electron microscopic analysis of different drug eluting stents after failed implantation: From nearly undamaged to major damaged polymers. *Catheter. Cardiovasc. Interv.* **2010**, *75*, 905–911. [CrossRef] [PubMed]
8. Penkov, O.V.; Khadem, M.; Lim, W.-S.; Kim, D.-E. A review of recent applications of atmospheric pressure plasma jets for materials processing. *J. Coatings Technol. Res.* **2015**, *12*, 225–235. [CrossRef]
9. Merche, D.; Vandencasteele, N.; Reniers, F. Atmospheric plasmas for thin film deposition: A critical review. *Thin Solid Films* **2012**, *520*, 4219–4236. [CrossRef]

10. Mertz, G.; Fouquet, T.; El-Ahrach, H.I.; Becker, C.; Phan, T.N.T.; Ziarelli, F.; Gigmes, D.; Ruch, D. Water Sensitive Coatings Deposited by Aerosol Assisted Atmospheric Plasma Process: Tailoring the Hydrolysis Rate by the Precursor Chemistry. *Plasma Process. Polym.* **2015**, *12*, 1293–1301. [CrossRef]
11. Mertz, G.; Fouquet, T.; Becker, C.; Ziarelli, F.; Ruch, D. A methacrylic anhydride difunctional precursor to produce a hydrolysis-sensitive coating by aerosol-assisted atmospheric plasma process. *Plasma Process. Polym.* **2014**, *11*, 728–733. [CrossRef]
12. Friedrich, J. Mechanisms of plasma polymerization-Reviewed from a chemical point of view. *Plasma Process. Polym.* **2011**, *8*, 783–802. [CrossRef]
13. O'Hare, L.-A.; O'Neill, L.; Goodwin, A.J. Anti-microbial coatings by agent entrapment in coatings deposited via atmospheric pressure plasma liquid deposition. *Surf. Interface Anal.* **2006**, *38*, 1519–1524. [CrossRef]
14. Da Ponte, G.; Sardella, E.; Fanelli, F.; Paulussen, S.; Favia, P. Atmospheric Pressure Plasma Deposition of Poly Lactic Acid-Like Coatings with Embedded Elastin. *Plasma Process. Polym.* **2014**, *11*, 345–352. [CrossRef]
15. Palumbo, F.; Porto, C.L.; Fracassi, F.; Favia, P. Recent advancements in the use of aerosol-assisted atmospheric pressure plasma deposition. *Coatings* **2020**, *10*, 440. [CrossRef]
16. Heyse, P.; Roeffaers, M.B.J.; Paulussen, S.; Hofkens, J.; Jacobs, P.A.; Sels, B.F. Protein Immobilization Using Discharges: A Route to a Straightforward Manufacture of Bioactive Films. *Plasma Process. Polym.* **2008**, *5*, 186–191. [CrossRef]
17. Heyse, P.; Van Hoeck, A.; Roeffaers, M.B.J.; Raffin, J.; Sto, T.; Lammertyn, J.; Verboven, P.; Jacobs, P.A.; Hofkens, J.; Paulussen, S.; et al. Exploration of Atmospheric Pressure Plasma Nanofilm Technology for Straightforward Bio-Active Coating Deposition: Enzymes, Plasmas and Polymers, an Elegant Synergy. *Plasma Process. Polym.* **2011**, *8*, 965–974. [CrossRef]
18. Palumbo, F.; Camporeale, G.; Yang, Y.; Wu, J.; Sardella, E.; Dilecce, G.; Calvano, C.D.; Quintieri, L.; Caputo, L.; Baruzzi, F.; et al. Direct Plasma Deposition of Lysozyme-Embedded Bio-Composite Thin Films. *Plasma Process. Polym.* **2015**, *12*, 1302–1310. [CrossRef]
19. Hsiao, C.; Wu, C.; Liu, Y.; Yang, Y.; Cheng, Y.; Palumbo, F.; Camporeale, G.; Favia, P.; Wu, J. Aerosol-Assisted Plasma Deposition of Biocomposite Coatings: Investigation of Processing Conditions on Coating Properties. *IEEE Trans. Plasma Sci.* **2016**, *44*, 3091–3098. [CrossRef]
20. Liu, Y.H.; Yang, C.H.; Lin, T.R.; Cheng, Y.C. Using aerosol-assisted atmospheric-pressure plasma to embed proteins onto a substrate in one step for biosensor fabrication. *Plasma Process. Polym.* **2018**, *15*, 1–10. [CrossRef]
21. Palumbo, F.; Treglia, A.; Lo Porto, C.; Fracassi, F.; Baruzzi, F.; Frache, G.; El Assad, D.; Pistillo, B.R.; Favia, P. Plasma-Deposited Nanocapsules Containing Coatings for Drug Delivery Applications. *ACS Appl. Mater. Interfaces* **2018**, *10*, 35516–35525. [CrossRef] [PubMed]
22. Lo Porto, C.; Palumbo, F.; Palazzo, G.; Favia, P. Direct plasma synthesis of nano-capsules loaded with antibiotics. *Polym. Chem.* **2017**, *8*, 1746–1749. [CrossRef]
23. Lo Porto, C.; Palumbo, F.; Buxadera-Palomero, J.; Canal, C.; Jelinek, P.; Zajickova, L.; Favia, P. On the plasma deposition of vancomycin-containing nano-capsules for drug-delivery applications. *Plasma Process. Polym.* **2018**, *15*, 1–11. [CrossRef]
24. Lo Porto, C.; Palumbo, F.; Fracassi, F.; Barucca, G.; Favia, P. On the formation of nanocapsules in aerosol-assisted atmospheric-pressure plasma. *Plasma Process. Polym.* **2019**, *16*, 1–7. [CrossRef]
25. Wang, L.; Lo Porto, C.; Palumbo, F.; Modic, M.; Cvelbar, U.; Ghobeira, R.; De Geyter, N.; De Vrieze, M.; Kos, Š.; Serša, G.; et al. Synthesis of antibacterial composite coating containing nanocapsules in an atmospheric pressure plasma. *Mater. Sci. Eng. C* **2021**, *119*. [CrossRef] [PubMed]
26. Bitar, R.; Cools, P.; De Geyter, N.; Morent, R. Acrylic acid plasma polymerization for biomedical use. *Appl. Surf. Sci.* **2018**, *448*, 168–185. [CrossRef]
27. Da Ponte, G.; Sardella, E.; Fanelli, F.; Van Hoeck, A.; d'Agostino, R.; Paulussen, S.; Favia, P. Atmospheric pressure plasma deposition of organic films of biomedical interest. *Surf. Coat. Technol.* **2011**, *205*, S525–S528. [CrossRef]
28. Da Ponte, G.; Sardella, E.; Fanelli, F.; D'Agostino, R.; Gristina, R.; Favia, P. Plasma deposition of PEO-like coatings with aerosol-assisted dielectric barrier discharges. *Plasma Process. Polym.* **2012**, *9*, 1176–1183. [CrossRef]
29. Yeo, Y.; Basaran, O.A.; Park, K. A new process for making reservoir-type microcapsules using ink-jet technology and interfacial phase separation. *J. Control. Release* **2003**, *93*, 161–173. [CrossRef] [PubMed]
30. Yeo, Y.; Park, K. A new microencapsulation method using an ultrasonic atomizer based on interfacial solvent exchange. *J. Control. Release* **2004**, *100*, 379–388. [CrossRef]
31. Pabari, R.M.; Sunderland, T.; Ramtoola, Z. Investigation of a Novel 3-Fluid Nozzle Spray Drying Technology for the Engineering of Multifunctional Layered Microparticles. *Expert Opin. Drug Deliv.* **2012**, *9*, 1463–1474. [CrossRef]
32. Kondo, K.; Niwa, T.; Danjo, K. Preparation of sustained-release coated particles by novel microencapsulation method using three-fluid nozzle spray drying technique. *Eur. J. Pharm. Sci.* **2014**, *51*, 11–19. [CrossRef] [PubMed]
33. Wan, F.; Maltesen, M.J.; Andersen, S.K.; Bjerregaard, S.; Foged, C.; Rantanen, J.; Yang, M. One-Step Production of Protein-Loaded PLGA Microparticles via Spray Drying Using 3-Fluid Nozzle. *Pharm. Res.* **2014**, *31*, 1967–1977. [CrossRef] [PubMed]
34. Petisco-Ferrero, S.; Sánchez-Ilárduya, M.B.; Díez, A.; Martín, L.; Meaurio Arrate, E.; Sarasua, J.R. Surface functionalization of an osteoconductive filler by plasma polymerization of poly(ϵ-caprolactone) and poly(acrylic acid) films. *Appl. Surf. Sci.* **2016**, *386*, 327–336. [CrossRef]

polymers MDPI

Article

Polyethyleneimine-Oleic Acid Micelles-Stabilized Palladium Nanoparticles as Highly Efficient Catalyst to Treat Pollutants with Enhanced Performance

Xiang Lai [1,†], Xuan Zhang [1,†], Shukai Li [1], Jie Zhang [1], Weifeng Lin [2] and Longgang Wang [1,*]

1 Key Laboratory of Applied Chemistry, Hebei Key Laboratory of Heavy Metal Deep-Remediation in Water and Resource Reuse, College of Environmental and Chemical Engineering, Yanshan University, Qinhuangdao 066004, China; lx0406x@163.com (X.L.); 15603395687@163.com (X.Z.); lishukai1998@163.com (S.L.); 19991641458@163.com (J.Z.)
2 Department of Molecular Chemistry and Materials Science, Weizmann Institute of Science, Rehovot 76100, Israel; lin.weifeng@weizmann.ac.il
* Correspondence: lgwang@ysu.edu.cn
† These authors contributed equally to this work.

Abstract: Water soluble organic molecular pollution endangers human life and health. It becomes necessary to develop highly stable noble metal nanoparticles without aggregation in solution to improve their catalytic performance in treating pollution. Polyethyleneimine (PEI)-based stable micelles have the potential to stabilize noble metal nanoparticles due to the positive charge of PEI. In this study, we synthesized the amphiphilic PEI-oleic acid molecule by acylation reaction. Amphiphilic PEI-oleic acid assembled into stable PEI-oleic acid micelles with a hydrodynamic diameter of about 196 nm and a zeta potential of about 34 mV. The PEI-oleic acid micelles-stabilized palladium nanoparticles (PO-PdNPs$_n$) were prepared by the reduction of sodium tetrachloropalladate using $NaBH_4$ and the palladium nanoparticles (PdNPs) were anchored in the hydrophilic layer of the micelles. The prepared PO-PdNPs$_n$ had a small size for PdNPs and good stability in solution. Noteworthily, PO-PdNPs$_{150}$ had the highest catalytic activity in reducing 4-nitrophenol (4-NP) (K_{nor} = 18.53 s^{-1}mM^{-1}) and oxidizing morin (K_{nor} = 143.57 s^{-1}M^{-1}) in aqueous solution than other previous catalysts. The enhanced property was attributed to the improving the stability of PdNPs by PEI-oleic acid micelles. The method described in this report has great potential to prepare many kinds of stable noble metal nanoparticles for treating aqueous pollution.

Keywords: polyethyleneimine; micelles; palladium; nanoparticles; catalytic

Citation: Lai, X.; Zhang, X.; Li, S.; Zhang, J.; Lin, W.; Wang, L. Polyethyleneimine-Oleic Acid Micelles-Stabilized Palladium Nanoparticles as Highly Efficient Catalyst to Treat Pollutants with Enhanced Performance. *Polymers* **2021**, *13*, 1890. https://doi.org/10.3390/polym13111890

Academic Editor: M. Ali Aboudzadeh

Received: 28 April 2021
Accepted: 2 June 2021
Published: 6 June 2021

1. Introduction

The quality of water is highly related with our health. Many countries have strictly controlled the emissions of various organic pollutants in water [1]. For example, 4-nitrophenol (4-NP) is highly toxic, but its reduced product 4-aminophenol (4-AP) is relatively low in toxicity and is a pharmaceutical intermediate. The catalytic reduction of 4-NP and similar phenol compounds is carried out on catalysts treated with $NaBH_4$. In addition, morin is a kind of polyphenol dye that belongs to flavonoid dyes. Morin has been used as a model matrix for the study of catalytic bleaching processes in laundry detergents [2,3]. Morin can be degraded by using nanoparticles with H_2O_2. These nanoparticles play an important role for the catalytic generation of reactive oxygen species from decomposition of H_2O_2. Thus, the efficiency of treatment of organic pollutants is highly dependent on the property of catalysts.

Many kinds of noble metal nanoparticles, such as platinum nanoparticles (PtNPs) [4], gold nanoparticles (AuNPs) [5,6], and palladium nanoparticles (PdNPs) [7–9], catalyze the reduction of 4-NP and the oxidation of morin. The catalytic activity of noble metal nanoparticles is highly dependent on the active atoms on their surface, which results in extremely high surface energy of the nanoparticles [10,11]. However, it is for this reason that the nanoparticles are easily agglomerated in the preparation and catalytic reaction process [12], which results in a significant decrease in the number of surface-active atoms and the catalytic activity. This problem limits noble metal nanoparticles in practical applications [13]. Researchers have developed a variety of stabilizers to prevent the coagulation of noble metal nanoparticles, thereby increasing their catalytic efficiency [14,15]. For example, Pitchaimani Veerakumar and his colleagues report a method for immobilizing PdNPs (Pd/NH_2-SiO_2) with PEI (Mw = 25,000) functionalized silica nanoparticles that have good separation properties and exhibit excellent catalytic performance [16]. As a kind of polymer, PEI has a good advantage as a stabilizer. Each unit with three atoms in the PEI skeleton has a nitrogen atom [17]. Since PEI has a positive charge, it generates electrostatic attraction with the negatively charged noble metal precursor [18], thereby stabilizing the noble metal nanoparticles. However, it is difficult for the low-molecule-weight PEI to stabilize the noble metal nanoparticles. The micelles formed by amphiphilic molecules based on the reaction of low-molecule-weight PEI with other hydrophobic molecules can be one method to stabilize noble metal nanoparticles and solve the shortcomings of low-molecule-weight PEI.

Herein, PEI-oleic acid micelle-stabilized PdNPs (PO-$PdNPs_n$) were prepared by using PEI-based micelles. The amphiphilic molecule consisted of low-molecule-weight PEI (Mw = 600) as hydrophilic moiety and oleic acid as hydrophobic moiety. PEI and oleic acid are also inexpensive and readily available. The prepared PdNPs were small in size and had a narrow size distribution. The PO-$PdNPs_n$ showed high stability and an enhanced catalytic efficiency to treat pollutants such as 4-NP and morin. The current work provides new ideas for the synthesis of noble metal nanoparticle catalysts with amphiphilic molecular micelles to treat pollutants.

2. Materials and Methods

2.1. Preparation of PEI-Oleic Acid Micelles

The synthesis of PEI-oleic acid refers to the method reported in the previous literature [17]. First, 0.60 g PEI, 0.28 g oleic acid, 0.40 g EDC·HCl, 0.27 g HOBt, and 6 mL of anhydrous dimethyl formamide were added into a flask and passed through with N_2 for 0.5 h. The obtained solution was dialyzed against methanol with a dialysis bag (MWCO = 500) after 1 day. Then, the PEI-oleic acid was obtained by the removal of methanol with the help of a rotary evaporator (RE-52A, Shanghai Yarong Biochemical Instrument Factory, Shanghai, China). PEI-oleic acid in methanol solution was added dropwise to water (methanol: H_2O = 1:9) at 25 °C for 10 min. The PEI-oleic acid micelles solution was obtained by dialysis.

2.2. Preparation of PO-$PdNPs_n$

Two mM Na_2PdCl_4 and the PEI-oleic acid micelles were mixed. The molar ratios of N atoms of the PEI-oleic acid micelles to Pd atoms of Na_2PdCl_4 were 75, 100, and 150, respectively. After 20 min, a 5-fold molar excess of $NaBH_4$ in 0.3 M NaOH was added. One M HCl was added to tune the pH to neutral after 20 min. The mixed solution reacted for 1 h to obtain PO-$PdNPs_n$ (n = 75, 100, 150).

2.3. Critical Micelles Concentration Measurement

The critical micelle concentration of the PEI-oleic acid micelles in aqueous solution was determined by a pyrene fluorescence probe [19]. Briefly, the fluorescence intensity values of 374 nm and 384 nm at the excitation wavelength of 334 nm were measured by fluorescence spectrophotometer. The CMC value of the PEI-oleic acid micelles in water was determined by the ratio of fluorescence intensity at 374 nm and 384 nm.

2.4. Catalytic Reaction on 4-NP

To study the degradation of 4-NP over time, 250 μL of 4-NP solution (600 μM), 1 mL of water, and 1 mL of fresh $NaBH_4$ solution (0.5 M) were added in a quartz cuvette (1 × 1 cm^2) in sequence at 25 °C. Then, 50 μL of PO-$PdNPs_{75}$ (8.35 μM) was added. UV–Vis spectra within 200–800 nm of the mixed solution were recorded every 3 min.

To explore the relationship between PO-$PdNPs_n$ and the catalytic rate, 250 μL of 4-NP (600 μM) solution, water, and 1 mL of fresh $NaBH_4$ solution (0.5 M) were added into a quartz cuvette followed by the addition 50 μL of PO-$PdNPs_n$ (n = 75, 100, 150), respectively. The absorbance at 400 nm was measured by UV-TU1810 (Beijing Purkinje General Instrument Co., Ltd., Beijing, China).

2.5. Catalytic Reaction on Morin

Eighty μL of morin in a carbonate buffer at pH 9.2 (3 mM), 1790 μL of carbonate buffer at pH 9.2, 50 μL of PO-$PdNPs_{75}$ (0.475 mM), and 80 μL of hydrogen peroxide (0.4 mM) were added to the cuvette. UV–Vis spectra within 200–800 nm of the mixed solution were recorded every 2 min. As the same time, the spectra of the control experiment were measured every 5 min.

The effects of different catalyst concentrations and different morin concentrations on the catalytic reaction were determined by UV-TU1810 (Beijing Purkinje General Instrument Co., Ltd., Beijing, China) at 403 nm. One mM morin, carbonate buffer, PO-$PdNPs_n$ (n = 75, 100, 150), and 0.4 M hydrogen peroxide were added to a cuvette. The final concentration of [N] in $PdNPs_n$ and hydrogen peroxide were 7.8 mM and 10 mM, respectively. The concentration of morin was from 0.5 to 1.25 mM. The next experiment had similar procedure. The concentration of [N] in the catalyst was fixed at 0.78–5.47 mM with 3 mM morin.

2.6. Statistical Analysis

The k_{app} data of the PO-$PdNPs_n$ of 4-NP and the morin treatment were analyzed by SPSS 25 to assess the statistical differences between the groups. $p < 0.05$ is considered statistically significant.

3. Results

3.1. Characterization of PEI-Oleic Acid Micelles

Acylation reaction was employed into the synthesis of the amphiphilic molecules of PEI-oleic acid, which is illustrated in Figure 1. In this reaction, PEI (Mw = 600) and oleic acid had the same amount, EDC·HCl and HOBt acted as coupling reagents in dry DMF at room temperature for 24 h in the acylation reaction. The prepared PEI-oleic acid amphiphilic molecule had PEI as the hydrophilic segment and oleic acid as the hydrophobic segment.

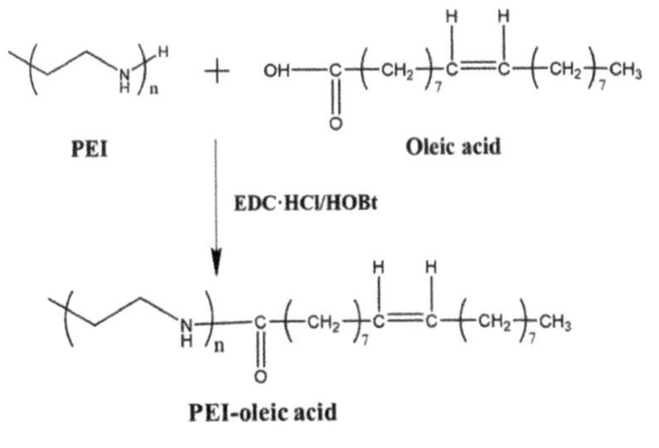

Figure 1. Synthesis of amphiphilic molecule PEI-oleic acid.

The PEI-oleic acid amphiphilic molecules self-assembled in water to form micelles, and the hydrophilic PEI segment acted as a shell layer of the micelles. The critical micelle concentration (CMC) of the PEI-oleic acid micelles was [N] = 0.12 mg/mL based on the fluorescence method using pyrene. The hydrodynamic diameter and zeta potential of PEI-oleic acid micelles were determined to be 196 nm and 34 mV by DLS (DLS, Malvern, Worcestershire, UK), respectively. The positive charge of the micelles was attributed to the hydrophilic PEI shell.

3.2. Characterization of PO-PdNPs$_n$

To prepare PO-PdNPs$_n$, Na_2PdCl_4 was mixed with PEI-oleic acid micelles and then further reduced with $NaBH_4$. As shown in Figure 2a, two characteristic absorption peaks at 305 and 420 nm were detected for the Na_2PdCl_4. After the reduction of Na_2PdCl_4 by $NaBH_4$ to produce PdNPs, a new absorption spectrum appeared, which indicated the formation of PdNPs. This was consistent with the results reported in the previous literature [20]. As the molar ratio of [N]:[Pd] ranged from 75 to 150, the color of solution gradually decreased as shown in Figure 2b. The UV–Vis spectra and color change of solution indicated that the PdNPs were successfully stabilized by the PEI-oleic acid micelles. The positively charged PEI shell played an important role in adsorbing and stabilizing PdNPs. As shown in Figure S1 and Table S1, the infrared spectroscopy has shown the successful preparation of PEI-oleic acid micelles and PO-PdNPs$_{75}$.

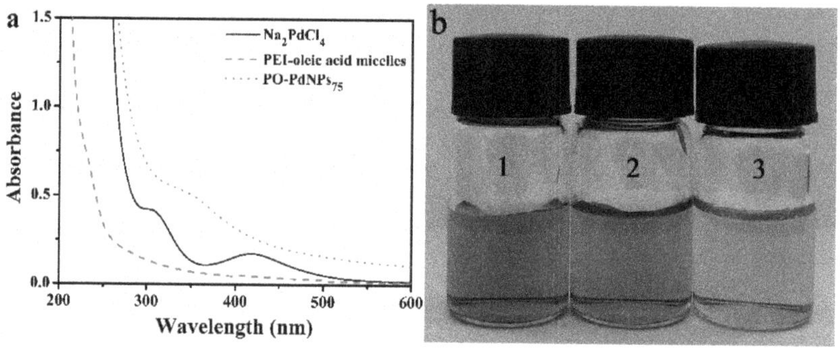

Figure 2. (**a**) UV–Vis spectra of Na_2PdCl_4, PEI-oleic acid micelles, and PO-PdNPs$_{75}$, (**b**) PO-PdNPs$_n$ with molar ratio of n = [N]: [Pd] = 75 (**b1**), 100 (**b2**), and 150 (**b3**), respectively.

The size of the PdNPs of PO-PdNPs$_n$ was measured by TEM (TEM, JEM-1230EX, Hitachi, Tokyo, Japan). Figure 3 displays the monodispersed characteristic of PO-PdNPs$_n$. The mean particle sizes of PdNPs within PO-PdNPs$_n$ at [N]:[Pd] = 75, 100, and 150 were 2.01 $\pm$ 0.30, 1.85 $\pm$ 0.25, and 1.67 $\pm$ 0.27 nm, respectively. The size of the PdNPs decreased with the increasing the molar ratio of [N]:[Pd]. In short, the prepared PdNPs of PO-PdNPs$_n$ had a small particle size and a narrow size distribution.

In order to determine the stability of the prepared PO-PdNPs$_n$ in aqueous solution, the hydrodynamic diameter and zeta potential were measured by DLS (DLS, Malvern, Worcestershire, UK) [21,22]. As shown in Figure 4a, the hydrodynamic diameters were about 53.89, 43.95, and 37.86 nm at a molar ratio of [N]:[Pd] = 75, 100, and 150, respectively. These hydrodynamic diameters of the prepared PO-PdNPs were smaller than that of the PEI-oleic acid micelles, which should be attributed to PdNPs located in the PEI-oleic acid micelles' shell. The measured hydrodynamic diameters of the prepared PO-PdNPs was larger than the TEM particle size measured in the dry state [23,24]. Figure 4b showed that the zeta potential of PO-PdNPs$_n$ was 21.5, 25.55, and 31.1 mV at a molar ratio of [N]:[Pd] = 75, 100, and 150, respectively. The positive charge on the surface of PO-PdNPs$_n$ was larger than 20 mV. Thus, the strong electrostatic repulsion was good for the high stability of PO-PdNPs$_n$ in aqueous solution. In summary, PO-PdNPs$_n$ had high stability in solution and big positive charge, so it kept from coagulation within one month.

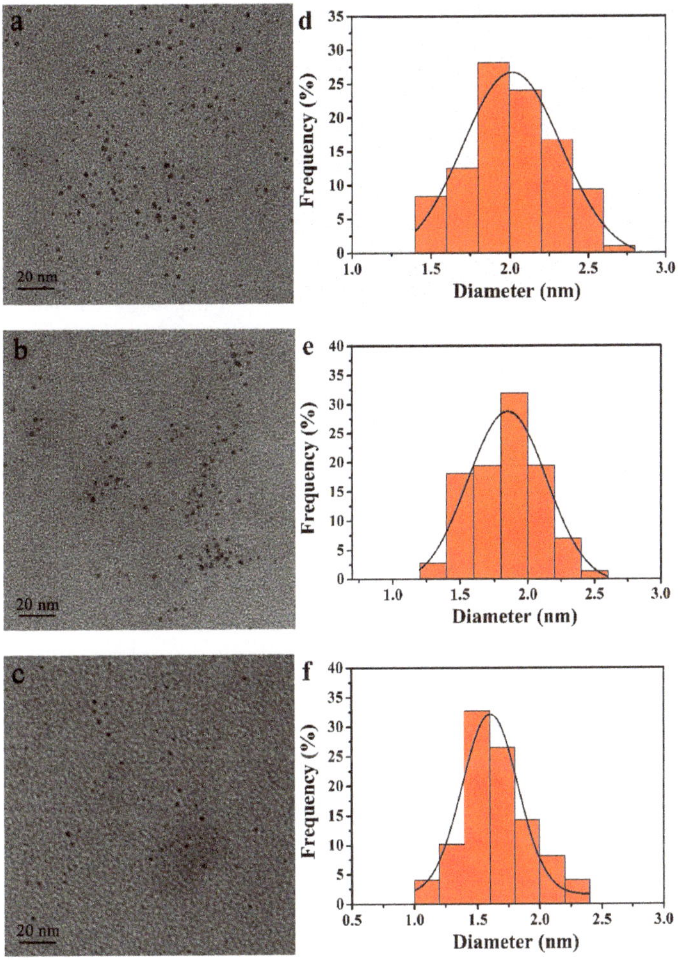

Figure 3. TEM images and size distribution analysis of the PdNPs of PO-PdNPsn: (**a**,**d**) n = 75, (**b**,**e**) n = 100, (**c**,**f**) n = 150.

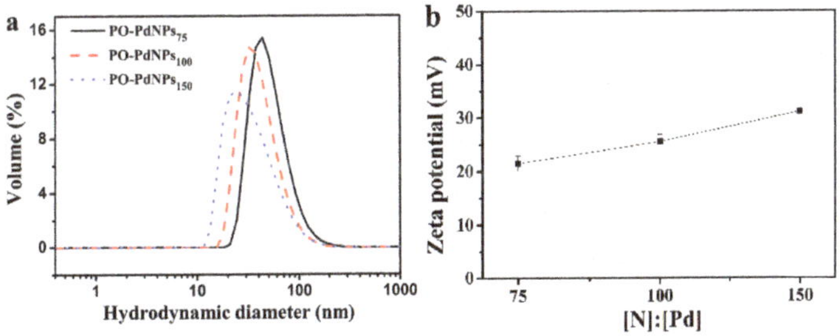

Figure 4. (**a**) Hydrodynamic diameter and (**b**) zeta potential of PO-PdNPs$_n$.

3.3. Catalytic Activity of PO-PdNPs$_n$ on 4-NP

The catalytic reduction of 4-NP was employed to evaluate the catalytic activity of PO-PdNPs$_n$. Due to the high toxicity of 4-NP and its pollution of water resources, it has attracted great attention. 4-AP is the reduction product of 4-NP. 4-AP has low toxicity and important applications in industry. It is important to study the method to transform 4-NP into 4-AP. The concentration of PdNPs in the solution of the reacting mixture was as low as 10^{-7} mM. Therefore, the absorption spectra of PdNPs almost had no effect on the absorbance of the mixed solution. The absorption peak of 4-NP existed at 317 nm. When the $NaBH_4$ solution was added, the color of the solution rapidly turned yellow, and the obvious absorption peak appeared at 400 nm, which corresponded to 4-hydroxyaminophenol [25]. In Figure 5a, it was found that the absorption peak at 400 nm decreased gradually, and the absorption peak at 310 nm increased. The results indicated the formation of 4-AP due to the consumption of 4-hydroxyaminophenol. After 12 min, the color of the solution became colorless, and the absorption peak at 400 nm decreased to 0.119 as shown in Figure 5b, indicating that 4-NP completely changed to 4-AP. The turnover frequency (TOF) was defined as the number of reactants converted per h. The TOF of PO-PdNPs$_{75}$ reached 1796.4 h^{-1} in this experiment.

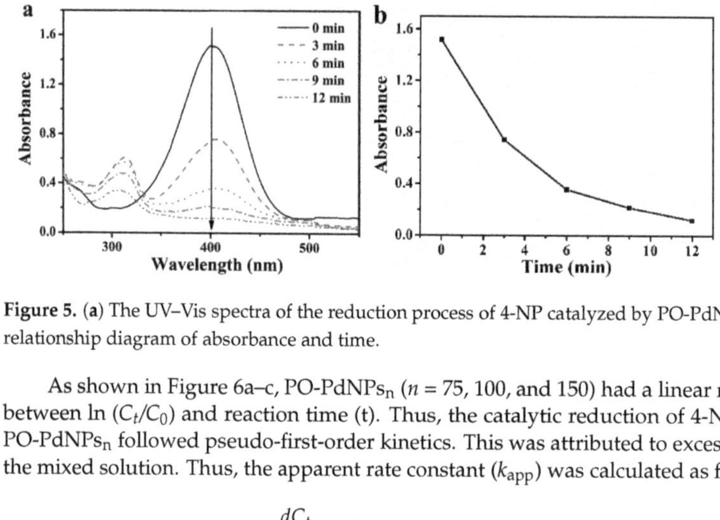

Figure 5. (**a**) The UV–Vis spectra of the reduction process of 4-NP catalyzed by PO-PdNPs$_{75}$, (**b**) the relationship diagram of absorbance and time.

As shown in Figure 6a–c, PO-PdNPs$_n$ (*n* = 75, 100, and 150) had a linear relationship between ln (C_t/C_0) and reaction time (t). Thus, the catalytic reduction of 4-NP by using PO-PdNPs$_n$ followed pseudo-first-order kinetics. This was attributed to excess $NaBH_4$ in the mixed solution. Thus, the apparent rate constant (k_{app}) was calculated as follows:

$$\frac{dC_t}{dt} = \ln(C_t/C_0) = -k_{app}t \tag{1}$$

The k_{app} values were calculated from the slope of the lines as shown in Figure 7. The corresponding k_{app} was shown in Figure 6d. The k_{app} value increased with increasing catalyst concentration of PO-PdNPs$_n$, indicating that the reduction rate was linear correlated with the concentration of PO-PdNPs$_n$. K_{nor} was used to compare the catalytic activity of different catalysts, K_{nor} was defined as the ratio of k_{app} to molar concentration of catalyst ($K_{nor} = k_{app}/C_{cat}$). PO-PdNPs$_{150}$ had the highest K_{nor} (18.53 $s^{-1}mM^{-1}$) in PO-PdNPs$_n$. Table 1 shows the K_{nor} and the TOF of the PO-PdNPs$_n$ and other catalysts. PO-PdNPs$_{150}$ had the highest K_{nor}, indicating that they had the highest activity. PdNPs were located in the micelles' shells, which may contribute to their high catalytic activity caused by the low mass transfer resistance of the substrates. In short, this catalytic reaction followed pseudo-first-order kinetics and PO-PdNPs$_{150}$ had the highest activity.

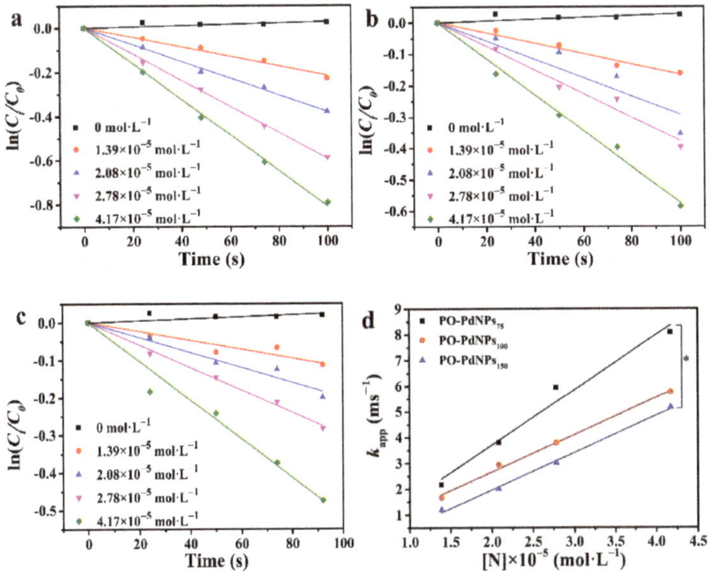

Figure 6. The catalytic process for $\ln(C_t/C_0)$ and time (**a**) PO-PdNPs$_{75}$, (**b**) PO-PdNPs$_{100}$, (**c**) PO-PdNPs$_{150}$, (**d**) the linear relationship between k_{app} and concentration of PO-PdNPs$_n$. * $p < 0.05$.

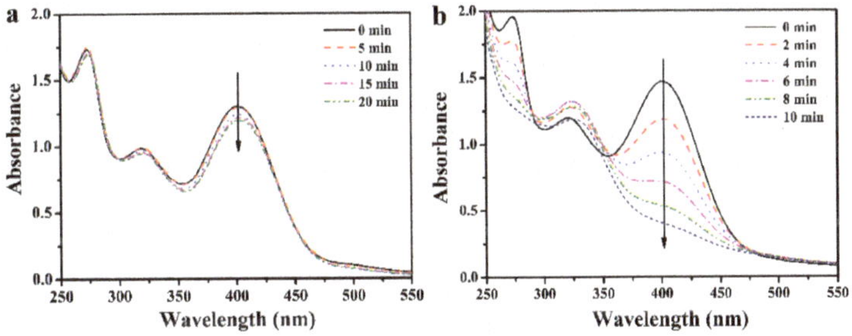

Figure 7. The UV–Vis spectra of the oxidation of morin (**a**) without the catalyst every 5 min and (**b**) with the PO-PdNPs$_{75}$ every 2 min.

Table 1. Comparison of K_{nor} and TOF of PO-PdNPs$_n$ with other Pd-based catalysts in 4-NP reduction.

Catalyst	K_{nor} ($s^{-1}mM^{-1}$)	TOF (h^{-1})	Reference
PdP/CNSs	1.4	504	[2]
Pd/Fe_3O_4@SiO_2@KCC-1	2.78	-	[26]
Pd/SBA-15	0.118	-	[27]
@Pd/CeO_2	-	1068	[28]
PO-PdNPs$_{75}$	-	1796	This work
PO-PdNPs$_{150}$	18.53	-	This work

3.4. Catalytic Activity of PO-PdNPs$_n$ on Morin

Morin is a polyphenol dye that has been used as a model to study the catalytic ability of precious metal nanoparticle catalysts. The catalytic activity and catalytic mechanism of PO-PdNPs$_n$ were evaluated with the addition of H_2O_2 [29,30]. In a carbonate buffer with pH 9.2, the maximum peak of the morin was $\lambda = 403$ nm. Figure 7a shows that without the addition of a catalyst, the maximum peak of morin hardly decreased within 20 min with H_2O_2. Figure 7b shows that after adding the catalyst PO-PdNPs$_{75}$, the absorbance at 403 nm decreased rapidly with time, while the absorbance at 325 nm gradually increased with time. This phenomenon indicated that the morin underwent catalyzed oxidation to benzofuranone [31]. After a longer time, the peak at 325 nm decreased again, indicating that benzofuranone was further oxidized [32,33]. Therefore, the experimental study of the reaction kinetics was carried out by controlling the reaction time.

To study the relationship between k_{app} and PO-PdNPs$_n$ concentration, the effect of changing the concentration of PO-PdNPs$_n$ on the catalytic reaction under the conditions of constant concentration of morin and H_2O_2 was measured. Figure 8a showed that the relationship between k_{app} and the catalyst concentration was linear. With the increase of catalyst concentration, k_{app} became higher. The three proportions of catalysts had the same trend as follows: PO-PdNPs$_{75}$ > PO-PdNPs$_{100}$ > PO-PdNPs$_{150}$. The K_{nor} of PO-PdNPs$_{75}$, PO-PdNPs$_{100}$, and PO-PdNPs$_{150}$ were 105.77, 129.49, and 143.57 $s^{-1}M^{-1}$, respectively, so the catalytic activity was increased according to the order of PO-PdNPs$_{75}$ < PO-PdNPs$_{100}$ < PO-PdNPs$_{150}$. This was because the particle size of PdNPs in PO-PdNPs$_n$ decreased in the order of PO-PdNPs$_{75}$ <PO-PdNPs$_{100}$ <PO-PdNPs$_{150}$. At the same concentration, the smaller particle size resulted in the larger specific surface area and the higher catalytic efficiency.

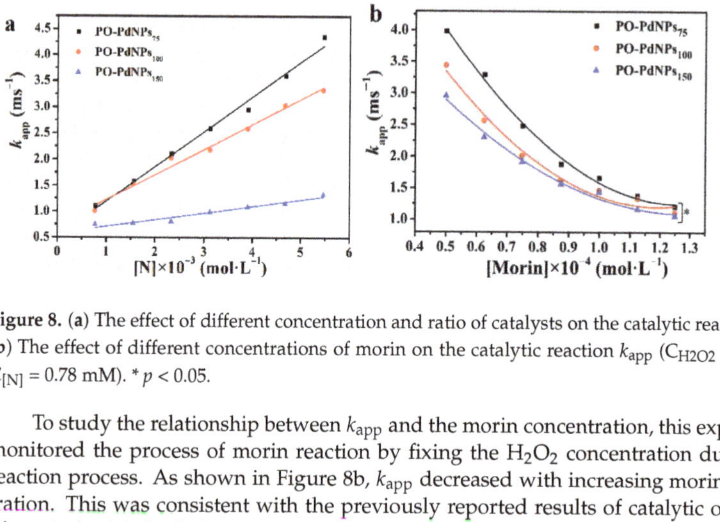

Figure 8. (**a**) The effect of different concentration and ratio of catalysts on the catalytic reaction k_{app}. (**b**) The effect of different concentrations of morin on the catalytic reaction k_{app} (C_{H2O2} = 10 mM, $C_{[N]}$ = 0.78 mM). * $p < 0.05$.

To study the relationship between k_{app} and the morin concentration, this experiment monitored the process of morin reaction by fixing the H_2O_2 concentration during the reaction process. As shown in Figure 8b, k_{app} decreased with increasing morin concentration. This was consistent with the previously reported results of catalytic oxidation of morin by MnOx [34]. Surface coverage was the primary factor of the reaction rate that was dependent on the different concentration and adsorption constant of morin and H_2O_2. Generally, during the reaction of morin, the adsorption constant K_{morin} is higher than K_{H2O2}. Therefore, compared with H_2O_2, morin is more easily adsorbed on the catalyst surface. As the concentration of morin increased, the active sites on the surface of noble metal nanoparticles are covered, resulting in a decrease in the active sites available for H_2O_2 adsorption, which gradually decreases the reaction rate when the concentration of morin increases.

3.5. Catalytic Mechanism

During the catalytic reduction of 4-NP, the Langmuir–Hinshelwood kinetics method was followed according to previous reports [35,36]. Similarly, the catalytic reduction of 4-NP by PO-PdNPs$_n$ should also meet the equation. 4-NP was quickly reduced to a stable intermediate 4-hydroxyaminophenol. The electrons from sodium borohydride and 4-hydroxyaminophenol ions combined on the surface of PdNPs to produce 4-AP ions, and then 4-AP ions were desorbed to form 4-AP molecules. In this process, the adsorption/desorption equilibria of all related compounds on the surface of PdNPs were rapidly achieved. In the oxidation process of morin by PO-$PdNPs_n$, the Langmuir–Hinshelwood kinetics method was also followed [29,37].

4. Conclusions

In conclusion, PEI-oleic acid micelles were prepared and used to stabilize PdNPs in aqueous solution. The size of PdNPs in the PO-$PdNPs_n$ was between 1.67 and 2.01 nm with a narrow size distribution. The PO-$PdNPs_n$ maintained high stability due to its zeta potentials larger than 20 mV. In addition, the prepared PO-$PdNPs_n$ effectively catalyzed the reduction of 4-NP to form 4-AP. PO-$PdNPs_n$ also had high catalytic efficiency for morin. The catalytic ability of PO-$PdNPs_n$ was higher than those of other catalysts, which could result from the small size and high stability of Pd NPs. The location of PdNPs in the PEI-oleic acid micelles was good for low mass transfer resistance. The prepared PO-$PdNPs_n$ has great potential applications in treating various organic contaminants in solution in the future.

Supplementary Materials: The following are available online at https://www.mdpi.com/article/10.3390/polym13111890/s1, Figure S1: FTIR spectra of PEI-oleic acid micelles and PO-$PdNPs_{75}$, Table S1: IR bands of the two compounds and their assignments.

Author Contributions: Conceptualization, L.W.; methodology, L.W.; investigation, X.Z.; writing—original draft preparation, X.L.; writing—review and editing, X.L., S.L., J.Z., W.L. and L.W.; visualization, X.Z.; supervision, L.W. All authors have read and agreed to the published version of the manuscript.

Funding: This research was funded by Natural Science Foundation of Hebei Province, grant number B2017203229, China Postdoctoral Science Foundation, grant number 2016M601284, National undergraduate Innovation and Entrepreneurship Training Program of Yanshan University, grant number S202010216074 and 202010216055.

Institutional Review Board Statement: Not applicable.

Informed Consent Statement: Not applicable.

Data Availability Statement: The data presented in this study are available on request from the corresponding author.

Conflicts of Interest: The authors declare no conflict of interest.

References

1. Dong, Z.; Le, X.; Dong, C.; Zhang, W.; Li, X.; Ma, J. Ni@Pd core–shell nanoparticles modified fibrous silica nanospheres as highly efficient and recoverable catalyst for reduction of 4-nitrophenol and hydrodechlorination of 4-chlorophenol. *Appl. Catal. B* **2015**, *162*, 372–380. [CrossRef]
2. Zhao, Z.; Ma, X.; Wang, X.; Ma, Y.; Liu, C.; Hang, H.; Zhang, Y.; Du, Y.; Ye, W. Synthesis of amorphous PdP nanoparticles supported on carbon nanospheres for 4-nitrophenol reduction in environmental applications. *Appl. Surf. Sci.* **2018**, *457*, 1009–1017. [CrossRef]
3. Ghosh, P.; Patwari, J.; Dasgupta, S. Complexation with human serum albumin facilitates sustained release of morin from polylactic-co-glycolic acid nanoparticles. *J. Phys. Chem. B* **2017**, *121*, 1758–1770. [CrossRef]
4. Zhang, X.-F.; Zhu, X.-Y.; Feng, J.-J.; Wang, A.-J. Solvothermal synthesis of N-doped graphene supported PtCo nanodendrites with highly catalytic activity for 4-nitrophenol reduction. *Appl. Surf. Sci.* **2018**, *428*, 798–808. [CrossRef]
5. Hao, Y.; Shao, X.; Li, B.; Hu, L.; Wang, T. Mesoporous TiO_2 nanofibers with controllable Au loadings for catalytic reduction of 4-nitrophenol. *Mater. Sci. Semicond. Process.* **2015**, *40*, 621–630. [CrossRef]
6. Zhang, H.; Xin, X.; Sun, J.; Zhao, L.; Shen, J.; Song, Z.; Yuan, S. Self-assembled chiral helical nanofibers by amphiphilic dipeptide derived from d- or l-threonine and application as a template for the synthesis of Au and Ag nanoparticles. *J. Colloid Interface Sci.* **2016**, *484*, 97–106. [CrossRef] [PubMed]
7. Duan, X.; Xiao, M.; Liang, S.; Zhang, Z.; Zeng, Y.; Xi, J.; Wang, S. Ultrafine palladium nanoparticles supported on nitrogen-doped carbon microtubes as a high-performance organocatalyst. *Carbon* **2017**, *119*, 326–331. [CrossRef]
8. Le, X.; Dong, Z.; Li, X.; Zhang, W.; Le, M.; Ma, J. Fibrous nano-silica supported palladium nanoparticles: An efficient catalyst for the reduction of 4-nitrophenol and hydrodechlorination of 4-chlorophenol under mild conditions. *Catal. Commun.* **2015**, *59*, 21–25. [CrossRef]

9. Deraedt, C.; Salmon, L.; Astruc, D. "Click" dendrimer-stabilized palladium nanoparticles as a green catalyst down to parts per million for efficient CC cross-coupling reactions and reduction of 4-nitrophenol. *Adv. Synth. Catal.* **2014**, *356*, 2525–2538. [CrossRef]
10. Fedorczyk, A.; Ratajczak, J.; Kuzmych, O.; Skompska, M. Kinetic studies of catalytic reduction of 4-nitrophenol with $NaBH_4$ by means of Au nanoparticles dispersed in a conducting polymer matrix. *J. Solid State Electrochem.* **2015**, *19*, 2849–2858. [CrossRef]
11. Ma, T.; Yang, W.; Liu, S.; Zhang, H.; Liang, F. A comparison reduction of 4-nitrophenol by gold nanospheres and gold nanostars. *Catalysts* **2017**, *7*, 38. [CrossRef]
12. Fan, L.; Ji, X.; Lin, G.; Liu, K.; Chen, S.; Ma, G.; Xue, W.; Zhang, X.; Wang, L. Green synthesis of stable platinum nanoclusters with enhanced peroxidase-like activity for sensitive detection of glucose and glutathione. *Microchem. J.* **2021**, *166*, 106–202. [CrossRef]
13. Li, N.; Zhao, P.; Astruc, D. Anisotropic gold nanoparticles: Synthesis, properties, applications, and toxicity. *Angew. Chem. Int. Ed. Engl.* **2014**, *53*, 1756–1789. [CrossRef]
14. Ye, W.; Yu, J.; Zhou, Y.; Gao, D.; Wang, D.; Wang, C.; Xue, D. Green synthesis of Pt–Au dendrimer-like nanoparticles supported on polydopamine-functionalized graphene and their high performance toward 4-nitrophenol reduction. *Appl. Catal. B* **2016**, *181*, 371–378. [CrossRef]
15. Wang, L.; Yang, Q.; Cui, Y.; Gao, D.; Kang, J.; Sun, H.; Zhu, L.; Chen, S. Highly stable and biocompatible dendrimer-encapsulated gold nanoparticle catalysts for the reduction of 4-nitrophenol. *New J. Chem.* **2017**, *41*, 8399–8406. [CrossRef]
16. Veerakumar, P.; Velayudham, M.; Lu, K.-L.; Rajagopal, S. Silica-supported PEI capped nanopalladium as potential catalyst in Suzuki, Heck and Sonogashira coupling reactions. *Appl. Catal. A* **2013**, *455*, 247–260. [CrossRef]
17. Wang, L.; Zhang, X.; Cui, Y.; Guo, X.; Chen, S.; Sun, H.; Gao, D.; Yang, Q.; Kang, J. Polyethyleneimine-oleic acid micelle-stabilized gold nanoparticles for reduction of 4-nitrophenol with enhanced performance. *Transit. Met. Chem.* **2020**, *45*, 31–39. [CrossRef]
18. Xu, M.; Qian, J.; Suo, A.; Liu, T.; Liu, X.; Wang, H. A reduction-dissociable PEG-b-PGAH-b-PEI triblock copolymer as a vehicle for targeted co-delivery of doxorubicin and P-gp siRNA. *Polym. Chem.* **2015**, *6*, 2445–2456. [CrossRef]
19. Wang, L.; Zhu, L.; Bernards, M.T.; Chen, S.; Sun, H.; Guo, X. Dendrimer-based biocompatible zwitterionic micelles for efficient cellular internalization and enhanced anti-tumor effects. *ACS Appl. Polym. Mater.* **2019**, *2*, 159–171. [CrossRef]
20. Wang, L.; Zhang, J.; Guo, X.; Chen, S.; Cui, Y.; Yu, Q.; Yang, L.; Sun, H.; Gao, D.; Xie, D. Highly stable and biocompatible zwitterionic dendrimer-encapsulated palladium nanoparticles that maintain their catalytic activity in bacterial solution. *New J. Chem.* **2018**, *42*, 19740–19748. [CrossRef]
21. Cui, T.; Li, S.; Chen, S.; Liang, Y.; Sun, H.; Wang, L. "Stealth" dendrimers with encapsulation of indocyanine green for photothermal and photodynamic therapy of cancer. *Int. J. Pharm.* **2021**, *600*, 120–502. [CrossRef] [PubMed]
22. Dong, L.; Li, R.; Wang, L.; Lan, X.; Sun, H.; Zhao, Y.; Wang, L. Green synthesis of platinum nanoclusters using lentinan for sensitively colorimetric detection of glucose. *Int. J. Biol. Macromol.* **2021**, *172*, 289–298. [CrossRef] [PubMed]
23. Ma, G.; Lin, W.; Wang, Z.; Zhang, J.; Qian, H.; Xu, L.; Yuan, Z.; Chen, S. Development of polypeptide-based zwitterionic amphiphilic micelles for nanodrug delivery. *J. Mater. Chem. B* **2016**, *4*, 5256–5264. [CrossRef] [PubMed]
24. Sun, H.; Chang, M.Y.Z.; Cheng, W.I.; Wang, Q.; Commisso, A.; Capeling, M.; Wu, Y.; Cheng, C. Biodegradable zwitterionic sulfobetaine polymer and its conjugate with paclitaxel for sustained drug delivery. *Acta Biomater.* **2017**, *64*, 290–300. [CrossRef] [PubMed]
25. Gu, S.; Kaiser, J.; Marzun, G.; Ott, A.; Lu, Y.; Ballauff, M.; Zaccone, A.; Barcikowski, S.; Wagener, P. Ligand-free gold nanoparticles as a reference material for kinetic modelling of catalytic reduction of 4-nitrophenol. *Catal. Lett.* **2015**, *145*, 1105–1112. [CrossRef]
26. Le, X.; Dong, Z.; Liu, Y.; Jin, Z.; Huy, T.-D.; Le, M.; Ma, J. Palladium nanoparticles immobilized on core–shell magnetic fibers as a highly efficient and recyclable heterogeneous catalyst for the reduction of 4-nitrophenol and Suzuki coupling reactions. *J. Mater. Chem. A* **2014**, *2*, 19696–19706. [CrossRef]
27. Morère, J.; Tenorio, M.J.; Torralvo, M.J.; Pando, C.; Renuncio, J.A.R.; Cabañas, A. Deposition of Pd into mesoporous silica SBA-15 using supercritical carbon dioxide. *J. Supercrit. Fluids* **2011**, *56*, 213–222. [CrossRef]
28. Liu, B.; Yu, S.; Wang, Q.; Hu, W.; Jing, P.; Liu, Y.; Jia, W.; Liu, Y.; Liu, L.; Zhang, J. Hollow mesoporous ceria nanoreactors with enhanced activity and stability for catalytic application. *Chem. Commun.* **2013**, *49*, 3757–3759. [CrossRef]
29. Ncube, P.; Hlabathe, T.; Meijboom, R. The preparation of well-defined dendrimer-encapsulated palladium and platinum nanoparticles and their catalytic evaluation in the oxidation of morin. *Appl. Surf. Sci.* **2015**, *357*, 1141–1149. [CrossRef]
30. Nemanashi, M.; Meijboom, R. Catalytic behavior of different sizes of dendrimer-encapsulated Au(n) nanoparticles in the oxidative degradation of morin with H_2O_2. *Langmuir* **2015**, *31*, 9041–9053. [CrossRef]
31. Ilunga, A.K.; Meijboom, R. Catalytic and kinetic investigation of the encapsulated random alloy (Pdn-Au110-n) nanoparticles. *Appl. Catal. B* **2016**, *189*, 86–98. [CrossRef]
32. Ndolomingo, M.J.; Meijboom, R. Kinetics of the catalytic oxidation of morin on γ-Al_2O_3 supported gold nanoparticles and determination of gold nanoparticles surface area and sizes by quantitative ligand adsorption. *Appl. Catal. B* **2016**, *199*, 142–154. [CrossRef]
33. Polzer, F.; Wunder, S.; Lu, Y.; Ballauff, M. Oxidation of an organic dye catalyzed by MnO_x nanoparticles. *J. Catal.* **2012**, *289*, 80–87. [CrossRef]
34. Xaba, M.S.; Meijboom, R. Kinetic and catalytic analysis of mesoporous Co_3O_4 on the oxidation of morin. *Appl. Surf. Sci.* **2017**, *423*, 53–62. [CrossRef]
35. Gu, S.; Wunder, S.; Lu, Y.; Ballauff, M.; Fenger, R.; Rademann, K.; Jaquet, B.; Zaccone, A. Kinetic analysis of the catalytic reduction of 4-nitrophenol by Metallic Nanoparticles. *J. Phys. Chem. C* **2014**, *118*, 18618–18625. [CrossRef]

36. Nasrollahzadeh, M.; Baran, T.; Baran, N.Y.; Sajjadi, M.; Tahsili, M.R.; Shokouhimehr, M. Pd nanocatalyst stabilized on amine-modified zeolite: Antibacterial and catalytic activities for environmental pollution remediation in aqueous medium. *Sep. Purif. Technol.* **2020**, *239*, 116–542. [CrossRef]
37. Xiao, H.; Wang, R.; Dong, L.; Cui, Y.; Chen, S.; Sun, H.; Ma, G.; Gao, D.; Wang, L. Biocompatible Dendrimer-Encapsulated Palladium Nanoparticles for Oxidation of Morin. *ACS Omega* **2019**, *4*, 18685–18691. [CrossRef]

MDPI
St. Alban-Anlage 66
4052 Basel
Switzerland
Tel. +41 61 683 77 34
Fax +41 61 302 89 18
www.mdpi.com

Polymers Editorial Office
E-mail: polymers@mdpi.com
www.mdpi.com/journal/polymers

www.ingramcontent.com/pod-product-compliance
Lightning Source LLC
LaVergne TN
LVHW070741100526
838202LV00013B/1278
* 9 7 8 3 0 3 6 5 3 7 7 6 4 *